Advances in Intelligent Systems and Computing

Volume 581

Series editor

Janusz Kacprzyk, Polish Academy of Sciences, Warsaw, Poland
e-mail: kacprzyk@ibspan.waw.pl

About this Series

The series "Advances in Intelligent Systems and Computing" contains publications on theory, applications, and design methods of Intelligent Systems and Intelligent Computing. Virtually all disciplines such as engineering, natural sciences, computer and information science, ICT, economics, business, e-commerce, environment, healthcare, life science are covered. The list of topics spans all the areas of modern intelligent systems and computing.

The publications within "Advances in Intelligent Systems and Computing" are primarily textbooks and proceedings of important conferences, symposia and congresses. They cover significant recent developments in the field, both of a foundational and applicable character. An important characteristic feature of the series is the short publication time and world-wide distribution. This permits a rapid and broad dissemination of research results.

Advisory Board

More information about this series at http://www.springer.com/series/11156

Vicenç Torra · Radko Mesiar
Bernard De Baets

Editors

Aggregation Functions in Theory and in Practice

 Springer

Editors
Vicenç Torra
School of Informatics
University of Skövde
Skövde
Sweden

Radko Mesiar
Department of Mathematics
and Descriptive Geometry
Slovak University of Technology
Bratislava
Slovakia

Bernard De Baets
KERMIT Research Unit
Knowledge-based Systems
Ghent University
Ghent
Belgium

ISSN 2194-5357 ISSN 2194-5365 (electronic)
Advances in Intelligent Systems and Computing
ISBN 978-3-319-59305-0 ISBN 978-3-319-59306-7 (eBook)
DOI 10.1007/978-3-319-59306-7

Library of Congress Control Number: 2017940833

Printed on acid-free paper

This Springer imprint is published by Springer Nature
The registered company is Springer International Publishing AG
The registered company address is: Gewerbestrasse 11, 6330 Cham, Switzerland

Preface

Aggregation functions are usually defined as those functions that are monotonic and that satisfy some boundary conditions. In particular settings, these conditions are relaxed. Aggregation functions are used for data fusion and decision-making. Examples of these functions include means, t-norms and t-conorms, uninorms and nullnorms, copulas and fuzzy integrals (e.g. the Choquet and Sugeno integrals). Besides the aggregation of real inputs, aggregation functions on general/particular lattices are also considered.

This volume collects the final revised manuscripts of 26 accepted contributions of participants to the 9th International Summer School on Aggregation Functions that took place in Skövde (Sweden) on 19–22 June 2017. Note that AGOP conferences are biannually organized by the working group AGOP of the EUSFLAT association, and it is the ninth in a series of AGOP summer schools, including AGOP 2001 (Oviedo, Spain), AGOP 2003 (Alcalá de Henares, Spain), AGOP 2005 (Lugano, Switzerland), AGOP 2007 (Gent, Belgium), AGOP 2009 (Palma de Mallorca, Spain), AGOP 2011 (Benevento, Italy), AGOP 2013 (Pamplona, Spain) and AGOP 2015 (Katowice, Poland). The volume also includes the abstracts of the invited talks and tutorials given in the School. All included contributions were reviewed by PC members and several external reviewers, and they include works from theory and fundamentals of aggregation functions to their use in applications. Together, they provide a good overview of recent trends in research on aggregation functions.

March 2017

Vicenç Torra
Radko Mesiar
Bernard De Baets

9th International Summer School on Aggregation Functions – AGOP 2017

General Chairs

Vicenç Torra	University of Skövde, Sweden
Radko Mesiar	Slovak University of Technology, Slovakia
Bernard De Baets	Ghent University, Belgium

Program Chairs

Vicenç Torra	University of Skövde, Sweden
Radko Mesiar	Slovak University of Technology, Slovakia
Bernard De Baets	Ghent University, Belgium

Program Committee

M. Baczyński, Poland
B. Bedregal, Brazil
G. Beliakov, Australia
H. Bustince, Spain
T. Calvo, Spain
J. Dombi, Hungary
J. Drewniak, Poland
P. Drygas, Poland
J. Dujmović, USA
F. Durante, Italy

J. Fernandez, Spain
J.L. Garcia-Lapresta, Spain
M. Grabisch, France
S. Greco, Italy
E. Herrera-Viedma, Spain
E. Indurain, Spain
B. Jayaram, India
M. Kalina, Slovakia
F. Karacal, Turkey
E.P. Klement, Austria

A. Kolesárová, Slovakia
J. Li, China
J.L. Marichal, Luxembourg
G. Mayor, Spain
A. Mesiarová-Zemánková,
    Slovak Republic
J. Montero, Spain
S. Montes, Spain
S. Massanet, Spain
Y. Narukawa, Japan

E. Pap, Serbia
I. Perfilieva, Czech Republic
H. Prade, France
A. Pradera, Spain
J.V. Riera, Spain
B. Teheux, France
V. Torra, Sweden
J. Torrens, Spain
R.R. Yager, USA

Local Organizing Committee Chair

Vicenç Torra   University of Skövde, Sweden

Local Organizing Committee

SAIL group

Additional Referees

Jimmy Devillet

Supporting Institutions

University of Skövde
The Catalan Association for Artificial Intelligence (ACIA)
The European Society for Fuzzy Logic and Technology (EUSFLAT)

Tutorials and Invited Talks

The Role of Aggregation Functions on Auctions

Beatriz López

University of Girona, Campus Montilivi, Girona, Spain
beatriz.lopez@udg.edu

Auctions are mechanisms for allocating resources (tasks or goods) among self-interested agents [3]. An auction consists in the following four steps:

1. Call for proposals: the auctioneer announces the resources to be committed
2. Bidding: the bidders express their preferences on the resources
3. Winner determination problem (WDP): the auctioneer decides which agents will have the resources
4. Payment: the winner bidders pay to the auctioneer.

This basic mechanism could have several instantiations depending on the role of the participants (forward if the auctioneer sells; reverse if the auctioneer buys), number of bidding sides (one-side when an agent can be either auctioneer or bidder; double-side when the agent can have both roles) bid composition (single attribute or multi-attribute), number of different resources involved (single item versus combinatorial), number of items considered (single-unit versus multi-unit) [1, 6]. Moreover, the strategic decision made in each step depends on the kind of resource being auctioned: static or dynamic (consumable, perishable), divisible or indivisible, controlled or uncontrolled (e.g. public goods) [6].

All of the agents make decisions in order to maximize their utility regarding the selling (auctioneer) or buying (bidders) of the resources, u(R),

- Auctioneer: $u(R) = p - V(R)$
- Bidder: $u(R) = V(R) - p$

where p is the payment made for the resources and $V(R)$ is a valuation function that measures the value of R for the agent. When a single resource is being sold, characterized by a single attribute as, for example, the price, the WDP is simple: take the maximum value. However, when there is a set of resources to be allocated or the number of attributes that characterizes the resources has some dimensionality, then the WDP requires a more complex $V(\cdot)$ function.

Mechanism design [3] is the study concerning on the definition of auction components, as $V(\cdot)$. Other issues include social welfare measures, and dealing with cheaters, among others. Social welfare assesses the quality of the allocation in a global perspective. In that regard, a social welfare measure aggregates either the utility, benefits, satisfaction, or other gratifications of the agents. In recurrent scenarios, in which auctions are repeated over time, an auctioneer could learn trust models regarding the cheating behaviour of agents that later on conditions the $V(\cdot)$ function too. In this talk, we analyse the use of aggregation functions in all of these issues [2, 8, 12].

First, from a design perspective, the set of requirements that the aggregation function should fulfil to be $V(\cdot)$ will be reviewed, so that the properties desired for the mechanism are guaranteed [8]. There are some frameworks that could help, as for example constraints, in order to verify the properties [11]. Two cases will be studied in detail: the application of aggregation functions on multi-attribute auctions [7, 10] and on combinatorial auctions [14] will be described. Second, the role of the aggregation function in social welfare measures will be presented [9]. Third, an example how the parameters of aggregation functions can be tuned thanks to trust methods will be provided [13].

Cases study will be provided in several application domains, including different types of resources: workflow resource allocation with energy constraints [14], wastewater management [4, 5] and e-services [6].

Acknowledgement. This work was supported by the University of Girona (grant number MPCUdG2016) and the Spanish MINECO (grant number DPI2013-47450-C21-R).

Work developed with the support of the research group SITES awarded with distinction by the Generalitat de Catalunya (SGR 2014-2016).

References

1. Chevaleyre, Y., Dunne, P.E., Endriss, U., Lang, J., Lemaître, M., Maudet, N., Padget, J., Phelps, S., Rodriguez-Aguilar, J.A., Sousa, P.: Issues in multiagent resource allocation. Informatica **30**(1) (2006)
2. Grabisch, M., Marichal, J.-L., Mesiar, R., Pap, E., Aggregation functions: means. Inf. Sci. **181**(1) (2011)
3. Leyton-Brown, K., Shoham, Y.: Mechanism design and auctions. In: Weiss, G. (ed.) Multiagent Systems, Chap. 7, 2nd edn., pp. 285–32. The MIT Press (2013)
4. Murillo, J., Busquets, D., Dalmau, J., López, B., Muñoz, V., Rodríguez-Roda, I.: Improving urban wastewater management through an auction-based management of discharges. Environ. Model. Softw. **26**(6), 689–696 (2011)
5. Murillo, J., Muñoz, V., Busquets, D., López, B.: Schedule coordination through egalitarian recurrent multi-unit combinatorial auctions. Appl. Intell. **34**(1), 47–63 (2011)
6. Murillo Espinar, J.: Egalitarian behaviour in multi unit combinatorial auctions, Ph.D. thesis, University of Girona (2010)
7. Pla, A.: Multi-attribute auctions: application to workflow management systems, Ph.D. thesis, University of Girona (2014)
8. Pla, A., Lopez, B., Murillo, J.: Multi criteria operators for multi-attribute auctions. In: Torra, V., Narukawa, Y., López, B., Villaret M. (eds,) MDAI 2012. LNCS, vol 7647. Springer, Heidelberg (2012)
9. Pla, A., López, B., Murillo, J.: Multi-dimensional fairness for auction-based resource allocation. Knowl.-Based Syst. **73**, 134–148 (2015)

10. Pla, A., López, B., Murillo, J., Maudet, N.: Multi-attribute auctions with different types of attributes: enacting properties in multi-attribute auctions. Expert Syst. Appl. **41**(10), 4829–4843 (2014)
11. Pla, B., López, A. Murillo, J.: How to demonstrate incentive compatibility in multi-attribute auctions. In: Gibert, K., Botti, V., Reig-Bolaño, R. (eds.) Artificial Intelligence Research and Development (CCIA). Frontiers in Artificial Intelligence and Applications, vol. 256, pp. 303–306 (2013)
12. Torra, V., Narukawa, Y.: Modeling Decisions: Information Fusion and Aggregation Operators. Springer, Heidelberg (2007)
13. Torrent-Fontbona, F., Pla, A., López, B.: New perspective of trust through multi-attribute auctions. In: Papers from the AAAI Workshop in Incentive, Trust in E-Communities, pp. 25–31 (2015)
14. Torrent-Fontbona, F., Pla, B., López, A.: Using multi-attribute combinatorial auctions for resource allocation. In: Müller, J.P., Weyrich, M., Bazzan, A.L.C. (eds.) MATES 2014. LNCS, vol. 8732, pp. 57–71. Springer, Heidelberg (2014)

Aggregation Operators in Information Retrieval

Gabriella Pasi

Information Retrieval Laboratory,
Università degli Studi di Milano-Bicocca, Milan, Italy

In the context of Information Retrieval, the issue of employing aggregation operators in various phases of the retrieval process has been extensively investigated in the literature. In particular, these approaches rely on the interpretation of Information Retrieval as a Multi-Criteria Decision Making (MCDM) problem, from various perspectives. The first, more straightforward perspective, is to interpret the overall IR process as a MCDM process aimed at selecting the best alternatives (documents) based on the assessment of the performance of multiple criteria (the keywords specified in a user's query). Another and strongly related perspective is to see the assessment of the overall relevance estimate of a document (still an alternative) to a query as the process of evaluating the performance of several relevance dimensions (e.g. topicality, novelty, recency), which in this case represent the criteria to be aggregated. Another process that may require the application of appropriate aggregation operators is the indexing process, when applied to structured documents. Metasearch constitutes another interesting task that can be seen as an instance of a Multi-Expert Decision Making (MEDM) problem, also strongly relying on the appropriate choice of an aggregation operator. By this task, a user query is separately evaluated by different search engines, each one providing its own relevance assessment of the considered documents. Metasearch aims to merge the ranked lists generated by the various search engines (experts) in response to a query, to the aim of providing a unique, consensual ranked list of results. A quite interesting aspect implied by the above interpretations of various phases of the IR process is that the choice of different aggregation operators can produce different results. In other words, the semantics of aggregation implies an interpretation of the affected process. For example, if considering the aggregation of different relevance assessments for a same query and the same documents, distinct rankings can be obtained by applying distinct aggregation strategies. Despite the potential impact of aggregation on the whole IR process, this aspect has not received the proper attention in the literature. Only recently, some approaches have appeared demonstrating the importance of this issue and its potential impact on the searching process. This lecture aims to shortly review the main contributions that in the literature have made use of aggregation operators in Information Retrieval.

Geometric Analysis on Cantor Sets and Trees

Jana Björn

Department of Mathematics, Linköping University, Linköping, Sweden

This is a joint work with A. Björn, J.T. Gill and N. Shanmugalingam [1]. We consider an infinite network represented by a weighted rooted tree which we equip with a metric and measure structure enabling first-order Sobolev spaces and harmonic and p-harmonic functions. This is a special case of a procedure called uniformization, see Bonk, Heinonen and Koskela [2]. The visual boundary of the tree at infinity is an ultrametric space and can be regarded as a Cantor type set, see Semmes [4, 5].

In this setting, we show that the trace of the Sobolev space is exactly a Besov space with an explicit smoothness exponent, cf. Bourdon and Pajot [3]. This, in particular, means that such Besov boundary data have harmonic extensions to the whole tree and it is possible to solve the Dirichlet and obstacle problems with such boundary data. These harmonic extensions can be seen as potentials or stationary flows in the network.

We also consider mappings between pairs of such trees and between their boundaries. It turns out that quasi-symmetries between two Cantor sets exactly extend to rough quasi-isometries between their generating trees, and vice versa.

References

1. Björn, A., Björn, J., Gill, J.T., Shanmugalingam, N.: Geometric analysis on Cantor sets and trees. J. Reine Angew. Math. (to apper). doi:10.1515/crelle-2014-0099, arXiv:1304.0566
2. Bonk, M., Heinonen, J., Koskela, P.: Uniformizing Gromov hyperbolic spaces, Astérisque **270** (2001)
3. Bourdon, M., Pajot, H.: Cohomologie l_p et espaces de Besov. J. Reine Angew. Math. **558**, 85–108 (2003)
4. Semmes, S.: An introduction to the geometry of ultrametric spaces (2007, preprint). arXiv:0711.0709
5. Semmes, S.: Cellular structures, quasisymmetric mappings, and spaces of homogeneous type (2007, preprint). arXiv:0711.1333

A Monometric-Based Approach to Data Aggregation

Bernard De Baets and Raúl Pérez-Fernández

KERMIT, Department of Mathematical Modelling,
Statistics and Bioinformatics, Ghent University,
Coupure links 653, 9000 Gent, Belgium
{bernard.debaets,raul.perezfernandez}@ugent.be

Abstract. Data aggregation is a common problem in many fields of application and is historically understood as a process of combining several real values into a single one. However, the aggregation of other types of structured data is lately receiving increasing attention [4]. Some examples are the aggregation of multidimensional data [5], the aggregation of rankings [6] and the aggregation of mappings [3].

As Yager described in his "general theory of information aggregation" [11], a natural approach to data aggregation is based on the search for the element minimizing some notion of "penalty". In the framework of real numbers, this penalty is usually provided with a well-founded semantic basis, for instance, with the requirement of the property of quasi-convexity in the second argument [10]. Some other examples can be found in [2, 12] or in the recent survey on the definition of penalty functions in data aggregation [1]. Unfortunately, outside the framework of real numbers, this well-founded semantic basis is usually disregarded.

This penalty-based approach to data aggregation is similar to that considered in social choice theory for the aggregation of rankings. The monometric rationalization of ranking rules [8] is the branch of social choice theory, where the process of aggregating several rankings is characterized as the minimization of the distance to a consensus state for some appropriate monometric. Formally, a monometric is a function satisfying the axioms of nonnegativity and coincidence of a distance function, while requiring the compatibility with a given betweenness relation [7, 9]. This monometric can be understood as a natural extension of a penalty function outside the framework of real values, where the well-founded semantic basis is provided by the compatibility with the chosen betweenness relation. In this contribution, monometrics and betweenness relations will be considered for different types of structured data (multidimensional data, maps, strings, compositional data, among others), leading to a natural expansion of the definition of a penalty function beyond its current confinement to real values.

References

1. Bustince, H., Beliakov, G., Dimuro, G.P., Bedregal, B., Mesiar, R.: On the definition of penalty functions in data aggregation. Fuzzy Sets, Syst. (accepted). http://dx.doi.org/10.1016/j.fss.2016.09.011
2. Calvo, T., Mesiar, R., Yager, R.R.: Quantitative weights and aggregation. IEEE Trans. Fuzzy Syst. **12**, 62–69 (2004)
3. De Miguel, L., Campóon, M.J., Candeal, J.C., Induráin, E., Paternain, D.: Pointwise aggregation of maps: Its structural functional equation and some applications to social choice theory. Fuzzy Sets Syst. (accepted). http://dx.doi.org/10.1016/j.fss.2016.05.010
4. Gagolewski, M.: Data Fusion. Theory, Methods and Applications, Institute of Computer Science. Polish Academy of Sciences, Warsaw (2015)
5. Gagolewski, M.: Penalty-based aggregation of multidimensional data. Fuzzy Sets, Syst. (accepted). http://dx.doi.org/10.1016/j.fss.2016.12.009
6. Kemeny, J.G.: Mathematics without numbers. Daedalus **88**(4), 577–591 (1959)
7. Pasch, M.: Vorlesungen über neuere Geometrie, vol. 23. Teubner, Leipzig, Berlin (1882)
8. Pérez-Fernández, R., Rademaker, M., Baets, B.: Monometrics and their role in the rationalisation of ranking rules. Inf. Fusion **34**, 16–27 (2017)
9. Pitcher, E., Smiley, M.F.: Transitivities of betweenness. Trans. Am. Math. Soc. **52**(1), 95–114 (1942)
10. Wilkin, T., Beliakov, G.: Weakly monotonic averaging functions. Int. J. Intell. Syst. **30**(2), 144–169 (2015)
11. Yager, R.R.: Toward a general theory of information aggregation. Inf. Sci. **68**, 191–206 (1993)
12. Yager, R.R., Rybalov, A.: Understanding the median as a fusion operator. Int. J. Gen. Syst. **26**, 239–263 (1997)

The Fusion of Uncertain Information: Principles and Examples of Merging Rules Across Uncertainty Theories

Didier Dubois

IRIT, Université de Toulouse, Toulouse, France
dubois@irit.fr

Abstract. We present basic principles for the fusion of incomplete or uncertain information items that should apply regardless of the formalism adopted for representing pieces of information coming from several sources. This formalism can be based on sets, logic, partial orders, possibility theory, belief functions or imprecise probabilities. The presented tutorial is based on past work performed especially with Henri Prade, Weiru Liu and Jianbing Ma, Ronald Yager.

Outline of the Presentation

Information fusion deals with extracting accurate knowledge from possibly conflicting pieces of information stemming from a set of sources, without introducing arbitrary precision [5]. It differs from belief revision [11] and preference aggregation [7]. Information fusion is useful in many areas ranging from databases [6] to image processing [4] and expert opinion aggregation [8]. The main reference for this presentation is the paper [12]. We propose a general notion of information item representing incomplete or uncertain information about the value of an entity of interest. Any kind of uncertain information is supposed to rank possible values in terms of relative plausibility, and explicitly point out impossible ones. Important issues affecting the results of the fusion process, such as the comparison of information items by their relative information content, the consistency of information items, as well as their mutual consistency, are discussed. For each representation setting, we write a version of the fusion postulates, present known fusion rules that obey them, and compare our postulates to existing ones proposed in the past and specific to the representation setting. In the crudest (Boolean) setting (where an information item is just defined as a set of possible values), we show that the understanding of a set in terms of most plausible values, or in terms of non-impossible ones matters for choosing a relevant fusion rule. In particular, in the latter case, our principles justify the old method of maximal consistent subsets [23], while the former is related to the fusion of logical bases [16, 17] that merges sets of preferred values. Then, we consider several formal settings for incomplete or uncertain information items, where our postulates are also instantiated: plausibility orderings [19], qualitative and quantitative possibility distributions [10, 14] and possibilistic knowledge bases [3, 18], the merging of probability distributions [8, 15, 20, 26], of belief functions [9, 13, 24, 25, 28] and of convex sets of probabilities [21, 27]. The aim of this work is to provide a unified picture of fusion rules across such various uncertainty representation settings. Finally, we discuss the connection with the Belnap approach [2] to inference under source-based inconsistent information, and discuss the possibility of non-destructive fusion methods that preserve the original information provided by the sources [1].

References

1. Assaghir, Z., Napoli, A., Kaytoue, M., Dubois, D., Numerical, P.H., Fusion, I.: Lattice of answers with supporting arguments. In: IEEE International Conference on Tools with Artificial Intelligence (ICTAI 2011), Boca Raton, pp. 621–628 (2011)
2. Belnap, N.D.: A useful four-valued logic. In: Dunn, J.M., Epstein, G. (eds.) Modern Uses of Multiple-Valued Logic, pp. 8–37. D. Reidel Publishing Company (1977)
3. Benferhat, S., Dubois, D., Kaci, S., Prade, H.: Possibilistic merging and distance-based fusion of propositional information. Ann. Math. Artif. Intell. 34(1–3), 217–252 (2002)
4. loch, I.: Fusion of image information under imprecision, uncertainty: numerical methods. In: Della Riccia, G., Lenz, H.-J., Kruse, R. (eds.) Data Fusion and Perception, CISM Courses and Lectures, vol. 431, pp. 135–168 (2001)
5. Bloch, I., Hunter, A. (eds.): Fusion: general concepts and characteristics. Int. J. Intell. Syst. 16(10–11), 1107–1136 (2001). Special Issue on Data and Knowledge Fusion
6. Cholvy, L., Moral, S.: Merging databases: problems and examples. Int. J. Intell. Syst. 16(10), 1193–1221 (2001)
7. Chopra, S., Ghose, A.K., Meyer, T.A.: Social choice theory, belief merging, and strategy-proofness. Inf. Fusion 7(1), 61–79 (2006)
8. Cooke, R.: Experts in Uncertainty. Oxford University Press, Oxford (1991)
9. Destercke, S., Dubois, D.: Idempotent conjunctive combination of belief functions: extending the minimum rule of possibility theory. Inf. Sci. 181(18), 3925–3945 (2011)
10. Destercke, S., Dubois, D., Chojnacki, E.: Possibilistic information fusion using maximal coherent subsets. IEEE Trans. Fuzzy Syst. 17(1), 79–92 (2009)
11. Dubois D.: Information fusion and revision in qualitative and quantitative settings. In: Liu, W. (eds.) ECSQARU 2011. LNCS, vol. 6717. Springer, Heidelberg (2011). doi:10.1007/978-3-642-22152-1_1
12. Dubois, D., Liu, W., Ma, J., Prade, H.: The basic principles of uncertain information fusion. An organised review of merging rules in different representation frameworks. Inf. Fusion 32, 12–39 (2016)
13. Dubois, D., Prade, H.: Representation and combination of uncertainty with belief functions and possibility measures. Comput. Intell. 4, 244–264 (1988)
14. Dubois, D., Prade, H., Yager, R.R. Merging fuzzy information. In: Bezdek, J.C., Dubois, D., Prade, H. (eds.) Fuzzy Sets in Approximate Reasoning and Information Systems. The Handbooks of Fuzzy Sets Series, pp. 335–401. Kluwer Academic Publishers, Dordrecht (1999)
15. Genest, C., Zidek, J.V.: Combining probability distributions: a critique and an annotated bibliography. Stat. Sci. 1(1), 114–135 (1986)

16. Konieczny, S., Pino-Pérez, R.: Logic based merging. J. Philos. Logic **40**(2), 239–270 (2011)
17. Liberatore, P., Schaerf, M.: Arbitration (or how to merge knowledge bases). IEEE Trans. Knowl. Data Eng. **10**(1), 76–90 (1998)
18. Liu, W., Qi, G., Bell, D.A.: Adaptive merging of prioritized knowledge bases. Fundamenta Informaticae **73**(3), 389–407 (2006)
19. Maynard-Reid-II, P., Shoham, Y.: Belief fusion: aggregating pedigreed belief states. J. Logic, Lang. Inf. **10**, 183–209 (2001)
20. McConway, K.J.: Marginalization and linear opinion pools. J. Am. Stat. Assoc. **76**(374), 410–414 (June 1981)
21. Moral, S., Sagrado, J.: Aggregation of Imprecise Probabilities. Aggregation and Fusion of Imperfect Information, pp. 162–188. Physica-Verlag, Heidelberg (1997)
22. Oussalah, M.: Study of some algebrical properties of adaptive combination rules. Fuzzy Sets Syst. **114**(3), 391–409 (2000)
23. Rescher, N., Manor, R.: On inference from inconsistent premises. Theor. Decis. **1**, 179–219 (1970)
24. Shafer, G.: A Mathematical Theory of Evidence. Princeton University Press (1976)
25. Smets, P.: Analyzing the combination of conflicting belief functions. Inf. Fusion **8**(4), 387–412 (2007)
26. Wagner, C., Lehrer, K.: Rational Consensus in Science and Society. D. Reidel, (1981)
27. Walley, P. The elicitation and aggregation of beliefs. Technical report, University of Warwick (1982)
28. Yager, R.: On the Dempster-Shafer framework and new combination rules. Inf. Sci. **41**, 93–138 (1987)

Aggregation of Multidimensional Data: A Review

Marek Gagolewski

Systems Research Institute, Polish Academy of Sciences, ul. Newelska 6,
01-447 Warsaw, Poland
Faculty of Mathematics and Information Science,
Warsaw University of Technology, ul. Koszykowa 75,
00-662 Warsaw, Poland
gagolews@ibspan.waw.pl

Abstract. Aggregation theory classically deals with functions to summarize a sequence of numeric values, e.g. in the unit interval, see [6, 7]. Since the notion of componentwise monotonicity plays a key role in many situations, there is an increasingly growing interest in methods that act on diverse ordered structures.

However, as far as the definition of a mean or an averaging function is concerned, see, e.g., [1, 2], the internality (or at least idempotence) property seems to be of a relatively higher importance than the monotonicity condition. In particular, the Bajraktarević means or the mode are among some well-known non-monotone means.

The concept of a penalty-based function was first investigated by Yager in [8] and then extended in numerous works, see, e.g., for a recent summary and a critical overview [3]. In such a framework, we are interested in minimizing the amount of "disagreement" between the inputs and the output being computed; the corresponding aggregation functions are at least idempotent and express many existing means in an intuitive and attractive way.

In this talk, I focus on the notion of penalty-based aggregation of sequences of points in $\mathbb{R}$, this time for some $d \geq 1$, see [4, 5]. I review three noteworthy subclasses of penalty functions: componentwise extensions of unidimensional ones, those constructed upon pairwise distances between observations, and those defined by measuring the so-called data depth. Then, I discuss their formal properties, which are particularly useful from the perspective of data analysis, e.g. different possible generalizations of internality or equivariances to various geometric transforms. I also point out the difficulties with extending some notions that are key in classical aggregation theory, like the monotonicity property.

References

1. Beliakov, G., Bustince, H., Calvo, T.: A Practical Guide to Averaging Functions. Springer, Heidelberg (2016)
2. Bullen, P.: Handbook of Means and Their Inequalities. Springer Science+Business Media, Dordrecht (2003)
3. Bustince, H., Beliakov, G., Dimuro, G.P., Bedregal, B., Mesiar, R.: On the definition of penalty functions in data aggregation. Fuzzy Sets Syst. (2016, in press). doi:10.1016/j.fss.2016.09.011
4. Gagolewski, M.: Data Fusion: Theory, Methods, and Applications. Institute of Computer Science, Polish Academy of Sciences, Warsaw, Poland (2015)
5. Gagolewski, M.: Penalty-based aggregation of multidimensional data. Fuzzy Sets Syst. (2016, in press). doi:10.1016/j.fss.2016.12.009
6. Grabisch, M., Marichal, J.L., Mesiar, R., Pap, E.: Aggregation Functions. Cambridge University Press (2009)
7. Torra, V., Narukawa, Y., Modeling Decisions: Information Fusion and Aggregation Operators. Springer-Verlag, Heidelberg (2007)
8. Yager, R.R.: Toward a general theory of information aggregation. Inf. Sci. **68**(3), 191–206 (1993)

Contents

Capacities, Survival Functions and Universal Integrals

Radko Mesiar and Andrea Stupňanová$^{(\boxtimes)}$

Faculty of Civil Engineering, Slovak University of Technology in Bratislava,
Radlinského 11, 810 05 Bratislava, Slovakia
{radko.mesiar,andrea.stupnanova}@stuba.sk

Abstract. Based on the equality of survival functions related to n-ary vectors and capacities on $X = \{1, \dots, n\}$, the equality of universal integrals $\mathbf{I}(\mu, \mathbf{x}) = \mathbf{I}(\mu, \mathbf{y})$ is discussed and studied. Some particular cases are highlighted, and a special stress is put on possibility and necessity measures. As a by-product, a new characterization of possibility (necessity) measures is introduced.

1 Introduction

In the classical probability theory, two random variables cannot be distinguished by means of their parameters (such as the expected value, variance, quantiles, etc.) whenever they coincide in distribution functions. Equivalently, the coincidence of survival functions ensures the coincidence in all parameters for the considered random variables. In non-additive measure and integral theory, in many cases we deal with the survival functions and the related integrals are then computed based on the resulting survival functions. Recall that for a given measure μ and variable f, the survival function $h_{\mu,f}$ expresses a fusion of information contained in μ and f in the form

$$h_{\mu,f}(t) = \mu(f \geq t).$$

By means of the survival function one can derive the Choquet integral [2] (as the Riemann integral from $h_{\mu,f}$), the Sugeno integral [10], etc. As a common framework for all such survival function-based integrals, the concept of universal integrals was proposed in [7]. Though the literature contains a lot of results concerning measures and integrals, see, e.g., handbook [8], monograph [11] or edited volume [6], a deeper study of coincidences of survival functions (and, consequently, of universal integrals) is missing. We did a first step in this direction in [1], where we have focused on possibility and necessity measures and related survival functions. The aim of this contribution is a deeper study of survival functions in connection with some particular properties of the considered measures.

The paper is organized as follows. In the next section, we recall capacities, survival functions and universal integrals, considering the finite universe $X = \{1, \dots, n\}$ and the domain $[0, 1]$ for measures and functions, which become

V. Torra et al. (eds.), *Aggregation Functions in Theory and in Practice*,
Advances in Intelligent Systems and Computing 581, DOI 10.1007/978-3-319-59306-7_1

n-dimensional vectors in our case. In Sect. 3, we discuss survival functions and their link to capacities with some particular properties. In Sect. 4, we bring several results connected to possibility and necessity measures. Finally, some concluding remarks are added.

2   Preliminaries

Throughout this paper, we fix $X = \{1, \ldots, n\}$ for some $n \geq 2$, and we consider the corresponding power set 2^X. Then any measurable function $f : X \to [0, 1]$ can be represented as an n-dimensional vector $\mathbf{x} = (x_1, \ldots, x_n) = (f(1), \ldots, f(n)) \in [0, 1]^n$.

A capacity $\mu : 2^X \to [0, 1]$ is a monotone set function satisfying two boundary conditions $\mu(\emptyset) = 0$ and $\mu(X) = 1$. We denote as $\mathscr{M}_n$ the set of all capacities on X.

Observe that an additive capacity μ, i.e., $\mu(A) + \mu(B) = \mu(A \cup B)$ for any disjoint subsets A and B of X, is a discrete probability measure, and then

$$\mu(A) = \sum_{i \in A} p_i, \qquad \text{where} \qquad p_i = \mu(\{i\}), \ i \in X.$$

If a capacity μ is maxitive, i.e., $\mu(A) \vee \mu(B) = \mu(A \cup B)$ for any subsets A, B of X, then μ is a discrete possibility measure, and then

$$\mu(A) = \bigvee_{i \in A} \pi_i, \qquad \text{where} \qquad \pi_i = \mu(\{i\}).$$

A dual (conjugate) $\mu^d : 2^X \to [0, 1]$ related to a capacity μ by $\mu^d(A) = 1 - \mu(X \setminus A)$ is also a capacity. Note that if μ is additive, then $\mu = \mu^d$.

If μ is a possibility measure, then its dual μ^d is called a necessity measure and it is given by

$$\mu^d = 1 - \bigvee_{i \notin A} \pi_i.$$

If a capacity μ depends on the cardinality of the considered set A only, then μ is called a symmetric capacity, and then there are constants

$$v_0 = 0 \leq v_1 \leq \cdots \leq v_n = 1 \qquad \text{so that} \qquad \mu(A) = v_{\text{card}(A)}.$$

For more details about capacities (sometimes called also as fuzzy measures) see [11] or [5].

Among several integrals introduced for capacities from $\mathscr{M}_n$ we recall only 3 of them:

the Choquet integral [2,5,11] $\mathbf{Ch}(\mu, \cdot) : [0, 1]^n \to [0, 1]$,

$$\mathbf{Ch}(\mu, \mathbf{x}) = \int_0^1 \mu(\{i \in X \mid x_i \geq t\}) \, dt = \sum_{i=1}^n x_{(i)} \left(\mu(A_{(i)}) - \mu(A_{(i+1)}) \right),$$

where $(\cdot) : X \to X$ is a permutation such that $x_{(1)} \leq \cdots \leq x_{(n)}$, $A_{(n+1)} = \emptyset$ and $A_{(i)} = \{(i), \ldots, (n)\}$ for $i = 1, \ldots, n$;
the Sugeno integral [5,10,11] $\mathbf{Su}(\mu, \cdot) : [0,1]^n \to [0,1]$,

$$\mathbf{Su}(\mu, \mathbf{x}) = \bigvee_{0 \leq t \leq 1} (\mu(\{i \in X \,|\, x_i \geq t\}) \wedge t) = \bigvee_{i=1}^n \left(x_{(i)} \wedge \mu(A_{(i)}) \right);$$

the Shilkret integral [5,9,11] $\mathbf{Sh}(\mu, \cdot) : [0,1]^n \to [0,1]$,

$$\mathbf{Sh}(\mu, \mathbf{x}) = \bigvee_{0 \leq t \leq 1} (\mu(\{i \in X \,|\, x_i \geq t\}) \cdot t) = \bigvee_{i=1}^n \left(x_{(i)} \cdot \mu(A_{(i)}) \right).$$

Recently, Klement et al. [7] have introduced the concept of universal integrals as a common framework for all 3 above mentioned integrals, covering also many other integrals known from the literature. This concept, when applied to the unit interval as a domain for measures and functions, is related to the notion of a semicopula $\otimes : [0,1]^2 \to [0,1]$, i.e., $\otimes$ is a monotone operation with neutral element 1. For more details on semicopulas see [4,5].

Definition 1. A mapping $\mathbf{I} : \mathcal{M}_n \times [0,1]^n \to [0,1]$ is called a universal integral (on $[0,1]^n$) whenever two next axioms are satisfied:

(I1) there is a semicopula $\otimes : [0,1]^2 \to [0,1]$ such that

$$\mathbf{I}(\mu, c \cdot 1_A) = c \otimes \mu(A) \text{ for any } \mu \in \mathcal{M}_n, \ c \in [0,1] \text{ and } A \subseteq X;$$

(I2) for any $(\mu, \mathbf{x}), (\eta, \mathbf{y}) \in \mathcal{M}_n \times [0,1]^n$ such that

$$h_{\mu,\mathbf{x}}(t) = \mu(\{i \in X \,|\, x_i \geq t\}) \geq h_{\eta,\mathbf{y}}(t) = \eta(\{i \in X \,|\, y_i \geq t\})$$

it holds $\mathbf{I}(\mu, \mathbf{x}) \leq \mathbf{I}(\eta, \mathbf{y})$.

Note that axiom (I2) ensures the non-decreasing monotonicity of the universal integral $\mathbf{I}$ in both components, and the coincidence of integrals $\mathbf{I}(\mu, \mathbf{x}) = \mathbf{I}(\eta, \mathbf{y})$ whenever the survival functions $h_{\mu,\mathbf{x}}$ and $h_{\eta,\mathbf{y}}$ coincide. Obviously, all 3 above mentioned integrals belong to the class of universal integrals. The Choquet and the Shilkret integrals are related to the product semicopula ".", while the Sugeno integral is related to the copula $\wedge$ (=min, the greatest semicopula).

3 Capacities and Coincidence of Survival Functions

As already mentioned, the coincidence of survival functions ensures the coincidence of all related universal integrals. This claim holds also from the opposite direction, i.e., $\mathbf{I}(\mu, \mathbf{x}) = \mathbf{I}(\eta, \mathbf{y})$ for any universal integral $\mathbf{I}$ only if $h_{\mu,\mathbf{x}} = h_{\eta,\mathbf{y}}$.

For any fixed $\mu \in \mathcal{M}_n$, we introduce an equivalence relation $\sim_\mu$ on $[0,1]^n$ given by

$$\mathbf{x} \sim_\mu \mathbf{y} \qquad \text{if and only if} \qquad h_{\mu,\mathbf{x}} = h_{\mu,\mathbf{y}},$$

and we denote by $\mathscr{H}(\mu, \mathbf{x})$ the μ-equivalence class containing the n-tuple $\mathbf{x} \in [0,1]^n$, i.e.,

$$\mathscr{H}(\mu, \mathbf{x}) = \{\mathbf{y} \in [0,1]^n | \, h_{\mu,\mathbf{y}} = h_{\mu,\mathbf{x}}\}.$$

In some particular cases, the knowledge of equivalence classes $\mathscr{H}(\mu, \mathbf{x})$, $\mathbf{x} \in [0,1]^n$, determines the capacity μ. This is, e.g., the case when

$$\mathscr{H}(\mu, \mathbf{x}) = \{\mathbf{y} \in [0,1]^n | \, y_i = x_i\} \text{ for some fixed } i \in \{1, \dots, n\},$$

and then $\mu = \delta_{\{i\}}$ is the Dirac measure, $\delta_{\{i\}}(A) = \begin{cases} 1 & \text{if } i \in A \\ 0 & \text{otherwise} \end{cases}$.

When considering dual capacities, it holds

$$\mathscr{H}(\mu^d, \mathbf{x}) = 1 - \mathscr{H}(\mu, 1 - \mathbf{x}) = \{\mathbf{y} \in [0,1]^n, h_{\mu,1-\mathbf{y}} = h_{\mu,1-\mathbf{x}}\},$$

where $h_{\mu,1-\mathbf{x}}(t) = \mu(\{i \, | \, 1 - x_i \geq t\})$.

We have also next general results.

Proposition 1. *Let* $\mu, \eta \in \mathscr{M}_n$ *satisfy* $\mathscr{H}(\mu, \mathbf{x}) = \mathscr{H}(\eta, \mathbf{x})$ *for all* $\mathbf{x} \in [0,1]^n$, *i.e.,* $\sim_\mu = \sim_\eta$. *Then there is a strictly monotone function* $\varphi : [0,1] \to [0,1]$, $\varphi(0) = 0$ *and* $\varphi(1) = 1$, *such that* $\eta = \varphi \circ \mu$.

Proposition 2. *Let* $\mu \in \mathscr{M}_n$ *satisfy* $\mathscr{H}(\mu, \mathbf{x}) = \{\mathbf{x}\}$ *for all* $\mathbf{x} \in [0,1]^n$. *Then* $\mathrm{card}(\mathrm{range}\,(\mu)) = 2^n$, *i.e.,* $\mu(A) \neq \mu(B)$ *whenever* $A \neq B$.

Proposition 3. *Let* $\mu \in \mathscr{M}_n$ *satisfy* $\mathscr{H}(\mu, \mathbf{x}) = \{\mathbf{x}_\sigma | \sigma \in \mathscr{P}_n\}$ *for all* $\mathbf{x} \in [0,1]^n$, *where* $\mathscr{P}_n$ *is the set of all permutations on* X, *and* $\mathbf{x}_\sigma = (x_{\sigma(1)}, \dots, x_{\sigma(n)})$. *Then* μ *is a symmetric capacity given by* $\mu(A) = v_{\mathrm{card}(A)}$, *where* $0 = v_0 < v_1 < \cdots < v_n = 1$.

Observe that if we replace the strict inequality in Proposition 3 by non-strict inequalities, i.e., $0 = v_0 \leq v_1 \leq \cdots \leq v_n = 1$, then we can only claim

$$\mathscr{H}(\mu, \mathbf{x}) \supseteq \{\mathbf{x}_\sigma | \sigma \in \mathscr{P}_n\}.$$

Similarly, if we consider a monotone function $\varphi : [0,1] \to [0,1]$, $\varphi(0) = 0$ and $\varphi(1) = 1$, which is not strictly monotone, compare Proposition 1, then we can conclude only $\mathscr{H}(\varphi \circ \mu | \mathbf{x}) \supseteq \mathscr{H}(\mu, \mathbf{x})$.

4 Survival Functions and Possibility and Necessity Measures

When considering a possibility (necessity) measure μ related to a possibility distribution π, we have several interesting results.

For a better clarification of survival function $h_{\mu,\mathbf{x}}$, observe first that $h_{\mu,\mathbf{x}}(t) = 1 = u_1$ for any $t \in [0, t_1]$, $t_1 = \bigvee_{\pi(i)=1} x_i$. Recall that $u_1 = \mu(X)$.

For $u_2 = \mu(X \setminus \{i \in X | x_i \leq t_1\}) = \mu(\{i \, | \, x_i > t_1\}) = \bigvee_{x_i > t_1} \pi(i)$, we have

- either $u_2 = 0$ and then $h_{\mu,\mathbf{x}}(t) = 0$ for any $t \in]t_1, 1]$,
- or $u_2 > 0$ and then $h_{\mu,\mathbf{x}}(t) = u_2$ for any $t \in]t_1, t_2]$, where $t_2 = \displaystyle\bigvee_{\pi(i)=u_2} x_i$.

By induction, we have

- $h_{\mu,\mathbf{x}}(t) = u_k > 0$ for any $t \in]t_{k-1}, t_k]$ for $k = 1, \ldots, r$, and
- $h_{\mu,\mathbf{x}}(t) = 0$ for any $t \in]t_r, 1]$,

where $t_0 = 0$, $u_k = \displaystyle\bigvee_{x_i > t_{k-1}} \pi(i)$, and

$$\diamond \text{ either } u_k = 0 \text{ and then } r = k - 1,$$

$$\diamond \text{ or } u_k > 0 \text{ and then } t_k = \bigvee_{\pi(i)=u_k} x_i.$$

Observe that if, by chance, an interval $]t, t]$ is considered, then the corresponding claim is always valid due to the fact that $]t, t] = \emptyset$.

Theorem 1. *Let $\mu \in \mathcal{M}_n$ be a possibility measure related to a possibility distribution π and let $\mathbf{x} \in [0, 1]^n$ be a score vector. Then $\mathbf{y} \in \mathcal{H}(\mu, \mathbf{x})$ if and only if*

$$\bigvee_{\pi(i)=u_j} y_i = t_j \text{ and } y_i \leq t_j \text{ whenever } \pi(i) > u_{j+1} \text{ for all } j = 1, \ldots, r.$$

The set $\mathcal{H}(\mu, \mathbf{x})$ is an upper semi-lattice with the top element $\mathbf{x}^\mu$ given by

$$x_i^\mu = \begin{cases} t_j & \text{whenever } u_j \geq \pi(i) > u_{j+1}, \quad j = 1, \ldots, r \\ 1 & \text{whenever } \pi(i) = 0 \end{cases},$$

for $i = 1, \ldots, n$.

For the strangest capacity $\mu^* \in \mathcal{M}_n$ (i.e., possibility measure with constant distribution function $\pi = 1$), it holds

$$\mathcal{H}(\mu^*, \mathbf{x}) = \left\{ \mathbf{x} \in [0, 1]^n \mid \bigvee_{i=1}^n y_i = \bigvee_{i=1}^n x_i \right\},$$

and thus, evidently, the maximal element of $\mathcal{H}(\mu^*, \mathbf{x})$ is given by

$$\mathbf{x}^{\mu^*} = \left(\bigvee_{i=1}^n x_i, \ldots, \bigvee_{i=1}^n x_i \right).$$

Note that the upper semi-lattice $\mathcal{H}(\mu^*, \mathbf{x})$ has for any $\mathbf{x} \neq \mathbf{0}$ exactly n minimal elements

$$\left(\bigvee_{i=1}^n x_i, 0, \ldots, 0 \right), \quad \ldots \quad , \left(0, \ldots, 0, \bigvee_{i=1}^n x_i \right).$$

Dubois and Rico [3] have introduced, for any $\mathbf{x} \in [0,1]^n$, a vector $\mathbf{x}^+ \in [0,1]^n$ related to a possibility distribution $\pi : X \to [0,1]$ by

$$x_i^+ = \bigvee_{\pi(j) \geq \pi(i)} x_j, \qquad i \in X$$

and they have proved that $\mathbf{Ch}(\mu, \mathbf{x}) = \mathbf{Ch}(\mu, \mathbf{x}^+)$, where $\mu \in \mathscr{M}_n$ is the possibility measure induced by π. Similarly, they have shown $\mathbf{Su}(\mu, \mathbf{x}) = \mathbf{Su}(\mu, \mathbf{x}^+)$.

It is not difficult to check that $\mathbf{x}^+ \in \mathscr{H}(\mu, \mathbf{x})$ and thus for any universal integral $\mathbf{I}$ it holds $\mathbf{I}(\mu, \mathbf{x}) = \mathbf{I}(\mu, \mathbf{x}^+)$. Note also that $\mathbf{x}^+$ is the top element of the class $\mathscr{H}(\mu, \mathbf{x})$ whenever $\pi(i) > 0$ for all $i \in X$.

For a permutation $\sigma : X \to X$ and $\mathbf{x} \in [0,1]^n$, we define a vector $\mathbf{x}^{\sigma,+}$ by

$$x^{\sigma,+}(i) = \bigvee_{j \leq \sigma^{-1}(i)} x_{\sigma(j)}, \qquad i \in X.$$

As an interesting characterization of possibility measures we have the next results. They are independent of any (universal) integral and they follow from the integral equivalence of $\mathbf{1}_A$ and $(\mathbf{1}_A)^{\sigma,+}$ (resp. $(\mathbf{1}_A)^{\sigma,-}$).

Theorem 2. *Let $\mu \in \mathscr{M}_n$ be a capacity. Then the following are equivalent:*

(i) μ is a possibility measure;
(ii) there is a permutation $\sigma : X \to X$ such that for any $A \subseteq X$,

$$(\mathbf{1}_A)^{\sigma,+} \in \mathscr{H}(\mu, \mathbf{1}_A),$$

where $\mathbf{1}_A$ is the characteristic function of A, $\mathbf{1}_A(i) = \begin{cases} 1 & \text{if } i \in A \\ 0 & \text{if } i \notin A \end{cases}$.

For the proof and some other details we recommend our paper [1].

Similar result can be formulated for necessity measures (this follows from the duality of possibility and necessity measures). We introduce only two of them.

Theorem 3. *Let $\mu \in \mathscr{M}_n$ be a necessity measure related to a possibility distribution $\pi : X \to [0,1]$. Then $\mathscr{H}(\mu, \mathbf{x})$ is a lower semilattice containing the vector $\mathbf{x}^- \in [0,1]^n$ given by*

$$x_i^- = \bigwedge_{\pi(j) \geq \pi(i)} x_j, \qquad i \in X.$$

If $\pi(i) > 0$ for all $i \in X$ then $\mathbf{x}^-$ is the bottom element of $\mathscr{H}(\mu, \mathbf{x})$.

Recall that the vector $\mathbf{x}^-$ was also introduced by Dubois and Rico [3] and that the above theorem ensuring $\mathbf{I}(\mu, \mathbf{x}) = \mathbf{I}(\mu, \mathbf{x}^-)$ for any universal integral $\mathbf{I}$ generalize the results of [3], where the equalities $\mathbf{Ch}(\mu, \mathbf{x}) = \mathbf{Ch}(\mu, \mathbf{x}^-)$ and

$\mathbf{Su}(\mu, \mathbf{x}) = \mathbf{Su}(\mu, \mathbf{x}^-)$ were shown. Similarly as the vector $\mathbf{x}^{\sigma,+}$, was introduced, we define a vector $\mathbf{x}^{\sigma,-}$ by

$$x_i^{\sigma,-} = \bigwedge_{j \leq \pi^{-1}(i)} x_{\sigma(j)}, \qquad i \in X.$$

Theorem 4. *Let $\mu \in \mathcal{M}_n$ be a capacity. Then the following are equivalent:*

(i) μ is a necessity measure;
(ii) there is a permutation $\sigma : X \to X$ such that for any $A \subseteq X$,

$$(\mathbf{1}_A)^{\sigma,-} \in \mathcal{H}(\mu, \mathbf{1}_A).$$

5 Concluding Remarks

Based on the equality of survival functions, we have introduced and studied equivalence classes $\mathcal{H}(\mu, \mathbf{x})$ of all score vectors $\mathbf{y} \in [0,1]^n$ satisfying $h_{\mu,\mathbf{y}} = h_{\mu,\mathbf{x}}$. Consequently, for each $\mathbf{y} \in \mathcal{H}(\mu, \mathbf{x})$ and for each universal integral $\mathbf{I}$, the equality $\mathbf{I}(\mu, \mathbf{y}) = \mathbf{I}(\mu, \mathbf{x})$ holds. We have introduced several original results dealing with equivalence classes $\mathcal{H}(\mu, \mathbf{x})$, with a particular stress on possibility and necessity measure. In that later case, recent characterizations of possibility and necessity measures of Dubois and Rico [3] considering the Choquet and the Sugeno integrals were generalized to cover all universal integrals.

Our approach allows to find new links between particular capacities and universal integrals. Consider, for example, an additive capacity $\mu \in \mathcal{M}_n$ such that $\mu(\{i\}) = \frac{2^{i-1}}{2^n-1}$. Then $\mu(A) \neq \mu(B)$ whenever $A \neq B$, and for any automorphism $\varphi : [0,1] \to [0,1]$, $\varphi \circ \mu \in \mathcal{M}_n$ is a capacity such that for any $\mathbf{x} \neq \mathbf{y}$ there is a universal integral such that $\mathbf{I}(\varphi \circ \mu, \mathbf{x}) \neq \mathbf{I}(\varphi \circ \mu, \mathbf{y})$.

As an interesting topic for the further study, we expect generalization of our results dealing with possibility/necessity measures to some particular subclasses of plausibility/belief functions. As another promising direction is the possibility to consider some other scale different from $[0,1]$, for example a finite ordinal (linguistic) scale.

Acknowledgements. The support of the grants APVV-14-0013, VEGA 1/0420/15 and VEGA 1/0682/16 is kindly announced.

References

1. Chen, T., Mesiar, R., Li, J., Stupňanová, A.: Possibility and necessity measures and integral equivalence. Int. J. Approx. Reason (in press). doi:10.1016/j.ijar.2017. 04.008
2. Choquet, G.: Theory of capacities. Ann. Inst. Fourier **5**, 131–295 (1953/54)
3. Dubois, D., Rico, A.: Axiomatisation of discrete fuzzy integrals with respect to possibility and necessity measures. In: Torra, V., et al. (eds.) MDAI 2016. LNAI, vol. 9880, pp. 94–106. Springer, Heidelberg (2016)

4. Durante, F., Sempi, C.: Semicopulæ. Kybernetika **41**(3), 315–328 (2005)
5. Grabisch, M., Marichal, J.L., Mesiar, R., Pap, E.: Aggregation Functions. Cambridge University Press, Cambridge (2009)
6. Grabisch, M., Murofushi, T., Sugeno, M. (eds.): Fuzzy Measures and Integrals. Studies in Fuzziness and Soft Computing, vol. 40. Physika-Verlag, Heidelberg (2000)
7. Klement, E.P., Mesiar, R., Pap, E.: A universal integral as common frame for Choquet and Sugeno integral. IEEE Trans. Fuzzy Syst. **18**, 178–187 (2010)
8. Pap, E. (ed.): Handbook of Measure Theory, vol. I, II. North-Holland, Amsterdam (2002)
9. Shilkret, N.: Maxitive measure and integration. Indag. Math. **33**, 109–116 (1971)
10. Sugeno, M.: Theory of fuzzy integrals and its applications. Ph.D. thesis, Tokyo Institute of Technology (1974)
11. Wang, Z., Klir, G.J.: Generalized Measure Theory. Springer, Heidelberg (2009)

Point-Interval-Valued Sets: Aggregation and Construction

Slavka Bodjanova[1] and Martin Kalina[2([⊠])]

[1] Department of Mathematics, Texas A&M University-Kingsville, MSC 172,
Kingsville, TX 78363, USA
kfsb000@tamuk.edu
[2] Department of Mathematics, Faculty of Civil Engineering,
Slovak University of Technology in Bratislava,
Radlinského 11, 810 05 Bratislava, Slovakia
kalina@math.sk

Abstract. The concept of a point-interval-valued set (PIV set) is proposed as a tool for summary characterization of data from a two-way table. A PIV set is an L-fuzzy set whose membership labels can be numbers as well as special subintervals from the unit interval. Two relations of partial order of PIV sets are introduced and corresponding operations of union and intersection are studied. Aggregation of PIV sets by bounded t-norms is suggested.

1 Introduction

In decision making, the choice of the best object (product, individual, method, etc.) with respect to a given set of criteria is often based on evaluation of the degree of compatibility of each object with respect to each criterion by a group of experts. Usually the degrees of compatibility are numbers from the unit interval, where 1 represents full compatibility, while 0 represents no compatibility. Then evaluations of objects from a set $X = \{x_1, \ldots, x_n\}$ by experts $E_1, \ldots, E_m$ can be described by a two-way table

$$\tau = \{\delta_{ij} : \delta_{ij} \in [0,1], i = 1, \ldots, n, j = 1, \ldots, m\}. \tag{1}$$

Numerous methods and aggregation functions for a summary characterization of τ were proposed, see e.g., [1–3,14] and references therein. In this contribution, a new summary characterization of τ is introduced as a mapping φ on X satisfying the following properties: if all experts agree that x_i has low compatibility with a criterion C, then $\varphi(x_i) \in [0, 0.5)$, if all experts agree that x_i has high compatibility with C, then $\varphi(x_i) \in (0.5, 1]$, and if some experts assume that there is a low compatibility between x_i and C while other think that the compatibility is high, then $\varphi(x_i) = [a, b]$, where $0 \le a \le 0.5 \le b \le 1, a \ne b$. A special consideration needs to be given to the evaluation $\delta_{ij} = 0.5$. In applications, 0.5 can be viewed as the largest small value or the smallest large value. This is the reason why 0.5 needs to be handled separately, as a middle value. Then the mapping φ

© Springer International Publishing AG 2018
V. Torra et al. (eds.), *Aggregation Functions in Theory and in Practice*,
Advances in Intelligent Systems and Computing 581, DOI 10.1007/978-3-319-59306-7_2

discussed above is a point-interval-valued set on X (PIV set) which is formally introduced in Definition 1.

Definition 1. Let X be a universe of discourse and $D = \{[a,b] : 0 \leq a \leq 0.5 \leq b \leq 1, a \neq b\}$. A point-interval-valued set (PIV set) on X is any mapping $\varphi : X \to [0,1] \cup D$.

Note that a fuzzy set $f : X \to [0,1]$, as well as a shadowed set on X introduced by Pedrycz [12] as any mapping $S : X \to \{0, [0,1], 1\}$, are examples of PIV sets (see also [13] for computing with shadowed sets). Further in this paper it is assumed that X is a finite set and $\Gamma(X), F(X)$ and $\Omega(X)$ denote the set of all PIV sets on X, the set of all fuzzy sets on X and the set of all shadowed sets on X, respectively.

Recall that in the theory of fuzzy sets [5,10,15] for $f, g \in F(X)$, operations of union and intersection are defined pointwise for all $x \in X$ by

$$(f \cup g)(x) = \max\{f(x), g(x)\}, \quad (f \cap g)(x) = \min\{f(x), g(x)\}. \tag{2}$$

In the case of shadowed sets $A, B \in \Omega(X)$, the union and the intersection are defined in Tables 1 and 2, respectively.

Table 1. Union of shadowed sets

$A \backslash B$	0	$[0,1]$	1
0	0	$[0,1]$	1
$[0,1]$	$[0,1]$	$[0,1]$	1
1	1	1	1

Table 2. Intersection of shadowed sets

$A \backslash B$	0	$[0,1]$	1
0	0	0	0
$[0,1]$	0	$[0,1]$	$[0,1]$
1	0	$[0,1]$	1

Operations in Tables 1 and 2 are isomorphic with the logic connectives in three-valued logic [7,11], in particular, Łukasiewicz logic.

The intent of this contribution is to accomplish the following tasks: First, to find operations of union and intersection of PIV sets which in the case of fuzzy sets correspond to the union and intersection of fuzzy sets, and, in the case of shadowed sets correspond to the union and intersection of shadowed sets. Then, to find operations of union and intersection of PIV sets such that the union and the intersection of membership labels of PIV sets expressed by intervals correspond to the union and intersection of intervals of real numbers. In applications, researchers can choose from a vast variety of aggregation functions [4,6]. Among them, triangular norms (t-norms) and triangular conorms (t-conomrs) are the most popular [8,9]. The next task of this contribution is to extend t-norms and t-conorms used in aggregation of fuzzy sets to t-norms and t-conorms suitable in aggregation of PIV sets. The final task is to propose a method for deriving a PIV set from data given by $\tau = \{\delta_{ij} : \delta_{ij} \in [0,1], i = 1, \ldots, n, j = 1, \ldots, m\}$. Note that majority of mathematical propositions in this contribution can be proved by simple algebraic manipulations. For this reason and because of the limited space, the proofs are omitted.

2   Operations on PIV Sets, Case I

The set of all possible labels of PIV sets on X is the set $\Lambda = [0,1] \cup D$. First, a relation of partial order on Λ will be introduced.

Definition 2. Consider $\lambda_1, \lambda_2 \in \Lambda$. Then λ_1 s-precedes λ_2, denoted by $\lambda_1 \leq_s \lambda_2$, if one of the following holds:

(1) $\lambda_1, \lambda_2 \in [0,1]$ and $\lambda_1 \leq \lambda_2$,
(2) $\lambda_1 \in [0,0.5)$ and $\lambda_2 \in D$,
(3) $\lambda_1 \in D$ and $\lambda_2 \in (0.5,1]$,
(4) $\lambda_1 = [b,c] \in D, \lambda_2 = [p,q] \in D$ and $(b \leq p, c \leq q)$,
(5) $\lambda_1 = [b,0.5] \in D, \lambda_2 = 0.5$,
(6) $\lambda_1 = 0.5, \lambda_2 = [0.5,q] \in D$.

Theorem 1. *The relation $\leq_s$ is reflexive, antisymmetric and transitive and for all $\lambda \in \Lambda$ we have $0 \leq_s \lambda \leq_s 1$.*

For $\lambda_1, \lambda_2 \in \Lambda$, there is exactly one $\lambda \in \Lambda$ such that with respect to the relation $\leq_s, \lambda = \inf(\lambda_1, \lambda_2) = \lambda_1 \wedge_s \lambda_2$. Operation $\wedge_s$ is defined in Table 3, where $u_1 = \min\{b,p\}, v_1 = \min\{c,q\}$,

$$u_2 = \begin{cases} [p,0.5] & \text{if } p < 0.5, \\ 0.5 & \text{if } p = 0.5, \end{cases} \qquad v_2 = \begin{cases} [b,0.5] & \text{if } b < 0.5, \\ 0.5 & \text{if } b = 0.5. \end{cases}$$

Table 3. Relation $\wedge_s$ on Λ

$\lambda_2 \backslash \lambda_1$	$a < 0.5$	$[b,c] \in D$	$d > 0.5$	0.5
$r < 0.5$	$\min\{a,r\}$	r	r	r
$[p,q] \in D$	a	$[u_1, v_1]$	$[p,q]$	u_2
$t > 0.5$	a	$[b,c]$	$\min\{t,d\}$	0.5
0.5	a	v_2	0.5	0.5

For $\lambda_1, \lambda_2 \in \Lambda$, there is exactly one $\lambda \in \Lambda$ such that with respect to the relation $\leq_s, \lambda = \sup(\lambda_1, \lambda_2) = \lambda_1 \vee_s \lambda_2$. Operation $\vee_s$ is defined in Table 4, where $u_3 = \max\{b,p\}, v_3 = \max\{c,q\}$,

$$u_4 = \begin{cases} [0.5,q] & \text{if } q > 0.5, \\ 0.5 & \text{if } q = 0.5, \end{cases} \qquad v_4 = \begin{cases} [0.5,c] & \text{if } c > 0.5, \\ 0.5 & \text{if } c = 0.5. \end{cases}$$

Corollary 1. $L_1 = (\Lambda, \leq_s, 0, 1)$ *is a bounded lattice.*

Recall that a mapping $f : X \to L$, where L is a lattice is called an L-fuzzy set on X.

Table 4. Relation $\vee_s$ on Λ

$\lambda_2 \backslash \lambda_1$	$a < 0.5$	$[b, c] \in D$	$d > 0.5$	0.5
$r < 0.5$	$\max\{a, r\}$	$[b, c]$	d	0.5
$[p, q] \in D$	$[p, q]$	$[u_3, v_3]$	d	u_4
$t > 0.5$	t	t	$\max\{t, d\}$	t
0.5	0.5	v_4	d	0.5

Corollary 2. *PIV sets on X are L_1-fuzzy sets on X.*

For PIV sets from $\Gamma(X)$, a relation of partial order can be defined pointwise, using the relation $\leq_s$.

Definition 3. Let $f, g \in \Gamma(X)$. Then f is s-included in g, denoted by $f \sqsubseteq_s g$, if for all $x \in X$,

$$f(x) \leq_s g(x). \tag{3}$$

Theorem 2. *PIV sets on X are partially ordered by the relation $\sqsubseteq_s$. For all $g \in \Gamma(X)$, $\emptyset \sqsubseteq_s g \sqsubseteq_s X$.*

The corresponding operations of s-intersection and s-union of PIV sets can be derived from operations $\wedge_s$ and $\vee_s$, respectively.

Definition 4. Let $f, g \in \Gamma(X)$. Then the s-intersection of f and g, denoted by $f \cap_s g$, is defined for all $x \in X$ by $f(x) \wedge_s g(x)$.

Observe that when f and g are fuzzy sets then $f \cap_s g = f \cap g$ given by (2). At the same time, when f and g are shadowed sets then $f \cap_s g = f \cap g$, where $\cap$ is defined in Table 2.

Theorem 3. *The following properties are satisfied for $g, h, k \in \Gamma(X)$:*

(1) commutativity: $g \cap_s h = h \cap_s g$,
(2) associativity: $g \cap_s (h \cap_s k) = (g \cap_s h) \cap_s k$,
(3) idempotency: $g \cap_s g = g$,
(4) boundary conditions: $g \cap_s \emptyset = \emptyset$ and $g \cap_s X = g$.

Based on the label $g(x)$ of a PIV set g defined on X, each object $x \in X$ can be classified as an object with low, high or neither low nor high (middle) compatibility with the concept described by g. Let $L(g) = \{x \in X : g(x) < 0.5\}$ be the low compatibility region of g, $M(g) = \{x \in X : g(x) \in D \cup \{0.5\}\}$ be the middle compatibility region of g and $H(g) = \{x \in X : g(x) > 0.5\}$ be the high compatibility region of g. The relationship between compatibility regions of PIV sets g, h and the compatibility regions of their s-intersection is given in Theorem 4.

Theorem 4. *Let $g, h \in \Gamma(X)$. Then*

(1) $L(g \cap_s h) = L(g) \cup L(h)$,
(2) $M(g) \cap M(h) \subset M(g \cap_s h) \subset M(g) \cup M(h)$,
(3) $H(g \cap_s h) = H(g) \cap H(h)$.

Note that $M(g \cap_s h) = X \setminus ((L(g) \cup L(h)) \cup (H(g) \cap H(h)))$.

Definition 5. *Let $f, g \in \Gamma(X)$. Then the s-union of f and g, denoted by $f \cup_s g$, is defined for all $x \in X$ by $f(x) \vee_s g(x)$.*

When f and g are fuzzy sets then $f \cup_s g = f \cup g$ given by (2). At the same time, when f and g are shadowed sets then $f \cup_s g = f \cup g$, where $\cup$ is defined in Table 1.

Theorem 5. *The following properties are satisfied for $g, h, k \in \Gamma(X)$:*

(1) commutativity: $g \cup_s h = h \cup_s g$,
(2) associativity: $g \cup_s (h \cup_s k) = (g \cup_s h) \cup_s k$,
(3) idempotency: $g \cup_s g = g$,
(4) boundary conditions: $g \cup_s \emptyset = g$ and $g \cup_s X = X$.

Theorem 6. *Let $g, h \in \Gamma(X)$. Then*

(1) $L(g \cup_s h) = L(g) \cap L(h)$,
(2) $M(g) \cap M(h) \subset M(g \cup_s h) \subset M(g) \cup M(h)$,
(3) $H(g \cup_s h) = H(g) \cup H(h)$.

Note that $M(g \cup_s h) = X \setminus ((L(g) \cap L(h)) \cup (H(g) \cup H(h)))$.

Theorem 7. *Let $g, h, k \in \Gamma(X)$. Then*

(1) $g \cap_s (h \cup_s k) = (g \cap_s h) \cup_s (g \cap_s k)$,
(2) $g \cup_s (h \cap_s k) = (g \cup_s h) \cap_s (g \cup_s k)$,
(3) $M(g \cap_s h) \cup M(g \cup_s h) = M(g) \cup M(h)$.

Corollary 3. *$(\Gamma(X), \subset_s, \emptyset, X)$ is a distributive bounded lattice.*

3   Operations on PIV Sets, Case II

In this section, another operation of partial order on $\Lambda = [0, 1] \cup D$ will be introduced and than extended to $\Gamma(X)$.

Definition 6. *Consider $\lambda_1, \lambda_2 \in \Lambda$. Then λ_1 p-precedes λ_2, denoted by $\lambda_1 \leq_p \lambda_2$, if one of the following holds:*

(1) $\lambda_1, \lambda_2 \in [0, 0.5]$ and $\lambda_1 \geq \lambda_2$,
(2) $\lambda_1, \lambda_2 \in [0.5, 1]$ and $\lambda_1 \leq \lambda_2$,
(3) $\lambda_1 \in [0, 1]$ and $[b, c] = \lambda_2 \in D$ and $b \leq \lambda_1 \leq c$,
(4) $\lambda_1 = [b, c] \in D$ and $\lambda_2 = [p, q] \in D$ and $p \leq b \leq c \leq q$.

Theorem 8. *The relation $\leq_p$ is reflexive, antisymmetric and transitive and for all $\lambda \in \Lambda$ we have $0.5 \leq_p \lambda \leq_p [0,1]$.*

Corollary 4. *The relation $\leq_p$ is partial order on Λ with the least element 0.5 and the largest element $[0,1]$.*

Note that the number 0.5 is the smallest element from all numbers in $[0,1]$ and the interval $[0,1]$ is the largest interval from all intervals in D.

For $\lambda_1, \lambda_2 \in \Lambda$, there is exactly one $\lambda \in \Lambda$ such that, with respect to the relation $\leq_p$, $\lambda = \inf(\lambda_1, \lambda_2) = \lambda_1 \wedge_p \lambda_2$. Operation $\wedge_p$ is defined in Table 5, where

$$u_1 = \begin{cases} [\max\{b,p\}, \min\{q,c\}] & \text{if } \max\{b,p\} < \min\{q,c\}, \\ 0.5 & \text{if } \max\{b,p\} = 0.5 = \min\{q,c\}. \end{cases}$$

Table 5. Relation $\wedge_p$ on Λ

$\lambda_2 \backslash \lambda_1$	$a \leq 0.5$	$[b,c] \in D$	$d \geq 0.5$
$r \leq 0.5$	$\max\{a,r\}$	$\max\{r,b\}$	0.5
$[p,q] \in D$	$\max\{a,p\}$	u_1	$\min\{d,q\}$
$t \geq 0.5$	0.5	$\min\{t,c\}$	$\min\{t,d\}$

For $\lambda_1, \lambda_2 \in \Lambda$, there is exactly one $\lambda \in \Lambda$ such that with respect to the relation $\leq_p$, $\lambda = \sup(\lambda_1, \lambda_2) = \lambda_1 \vee_p \lambda_2$. Operation $\vee_p$ is defined in Table 6, where $u_2 = \min\{b,p\}, v_2 = \max\{c,q\}$ and

$$u_3 = \begin{cases} [r,d] & \text{if } r \neq d, \\ 0.5 & \text{if } r = d, \end{cases} \qquad v_3 = \begin{cases} [a,t] & \text{if } a \neq t, \\ 0.5 & \text{if } a = t. \end{cases}$$

Table 6. Relation $\vee_p$ on Λ

$\lambda_2 \backslash \lambda_1$	$a \leq 0.5$	$[b,c] \in D$	$d \geq 0.5$
$r \leq 0.5$	$\min\{a,r\}$	$[\min\{r,b\},c]$	u_3
$[p,q] \in D$	$[\min\{a,p\},q]$	$[u_2,v_2]$	$[p,\max\{d,q\}]$
$t \geq 0.5$	v_3	$[b,\max\{c,t\}]$	$\max\{t,d\}$

Corollary 5. $L_2 = (\Lambda, \leq_p, 0.5, [0,1])$ *is a bounded lattice.*

For PIV sets from $\Gamma(X)$, a relation of partial order can be defined pointwise, using the relation $\leq_p$.

Definition 7. Let $f, g \in \Gamma(X)$. Then f is p-included in g, denoted by $f \subset_p g$, if for all $x \in X$,

$$f(x) \leq_p g(x). \tag{4}$$

Theorem 9. *PIV sets on X are partially ordered by the relation $\subset_p$. For all $g \in \Gamma(X)$,*

$$0.5 \subset_p g \subset_p [0,1].$$

Definition 8. *Let $f, g \in \Gamma(X)$. Then the p-intersection of f and g, denoted by $f \cap_p g$, is defined for all $x \in X$ by*

$$f(x) \wedge_p g(x). \tag{5}$$

Note that when $f(x) = [b,c] \in D$ and $g(x) = [r,q] \in D$ then $f(x) \wedge_p g(x) = [b,c] \cap [r,q]$.

Theorem 10. *The following properties are satisfied for $g, h, k \in \Gamma(X)$:*

(1) commutativity: $g \cap_p h = h \cap_p g$,
(2) associativity: $g \cap_p (h \cap_p k) = (g \cap_p h) \cap_p k$,
(3) idempotency: $g \cap_p g = g$,
(4) boundary conditions: $g \cap_p 0.5 = 0.5$ and $g \cap_p [0,1] = g$.

Then compatibility regions of $g \in \Gamma(X)$ are: the low compatibility region $L(g) = \{x \in X : g(x) < 0.5\}$, the high compatibility region $H(g) = \{x \in X : g(x) > 0.5\}$ and the middle compatibility region $M(g) = X \setminus (L(g) \cup H(g))$ consists of two parts, $M_1(g) = \{x \in X : g(x) = 0.5\}$ and $M_2 = \{x \in X : g(x) \in D\}$.

Theorem 11. *Let $g, h \in \Gamma(X)$. Then*

(1) $L(g) \cap L(h) \subset L(g \cap_p h) \subset L(g) \cup L(h)$,
(2) $H(g) \cap H(h) \subset H(g \cap_p h) \subset H(g) \cup H(h)$,
(3) $M_1(g) \cap M_1(h) \subset M_1(g \cap_p h) \subset M_1(g) \cup M_1(h)$,
(4) $M_2(g) \cap M_2(h) = M_2(g \cap_p h)$.

Definition 9. *Let $f, g \in \Gamma(X)$. Then the p-union of f and g, denoted by $f \cup_p g$, is defined for all $x \in X$ by*

$$f(x) \vee_p g(x). \tag{6}$$

Note that when $f(x) = [b,c] \in D$ and $g(x) = [r,q] \in D$ then $f(x) \vee_p g(x) = [b,c] \cup [r,q]$.

Theorem 12. *The following properties are satisfied for $g, h, k \in \Gamma(X)$:*

(1) commutativity: $g \cup_p h = h \cup_p g$,
(2) associativity: $g \cup_p (h \cup_p k) = (g \cup_p h) \cup_p k$,
(3) idempotency: $g \cap_p g = g$,
(4) boundary conditions: $g \cup_p 0.5 = g$ and $g \cup_p [0,1] = [0,1]$.

Theorem 13. *Let $g, h \in \Gamma(X)$. Then*

(1) $L(g \cup_p h) = L(g) \cap L(h)$,
(2) $H(g \cup_p h) = H(g) \cap H(h)$,
(3) $M_1(g \cup_p h) = M_1(g) \cap M_1(h)$,
(4) $M_2(g \cup_p h) \supset M_2(g) \cap M_2(h)$.

Theorem 14. *Let* $g, h, k \in \Gamma(X)$. *Then*

(1) $g \cap_p (h \cup_p k) = (g \cap_p h) \cup_p (g \cap_p k)$,
(2) $g \cup_p (h \cap_p k) = (g \cup_p h) \cap_p (g \cup_p k)$.

Corollary 6. $(\Gamma(X), \subset_p, 0.5, [0, 1])$ *is a distributive bounded lattice.*

In applications, the primary use of PIV sets is to characterize compatibility of objects from X with a concept defined on X by labels small, large or neither small nor large, described by numbers or intervals. The preference of researchers for s-inclusion or p-inclusion (and corresponding aggregation functions) of PIV sets depends on properties of PIV sets on lattices $(\Gamma(X), \subset_s, \emptyset, X)$ and $(\Gamma(X), \subset_p, 0.5, [0, 1])$.

4   Aggregation of PIV Sets by T-norms

T-norms and t-conorms applied to aggregation of fuzzy sets can be extended to t-norms and t-conorms on the lattice $(\Gamma(X), \subset_s, \emptyset, X)$ or on the lattice $(\Gamma(X), \subset_p, 0.5, [0, 1])$. An extension based on the notion of the bounded t-norm and the bounded t-conorm is proposed in this section.

Definition 10. Let T be a t-norm on $[0, 1]$ and $\delta \in [0, 1]$. Then the δ-bounded t-norm T is the mapping $T_{(\delta)} : [\delta, 1] \times [\delta, 1] \to [\delta, 1]$ such that for all $(x, y) \in [\delta, 1]^2$

$$T_{(\delta)}(x, y) = \max\{\delta, T(x, y)\}. \tag{7}$$

Theorem 15. $T_{(\delta)}$ *is a commutative, associative and nondecreasing function satisfying*

$$T_{(\delta)}(x, 1) = x \quad and \quad T_{(\delta)}(x, \delta) = \delta \tag{8}$$

for all $x \in [\delta, 1]$.

Any t-norm T is the δ-bounded t-norm T when $\delta = 0$. Obviously, $T_{min,(\delta)} = T_{min}$ for all $\delta \in [0, 1]$.

Definition 11. Let S be a t-conorm on $[0, 1]$ and $\delta \in [0, 1]$. Then the δ-bounded t-conorm S is the mapping $S_{(\delta)} : [0, \delta] \times [0, \delta] \to [0, \delta]$ such that for all $(x, y) \in [0, \delta]^2$

$$S_{(\delta)}(x, y) = \min\{\delta, S(x, y)\}. \tag{9}$$

Theorem 16. $S_{(\delta)}$ *is a commutative, associative and nondecreasing function satisfying*

$$S_{(\delta)}(x, 0) = x \quad and \quad S_{(\delta)}(x, \delta) = \delta \tag{10}$$

for all $x \in [0, \delta]$.

Any t-conorm S is the δ-bounded t-conorm S when $\delta = 1$. Obviously, $S_{max,(\delta)} = S_{max}$ for all $\delta \in [0, 1]$.

Theorem 17. *Consider a t-norm T on $[0,1]$ and the lattice $(\Gamma(X), \subset_s, \emptyset, X)$. Let $\varphi_T : \Gamma(X) \times \Gamma(X) \to \Gamma(X)$ be defined for all $f, g \in \Gamma(X)$ in Table 7, where*

$$u_1 = \begin{cases} [T(b,p), T_{(0.5)}(c,q)] & \text{if } T(b,p) < T_{(0.5)}(c,q), \\ 0.5 & \text{if } T(b,p) = T_{(0.5)}(c,q) = 0.5, \end{cases}$$

$$u_2(s) = \begin{cases} [T(s,0.5), 0.5] & \text{if } T(s,0.5) < 0.5, \\ 0.5 & \text{if } T(s,0.5) = 0.5. \end{cases}$$

Then φ_T is a t-norm on $(\Gamma(X), \subset_s, \emptyset, X)$,

Table 7. T-norm on $\Gamma(X)$, Case I

$g \backslash f$	$f(x) < 0.5$	$f(x) = [b,c]$	$f(x) > 0.5$	0.5
$g(x) < 0.5$	$T(f(x), g(x))$	$g(x)$	$g(x)$	$g(x)$
$g(x) = [p,q]$	$f(x)$	u_1	$g(x)$	$u_2(p)$
$g(x) > 0.5$	$f(x)$	$f(x)$	$T_{(0.5)}(f(x), g(x))$	0.5
0.5	$f(x)$	$u_2(b)$	0.5	0.5

When $T = T_{min}$ then $\varphi_T(f,g) = f \cap_s g$.

Theorem 18. *Consider a t-conorm S on $[0,1]$ and the lattice $(\Gamma(X), \subset_s, \emptyset, X)$. Let $\varphi_S : \Gamma(X) \times \Gamma(X) \to \Gamma(X)$ be defined for all $f, g \in \Gamma(X)$ in Table 8, where*

$$u_3 = \begin{cases} [S_{(0.5)}(b,p), S(c,q)] & \text{if } S_{(0.5)}(b,p) < S(c,q), \\ 0.5 & \text{if } S_{(0.5)}(b,p) = S(c,q) = 0.5, \end{cases}$$

$$u_4(s) = \begin{cases} [0.5, S(s,0.5)] & \text{if } S(s,0.5) > 0.5, \\ 0.5 & \text{if } S(s,0.5) = 0.5. \end{cases}$$

Then φ_S is a t-conorm on $(\Gamma(X), \subset_s, \emptyset, X)$.

Table 8. T-conorm on $\Gamma(X)$, Case I

$g \backslash f$	$f(x) < 0.5$	$f(x) = [b,c]$	$f(x) > 0.5$	0.5
$g(x) < 0.5$	$S_{(0.5)}(f(x), g(x))$	$f(x)$	$f(x)$	0.5
$g(x) = [p,q]$	$g(x)$	u_3	$f(x)$	$u_4(q)$
$g(x) > 0.5$	$g(x)$	$g(x)$	$S(f(x), g(x))$	$g(x)$
0.5	0.5	$u_4(c)$	$f(x)$	0.5

When $S = S_{max}$ then $\varphi_S(f,g) = f \cup_s g$.

Theorem 19. *Consider a t-norm T on $[0,1]$, its dual t-conorm S and the lattice $(\Gamma(X), \subset_p, 0.5, [0,1])$. Let $\rho_T : \Gamma(X) \times \Gamma(X) \to \Gamma(X)$ be defined for all $f, g \in \Gamma(X)$ in Table 9, where*

$$u_1 = \begin{cases} [S_{(0.5)}(b,p), T_{(0.5)}(q,c)] & \text{if } S_{0.5}(b,p) < T_{(0.5)}(q,c), \\ 0.5 & \text{if } S_{0.5}(b,p) = 0.5 = T_{(0.5)}(q,c). \end{cases}$$

Then ρ_T is a t-norm on $(\Gamma(X), \subset_p, 0.5, [0,1])$.

Table 9. T-norm on $\Gamma(X)$, Case II

$g\backslash f$	$f(x) \leq 0.5$	$f(x) = [b,c]$	$f(x) \geq 0.5$
$g(x) \leq 0.5$	$S_{(0.5)}(f(x), g(x))$	$S_{(0.5)}(g(x), b)$	0.5
$g(x) = [p,q]$	$S_{(0.5)}(f(x), p)$	u_1	$T_{(0.5)}(f(x), q)$
$g(x) \geq 0.5$	0.5	$T_{(0.5)}(g(x), c)$	$T_{(0.5)}(f(x), g(x))$

When $T = T_{min}$ then $\rho_T(f,g) = f \cap_p g$.

Theorem 20. *Consider a t-conorm S on $[0,1]$ and the lattice $(\Gamma(X), \subset_p, 0.5, [0,1])$. Let $\rho_S : \Gamma(X) \times \Gamma(X) \to \Gamma(X)$ be defined for all $f, g \in \Gamma(X)$ in Table 10, where*

Table 10. T-conorm on $\Gamma(X)$, Case II

$g\backslash f$	$f(x) \leq 0.5$	$f(x) = [b,c]$	$f(x) \geq 0.5$
$g(x) \leq 0.5$	$T(f(x), g(x))$	$[T(g(x), b), c]$	u_3
$g(x) = [p,q]$	$[T(f(x), p), q]$	$[u_2, v_2]$	$[p, S(f(x), q)]$
$g(x) \geq 0.5$	u_3	$[b, S(c, g(x))]$	$S(f(x), g(x))$

$$u_2 = T(b,p), \quad v_2 = S(c,q), \quad u_3 = \begin{cases} [f(x), g(x)] & \text{if } f(x) \neq g(x), \\ 0.5 & \text{if } f(x) = g(x). \end{cases}$$

Then ρ_S is a t-conorm on $(\Gamma(X), \subset_p, 0.5, [0,1])$.

When $S = S_{max}$ then $\rho_S(f,g) = f \cup_s g$.

5 Construction of PIV Sets

Different methods for construction of PIV sets from available data can be proposed. A simple approach based on a preselected aggregation function Ag is outlined below.

Assume a fuzzy set f on a finite universal set U and an aggregation function Ag such that for $(\delta_1, \ldots, \delta_n) \in [0,1]^n$

$$\min(\delta_1, \ldots, \delta_n) \leq Ag(\delta_1, \ldots, \delta_n) \leq \max(\delta_1, \ldots, \delta_n).$$

Let $W_1(f) = \{u \in U : f(u) \leq 0.5\}, W_2(f) = \{u \in U : f(u) \geq 0.5\}, a = Ag(f(u), u \in W_1(f))$ and $b = Ag(f(u), u \in W_2(f))$. Then the summary characterization of f based on Ag can be expressed by

$$S_{Ag}(f) = \begin{cases} a & \text{if } W_2(f) = \emptyset, \\ b & \text{if } W_1(f) = \emptyset, \\ [a, b] & \text{if } W_1(f) \neq \emptyset, W_2(f) \neq \emptyset \text{ and } a \neq b, \\ 0.5 & \text{if } a = b = 0.5. \end{cases} \tag{11}$$

The task is to find a summary characterization of data τ (see formula (1)). In our illustrative example τ may represent evaluations of objects from X by experts from $E = \{E_j, j = 1, \ldots, m\}$. Then the following method can be used:

Method 1:

Step 1: For each $i = 1, \ldots, n$, construct fuzzy set f_i on E such that for $E_j \in E$: $f_i(E_j) = \delta_{ij}$. Then choose aggregation function Ag and create $S_{Ag}(f_i)$.
Step 2: Construct PIV set φ on X such that for $x_i \in X, \varphi(x_i) = S_{Ag}(f_i)$.

Example 1. Suppose that data in Table 11 represent evaluations of objects $x_1, \ldots, x_5$ by experts $E_1, \ldots, E_6$ with respect to criterion C_1. Let Ag be the arithmetic mean. Then, using Method 1, summary characterizations of Table 11 can be obtained by PIV set φ_{C1} with labels presented in Table 12.

Table 11. Evaluations with respect to criterion C1

X	x_1	x_2	x_3	x_4	x_5
E_1	0.5	0.8	1.0	0.2	0.7
E_2	0.3	0.4	0.9	0.0	0.6
E_3	0.1	0.6	0.8	0.1	0.4
E_4	0.2	0.5	0.7	0.1	0.4
E_5	0.1	0.7	0.8	0.3	0.3
E_6	0.4	0.4	0.9	0.4	0.8

Table 12. Summary of evaluations with respect to criterion C1

X	x_1	x_2	x_3	x_4	x_5
φ_{C1}	[0.267, 0.5]	[0.433, 0.65]	0.85	0.183	[0.367, 0.7]

6   Conclusion

PIV sets allow evaluation of compatibility of objects from a set X with a concept defined on X in terms (labels) that can be interpreted as either small or large (described by either a small or a large number from the unit interval), or neither small nor large (described by 0.5 or a subinterval from the unit interval covering 0.5). Description of the label "neither small nor large" by an interval gives researchers more flexibility in decision making under uncertainty. A method for a simple summary characterization of a family of finite fuzzy sets by a PIV set was suggested.

From the theoretical point of view, relations s-inclusion and p-inclusion of PIV sets were introduced and corresponding operations of union and intersection were defined. A restriction of t-norms (t-conomrs) to bounded t-norms (t-conorms) was proposed and used in construction of t-norms (t-conorms) suitable for aggregation of PIV sets.

Acknowledgements. The work of Martin Kalina has been supported from the Science and Technology Assistance Agency under contract No. APVV-14-0013, and from the VEGA grant agency, grant No. 2/0069/16.

References

1. Aliev, R., Pedrycz, W., Fazlollahi, B., Huseynov, O.H., Alizadeh, A.V.: Fuzzy logic-based generalized decision theory with imperfect information. Inf. Sci. **189**, 18–42 (2012)
2. Beliakov, G., Calvo, T., Pradera, A.: Absorbent tuples of aggregation operators. Fuzzy Sets Syst. **158**, 1675–1691 (2007)
3. Beliakov, G., Dujmović, J.: Extension of bivariate means to weighted means of several arguments by using binary trees. Inf. Sci. **331**, 137–147 (2016)
4. Calvo, T., Mayor, G., Mesiar, R. (eds.): Aggregation Operators. Physica-Verlag, Heidelberg (2002)
5. Dubois, D., Prade, H. (eds.): Fundamentals of Fuzzy Sets. Kluwer, Dordrecht (2000)
6. Grabisch, M., Pap, V., Marichal, J.L., Mesiar, R.: Aggregation Functions. University Press, Cambridge (2009)
7. Hájek, P.: Mathematics of Fuzzy Logic. Kluwer, Dordrecht (1998)
8. Klement, E.P., Mesiar, R., Pap, E.: Triangular Norms. Kluwer, Dordrecht (2000)
9. Klement, E.P., Mesiar, R. (eds.): Logical, Algebraic, Analytic and Probabilistic Aspects of Triangular Norms. Elsevier, Amsterdam (2005)
10. Klir, G.J., Yuan, B.: Fuzzy Sets and Fuzzy Logic: Theory and Applications. Prentice Hall, Upper Saddle River (1995)
11. Łukasiewicz, J.: O logice trojwartosciowej. Ruch filozoficzny **5**, 170–171 (1920). (in Polish)
12. Pedrycz, W.: Shadowed sets: representing and processing fuzzy sets. IEEE Trans. Syst. Man Cybern. (Part B) **28**, 103–109 (1998)
13. Pedrycz, W., Vukovich, G.: Granular computing with shadowed sets. Int. J. Intell. Syst. **17**, 173–197 (2002)
14. Yager, R.R., Alajlan, N.: On the measure based formulation of multi-criteria decision functions. Inf. Sci. **370–371**, 256–269 (2016)
15. Zadeh, L.: Fuzzy sets. Inf. Control **8**, 338–353 (1965)

On Some Applications of Williamson's Transform in Copula Theory

Tomáš Bacigál[(✉)]

Faculty of Civil Engineering, Slovak University of Technology in Bratislava,
Radlinského 11, 810 05 Bratislava, Slovakia
bacigal@math.sk

Abstract. We show several interesting examples of connection between distribution of a positively valued random variable and an Archimedean copula through Williamson's transformation (and Laplace transform), especially when arranged in a sequence. Naturally, there appears a question: how can we use statistical properties of distance functions to draw statistical properties of copulas, and vice versa? This question is formulated in two open problems.

Keywords: Archimedean copula · Williamson's transform · Laplace transform

1 Introduction

Copulas [5,14,17] are particular functions describing the dependence stucture of random vectors. Not going into details, recall that one of the prominent copula classes important for numerous applications is the class of Archimedean copulas. Formally, for $n \geq 2$, a function $C \colon [0,1]^n \to [0,1]$ is an n-ary Archimedean copula whenever it is a Post associative n-ary copula (i.e., for any $(x_1, \ldots, x_{2n-1}) \in [0,1]^{2n-1}$ it holds $C\left(C(x_1, \ldots, x_n), x_{n+1}, \ldots, x_{2n-1}\right) = C\left(x_1, C(x_2, \ldots, x_{n+1}), x_{n+2}, \ldots, x_{2n-1}\right) = \ldots = C\left(x_1, \ldots, x_{n-1}, C(x_n, \ldots, x_{2n-1})\right)$ and $C(x, \ldots, x) < x$ for any $x \in]0,1[$, see [18]. Due to [9] we have next representation of n-ary Archimedean copulas.

Theorem 1. *Let $f \colon [0,1] \to [0,\infty]$ be a continuous strictly decreasing function such that $f(1) = 0$ (i.e., f is an additive generator of a continuous Archimedean t-norm, see [7]). Then the n-ary function $C \colon [0,1]^n \to [0,1]$ given by*

$$C(x_1, \ldots, x_n) = f^{(-1)} \left(\sum_{i=1}^{n} f(x_i) \right). \tag{1}$$

(where $f^{(-1)} \colon [0,\infty] \to [0,1]$ given by $f^{(-1)}(u) = f^{-1}\big(\min(u, f(0))\big)$ is the pseudo-inverse of f) is an n-ary copula if and only if the function $g \colon [-\infty, 0] \to [0,1]$ given by $g(u) = f^{(-1)}(-u)$ is $(n-2)$-times differentiable with non-negative derivatives $g', \ldots, g^{(n-2)}$ on $] -\infty, 0[$ (or equivalently, $(-1)^n (f^{(-1)})^{(n)}(u) \geq 0$), and $g^{(n-2)}$ is a convex function (Fig. 1).

© Springer International Publishing AG 2018
V. Torra et al. (eds.), *Aggregation Functions in Theory and in Practice*,
Advances in Intelligent Systems and Computing 581, DOI 10.1007/978-3-319-59306-7_3

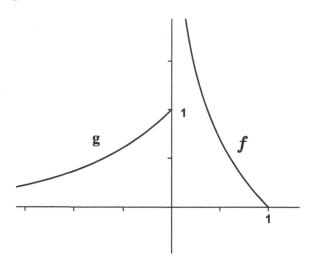

Fig. 1. Illustration of a generator f and its corresponding function g

We denote by $\mathcal{F}_n$ the class of all additive generators that generate n-ary copulas as characterized in Theorem 1.

Additive generators, which generate an n-ary copula for any $n \geq 2$, are called universal generators. Due to Theorem 1, we have the next result, see [6, 9].

Corollary 1. *Let $f: [0,1] \to [0,\infty]$ be an additive generator of a binary copula $C: [0,1]^2 \to [0,1]$. Then the n-ary extension $C: [0,1]^n \to [0,1]$ given by (1) is an n-ary copula for each $n \geq 2$ if and only if the function $g: [-\infty, 0] \to [0,1]$ given by $g(u) = f^{(-1)}(-u)$ is absolutely monotone, i.e., $g^{(k)}$ exists and is non-negative for each $k \in N = \{1, 2, \ldots\}$.*

The class of all universal additive generators will be denoted by $\mathcal{F}_\infty$. It is not difficult to check that $\mathcal{F}_2 \supset \mathcal{F}_3 \supset \ldots \supset \mathcal{F}_\infty$.

For any $n \geq 2$, there is an important link between the additive generators of n-ary Archimedean copulas and distance functions $F: [0,\infty[\to [0,1]$, i.e., distribution functions of positive random variables restricted to $[0,\infty[$. Observe that then $F(0) = 0$, F is monotone non-decreasing right-continuous and $\lim_{x\to\infty} F(x) = 1$. We denote the class of all distance functions as $\mathcal{D}$.

Based on the results of Williamson [19], we recall the next important result.

Theorem 2 (McNeil and Nešlehová [9], Corollary 3.1). *The following claims are equivalent for an arbitrary $n \in \{2, 3, \ldots\}$:*

(i) $f \in \mathcal{F}_n$

(ii) Under the notation of Theorem 1, the function $F: [0,\infty[\to [0,1]$ given by $F(0) = 0$ and for $x > 0$,

$$F(x) = 1 - \sum_{k=0}^{n-2} \frac{(-1)^k x^k (f^{(-1)})^{(k)}(x)}{k!} - \frac{(-1)^{n-1} x^{n-1} (f^{(-1)})_+^{(n-1)}(x)}{(n-1)!}, \quad (2)$$

is a distance function from $\mathcal{D}$, where $\cdot_{+}^{(n-1)}$ denotes the right-derivative of order $n-1$.

Note that due to [19], if F is a positive distance function, i.e., a distribution function of a positive random variable X, then for a fixed $n \in \{2, 3, \ldots\}$ the Williamson n-transform provides an inverse transformation to (2),

$$f^{(-1)}(x) = \int_x^\infty \left(1 - \frac{x}{t}\right)^{n-1} dF(t) = \begin{cases} \max\left(0, E\left[1 - \frac{x}{X}\right]^{n-1}\right), & x > 0 \\ 1 - F(0), & x = 0, \end{cases} \tag{3}$$

where $x \in [0, \infty[$ and $f^{(-1)}(\infty) = 0$.

Note that a similar relationship can be shown between additive generators from $\mathcal{F}_\infty$ and positive distance functions, based on the Laplace transform, i.e.

$$f^{(-1)}(x) = \int_0^\infty e^{-xt} dF(t). \tag{4}$$

For more and interesting details we recommend [9].

Note that if $F_X \in \mathcal{D}$ is a distance function linked to a positive random vector X, then for any positive real constant c, also $F_{cX} \in \mathcal{D}$, and for the related additive generators (independently of $n \geq 2$), $f_{cX}(x) = c f_X(x)$. However, both f_X and f_{cX} generate the same (n-ary) Archimedean copula.

The aim of this paper is to discuss some applications of the introduced link between additive generators and distance functions in the copula theory. The paper is organised as follows. In Sect. 2, some examples are given. In Sect. 3, we introduce and discuss particular sequences of additive generators (distance functions) related to a fixed distance function (additive generator). In Sect. 4, we open several interesting problems dealing with relations between classes $\mathcal{F}_n$ and $\mathcal{F}_m$ for $n \neq m$ and between distance functions related to classes $\mathcal{F}_n$ and $\mathcal{F}_m$, respectively. Finally, some concluding remarks are given.

2   Examples

Example 1. Let F be equal to a Dirac function δ_a focused at point $a > 0$,

$$F(x) = \delta_a(x) = \begin{cases} 0 & x < a \\ 1 & a \leq x \end{cases},$$

then, as is also shown in [9], by the Williamson n-transform we get generator $f_n(x) = a\left(1 - x^{\frac{1}{n-1}}\right)$ of the weakest n-dimensional Archimedean copula, i.e., the non-strict Clayton copula with parameter $\lambda = \frac{-1}{n-1}$, see Fig. 2. By rescaling generator to $\tilde{f}_n(x) = \frac{f(x)}{f(1/2)}$, $x \in [0, 1]$, the copula would not change, yet such a generator is fixed to the value $\tilde{f}_n(\frac{1}{2}) = 1$.

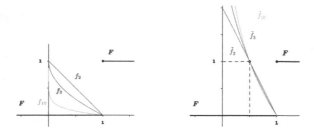

Fig. 2. Dirac function F, the corresponding generators f_n for different n and rescaled generators $\tilde{f}_n$.

Example 2. Let F be a uniform probability distribution function

$$F(x) = \begin{cases} 0 & x < 0 \\ x & 0 \le x < 1 \\ 1 & 1 \le x \end{cases}.$$

Then for dimension $n = 2$ we get

$$f_2^{(-1)}(x) = \int_x^\infty \left(1 - \frac{x}{t}\right)^{2-1} F'(t)dt = \begin{cases} \int_x^1 \left(1 - \frac{x}{t}\right) dt & 0 \le x < 1 \\ \int_x^\infty \left(1 - \frac{x}{t}\right) 0\, dt & 1 \le x \end{cases} =$$

$$= \begin{cases} [t - x \log t]_x^1 = 1 - x + x \log(x) & 0 \le x < 1 \\ 0 & 1 \le x \end{cases}$$

(where F' denotes the density related to F) from which the corresponding generator can be obtained only numerically, and so is the case also with the higher dimensions, e.g.,

$$f_3^{(-1)}(x) = \begin{cases} 1 + 2x \log x - x^2 & 0 \le x < 1 \\ 0 & 1 \le x \end{cases}.$$

We continue with the examples of constructing generators of non-strict Archimedean copulas while restricting the support of univariate distribution in the unit interval. By applying a suitable increasing transformation (such as power function) to a positive distance function on $[0,1]$ we obtain a new distribution.

Example 3. Consider a positive distance function $F(x) = min(1, x^2)$ and the corresponding density $F'(x) = 2x$ on $[0,1]$. Then

$$f_2^{(-1)}(x) = \int_x^\infty \left(1 - \frac{x}{t}\right)^{2-1} dF(t) = \begin{cases} \int_x^1 (t - x) \frac{2t}{t} dt = (1 - x)^2 & 0 \le x \le 1 \\ 0 & 1 < x \end{cases} =$$

$$= \max(1 - x, 0)^2.$$

Then the generator $f_2(x) = 1 - \sqrt{x}$, $x \in [0,1]$, is the generator of Clayton copula for parameter $\lambda = -\frac{1}{2}$. Nevertheless, in higher dimensions, $n \geq 3$, the generator has no closed form, e.g., $f_3^{(-1)}(x) = 1 - 4x + x^2(3 - 2\log x)$ for $x \in [0,1]$ and 0 otherwise (Fig. 3).

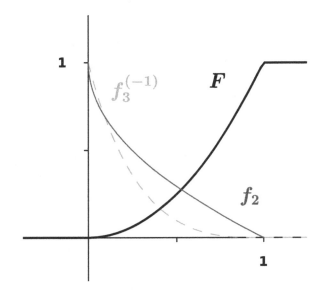

Fig. 3. Illustration of Example 3 with $a = 1$

Example 4. For any distance function $F \in \mathcal{D}$ related to a random variable X a shifted random variable $a + X$, $a \geq 0$, generates a distance function $F_a \in \mathcal{D}$ given by

$$F_a(x) = \begin{cases} 0 & x \leq a \\ F(x - a) & \text{otherwise} \end{cases}.$$

This observation allows to introduce parametric families of n-ary Archimedean copulas. Continuing in Example 2, distance functions F_a are just distribution functions of random variables uniformly distributed on $[a, a+1]$ and the related pseudo-inverses of additive generators are given by

$$f_2^{(-1)}(x) = \int_x^\infty \left(1 - \frac{x}{t}\right)^{2-1} F'(t) dt = \begin{cases} \int_a^{a+1} \left(1 - \frac{x}{t}\right) dt & x < a \\ \int_x^{a+1} \left(1 - \frac{x}{t}\right) dt & a \leq x < a+1 \\ \int_x^\infty \left(1 - \frac{x}{t}\right) 0 dt & a+1 \leq x \end{cases} =$$

$$= \begin{cases} [t - x\log t]_a^{a+1} = 1 - x\log\left(\frac{a+1}{a}\right) & x < a \\ [t - x\log t]_x^{a+1} = a + 1 - x - x\log\left(\frac{a+1}{x}\right) & a \leq x < a+1 \\ 0 & a+1 \leq x \end{cases}$$

and

$$f_3^{(-1)}(x) = \begin{cases} 1 - 2x \log\left(\frac{a+1}{a}\right) + \frac{x^2}{a(a+1)}, & x < a \\ a + 1 - 2x \log\left(\frac{a+1}{x}\right) - \frac{x^2}{a+1} & a \le x < a+1 \\ 0 & a+1 \le x \end{cases}.$$

displayed in Fig. 4.

Remark 1. Note that considering a random variable X uniformly distributed on $[a,b] \in [0, \infty[$, the random variable $Y = \frac{X}{b-a}$ is uniformly distributed on $[\frac{a}{b-a}, \frac{b}{b-a}] = [c, c+1]$ with $c = \frac{a}{b-a}$. Hence the additive generators of Archimedean copulas discussed in Example 4 covers all cases related to uniformly distributed random variables.

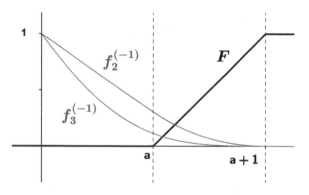

Fig. 4. Uniform $U(a, a+1)$ probability distribution function F and pseudo-inverses of the corresponding generators f_n.

Example 5. Generalizing Examples 2 and 3 such that $F(x) = \min(1, x^p)$, $p \in]0, \infty[$, we get

$$f_2^{(-1)}(x) = \begin{cases} 1 - \frac{px - x^p}{p-1} & 0 \le x \le 1 \\ 0 & 1 < x \end{cases} \quad \text{for } p \ne 1,$$

(with special case for $p = 1$ given in Example 2) whose corresponding generator for most values of p can be obtained only numerically, and the copulas C_p it generates span from M ($p \to 0$) to W ($p \to \infty$) excluding Π. Kendall's correlation coefficient as a function of parameter p can be expressed in the closed form $\tau_2(p) = 1 - 4\frac{p}{2(p+1)}$. Higher order Williamson transforms, e.g.,

$$f_3^{(-1)}(x) = \begin{cases} 1 - \frac{2x^p - p(p-1)x^2 + 2p(p-2)x}{(p-1)(p-2)} & 0 \le x \le 1 \\ 0 & 1 < x \end{cases} \quad \text{for } p \ne 1, 2,$$

(with special cases for $p = 1, 2$ given in Examples 2 and 3, respectively) neither provide convenience of generator in closed form, nor the full span of dependence range, e.g. $\tau_3(p) = 1 - 4\frac{p(p+3)}{3(p+1)(p+2)}$, see Fig. 5.

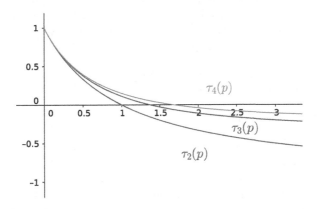

Fig. 5. Kendall's tau τ_n related to copula family generated by Williamson's transform of $F(x) = \min(1, x^p)$ distance function.

3   Williamson's Transforms and Sequences of Additive Generators/Distance Functions

Example 6. Take a generator of the product copula $f(x) = -\frac{1}{p} \log x$ with constant $p > 0$ and inverse $f^{-1}(x) = \exp(-px)$. From (2) for $n = 2$ we get $F(x) = 1 - \exp(-px)(1 - px)$. By comparing the density $\frac{\partial F(x)}{\partial x} = p^2 x \exp(-px)$ and the convolution of two exponential distribution $\mathcal{D}_p$ densities with parameter $p > 0$, $\int_0^x p \exp(-pt) p \exp(-p(x - t)) dt = p^2 x \exp(-px)$ it becomes clear that the resulting distribution is a distribution of the random variable $Y = X_1 + X_2$, where $X_1, X_2 \sim \mathcal{D}_p$ are independent (and identically distributed) random variables. The relation holds for any $n \geq 2$, thus (2) yields a cumulative distribution function of the sum of i.i.d. random variables $X_1, \ldots, X_n \sim \mathcal{D}_p$,
$F_{X_1 + \ldots + X_n}(x) = 1 - \exp(-px) \sum_{i=1}^n \frac{(px)^{i-1}}{(i-1)!}$ with $p > 0$ which defines the Erlang distribution with rate parameter p and shape parameter n.

Summarizing, we see that the sequence $(F_n)_{n=1}^\infty$ of the Erlang distribution functions (with either fixed or variable parameter p) is related via Williamson's transforms with the product copula. Observe that when considering the Laplace transform (4), then the product copula is related to Dirac function δ_p, $p \in]0, \infty[$.

Similarly, one can consider any other Archimedean copula for which each n-ary version is an n-ary copula, i.e., possessing a universal additive generator.

Example 7. Consider the generator of the Ali-Mikhail-Haq copula $f(x) = \frac{1}{x} - 1$ corresponding to the parameter $\lambda = 1$ and denote by F_n, $n = 2, 3, \ldots$, a positive distance function related to f through (2). Then $F_n(x) = 1 - \frac{1}{1+x} - \frac{x}{(1+x)^2} - \cdots - \frac{x^{n-1}}{(1+x)^n} = \left(\frac{x}{1+x}\right)^n$ which can be viewed as a parametric subfamily of all positive valued distribution functions F_p with any positive parameter p.

Observe that when considering the Laplace transform (4), then the discussed Ali-Mikhail-Haq copula is related to the exponential distribution with the distance function $F(x) = 1 - e^{-\lambda x}$, $\lambda > 0$

On the other hand, fixing a distance function F, one can introduce related n-ary copulas (universal copula) by means of (3) (of (4)).

Example 8. Starting with positive distance function of

- discrete random variable with probability mass concentrated in $\lambda > 0$, i.e. Dirac function $F(x) = 0$ for $x < \lambda$ and 1 otherwise, then the sequence from Example 1 is completed by the Laplace transform (4) that leads through $f^{-1}(x) = \exp(\lambda x)$ to the product copula Π.
- exponential distribution $F(x) = 1 - \exp(-\lambda x)$, $\lambda > 0$, by (4) we get $f^{-1}(x) = \frac{\lambda}{x+\lambda}$ and $f(x) = \lambda \left(\frac{1}{x} - 1\right)$ which generates the same copula (Clayton and Ali-Mikhail-Haq copula, both with parameter equal to 1) regardless of the choice of λ.
- distribution from Example 5, that is $F(x) = \min(1, x^p)$, $p \in]0, \infty[$ with Kendall's tau for $n = 1, 2, 3$ shown on Fig. 5, although no explicit form of universal generator inverse can be drawn, one can observe sequence of the lower bounds for Kendall's correlation coefficient, $\{\inf [\tau_n(p)]\}_{n=2}^{\infty} = \{-1, -\frac{1}{3}, -\frac{1}{5}, -\frac{1}{7}, \ldots\} = \left\{-\frac{1}{2n-3}\right\}_{n=2}^{\infty}$.

4   Some Open Problems

In Sect. 3 we have indicated some interesting consequences of the discussed links between additive generators of Archimedean copulas (of dimension $n = 2, 3, \ldots$ and universal) and distance functions via Williamson's transforms. Now we formulate some interesting arisen open problems explicitly.

Problem 1. Are there some statistical links between Archimedean copulas of dimensions n and m, $n \neq m$, related to the same distance function? Recall, for example, that fixing $F = \delta_p$ for some $p \in]0, \infty[$, the corresponding n-ary (universal) Archimedean copulas are the smallest n-ary (universal) Archimedean copulas.

Problem 2. For any fixed $n, m \geq 2$, $n \neq m$, one can define a transform $\varphi_{n,m} : \mathcal{D} \to \mathcal{D}$ on distance functions obtained as follows: for a fixed distance function $F \in \mathcal{D}$ one can define by means of (3) a pseudo-inverse $f_n^{(-1)}$ considering n in transform (3). Then, taking into account that $f_n^{(-1)}$ generates an

m-dimensional Archimedean copula for any $2 \leq m < n$, applying the transform (2), a new distance function $F_{n,m}$ is obtained. Now we put $\varphi_{n,m}(F) = F_{n,m}$. It is not difficult to check that if $2 \leq k < m < n$, then $\varphi_{m,k} \circ \varphi_{n,m} = \varphi_{n,k}$. Are there some interesting properties of transforms $\varphi_{n,m}$? For example, does this transform preserve the expected value, $E(F) = E(F_{n,m})$?

5   Conclusion

We have shown several interesting examples of connection between distributions of positively valued random variables (represented by distance function) and Archimedean copulas (represented by generator and it's inverse) through Williamson's transformation and Laplace transform, especially when arranged in a sequence. For instance, Williamson's n-transform ($n = 2, 3, \ldots$) links the product copula with distribution of sum of n exponentially distributed independent random variables while the Laplace transformation links it to the most elementary distance function, the Dirac function. Naturally there appears a question: how can we use statistical properties of distance functions to draw statistical properties of copulas, and vice versa? This question was itemized into two open problems, but surely more such problems could be formulated.

Acknowledgement. The work on this paper was supported by grant APVV-14-0013.

References

1. Bacigál, T., Juráňová, M., Mesiar, R.: On some new constructions of Archimedean copulas and applications to fitting problems. Neural Netw. World **20**(1), 81 (2010)
2. Bacigál, T., Mesiar, R., Najjari, V.: Generators of copulas and aggregation. Inf. Sci., submitted
3. Charpentier, A., Segers, J.: Convergence of Archimedean copulas. Stat. Probab. Lett. **78**(4), 412–419 (2008)
4. Jágr, V., Komorníková, M., Mesiar, R.: Conditioning stable copulas. Neural Netw. World **20**(1), 69–79 (2010)
5. Joe, H.: Multivariate Models and Dependence Concepts. Chapman and Hall, London (1997)
6. Kimberling, C.H.: A probabilistic interpretation of complete monotonicity. Aequationes Math. **10**(2), 152–164 (1974)
7. Klement, E.P., Mesiar, R., Pap, E.: Triangular norms. Kluwer Academic Publishers, Dodrecht (2000)
8. Klement, E.P., Mesiar, R., Pap, E.: Transformations of copulas. Kybernetika **41**(4), 425–434 (2005)
9. McNeil, A.J., Nešlehová, J.: Multivariate Archimedean copulas, d-Monotone functions and l_1-norm symmetric distributions. Ann. Stat. **37**(5B), 3059–3097 (2009)
10. Menger, K.: Statistical metrics. Proc. Nat. Acad. Sci. U.S.A. **28**(12), 535 (1942)
11. Mesiar, R.: On the pointwise convergence of continuous Archimedean t-norms and the convergence of their generators. BUSEFAL **75**, 39–45 (1998)
12. Michiels, F., De Schepper, A.: How to improve the fit of Archimedean copulas by means of transforms. Stat. Pap. **53**(2), 345–355 (2012)

13. Moynihan, R.: On τ_T semigroups of probability distributions II. Aequationes Math. **17**, 19–40 (1978)
14. Nelsen, R.B.: An Introduction to Copulas. Springer, New York (2006)
15. Rényi, A.: Wahrscheinlichketsrechnung mit einem Anhang über Informationstheorie. Deutsche Verlag der Wissenschaften, Berlin (1962)
16. Schweizer, B., Sklar, A.: Probabilistic Metric Spaces. Courier Dover Publications, Mineola (1983)
17. Sklar, A.: Fonctions de répartition à n dimensions et leurs marges. Publ. Inst. Statist. Univ. Paris **8**, 229–231 (1959)
18. Stupňanová, A., Kolesárová, A.: Associative n-dimensional copulas. Kybernetika **47**(1), 93–99 (2011)
19. Williamson, R.E.: Multiply monotone functions and their Laplace transforms. Duke Math. J. **23**, 189–207 (1956)

Some Remarks on Idempotent Nullnorms on Bounded Lattices

Gül Deniz Çaylı$^{(\boxtimes)}$ and Funda Karaçal

Department of Mathematics, Faculty of Sciences, Karadeniz Technical University,
61080 Trabzon, Turkey
guldeniz.cayli@ktu.edu.tr, fkaracal@yahoo.com

Abstract. Nullnorms are generalizations of triangular norms (t-norms) and triangular conorms (t-conorms) with a zero element to be an arbitrary point from an arbitrary bounded lattice. In this paper, we study on the existence of idempotent nullnorms on bounded lattices. We show that there exists unique idempotent nullnorm on an arbitrary distributive bounded lattice. We prove that an idempotent nullnorm may not always exist on every bounded lattice. Furthermore, we propose the construction method to obtain idempotent nullnorms on a bounded lattice under additional assumptions on given zero element. As by-product of this method, we see that it is in existence an idempotent nullnorm on non-distributive bounded lattices.

Keywords: Bounded lattice · Idempotent nullnorm · Zero element · Nullnorm

1 Introduction

T-norms with 1 as neutral element and t-conorms with 0 as neutral element have been introduced by Schweizer and Sklar in [24]. These operators have been extensively used in many applications in fuzzy set theory, fuzzy logics, multicriteria decision support and several branches of information sciences. For more details on t-norms, we refer to [1,2,17,18,22]. Subsequently, the concept of nullnorms and t-operators has been introduced in [4,19]. It has been stated that nullnorms and t-operators are equivalent in [20]. The nullnorms on unit interval have been also studied by many authors in other papers [7,10,23,25]. Nullnorms on the real unit interval as generalizations of t-norms and t-conorms admit a zero element a to be an arbitrary point from $[0, 1]$ and have to satisfy an additional condition. In case of $a = 1$, we obtain t-conorms and in case of $a = 0$, we obtain t-norms. Nullnorms have been found to be a useful tool in many different fields, such as the expert systems, neural networks, fuzzy logics in [21]. Moreover, they have to be used as aggregators in fuzzy logic maintain as many logical properties as possible.

Karaçal et al. [15] have studied nullnorms on bounded lattices. They have proved the existence of nullnorms with the zero element a for arbitrary element

© Springer International Publishing AG 2018
V. Torra et al. (eds.), *Aggregation Functions in Theory and in Practice*,
Advances in Intelligent Systems and Computing 581, DOI 10.1007/978-3-319-59306-7_4

$a \in L\backslash\{0,1\}$ with underlying t-norms and t-conorms on an arbitrary bounded lattice L. As a by-product, existence of the smallest nullnorm and of the greatest nullnorm has been shown. Moreover, Ince, Karaçal and Mesiar [13] have demonstrated the presence of idempotent nullnorms on a distributive bounded lattice L for any element $a \in L\backslash\{0,1\}$ playing the role of a zero element.

In this paper, we study idempotent nullnorms on bounded lattices. We prove that there is no idempotent nullnorm on a distributive bounded lattice L different from the proposal in [13]. Considering an arbitrary bounded lattice L, we show that there may not always exist an idempotent nullnorm V on L with the zero element $a \in L\backslash\{0,1\}$. Moreover, we introduce the method of constructing idempotent nullnorms on a bounded lattice L with given zero element $a \in L\backslash\{0,1\}$, if there is unique element in L incomparable with a.

The paper is organized as follows. After some preliminaries concerning nullnorms on bounded lattices, in Sect. 3 we discuss the existence of idempotent nullnorms on an arbitrary bounded lattice L with fixed zero element $a \in L\backslash\{0,1\}$. Next, it is given that the new method for constructing idempotent nullnorms on bounded lattices under additional assumption on $a \in L\backslash\{0,1\}$ which is considered as zero element. Finally, some examples and concluding remarks are added.

2   Preliminaries

In this section, some preliminaries concerning bounded lattices, t-norms, t-conorms and nullnorms on them are recalled.

Definition 1 [3]. A lattice $(L, \leqslant)$ is bounded if L has top and bottom elements, which are denoted as 1 and 0, respectively, that is, there exist two elements $1, 0 \in L$ such that $0 \leqslant x \leqslant 1$, for all $x \in L$.

Definition 2 [3]. A lattice $(L, \leqslant)$ is distributive lattice if the following two equivalent conditions hold:

(i) $x \wedge (y \vee z) = (x \wedge y) \vee (x \wedge z)$ for all $x, y, z \in L$.
(ii) $x \vee (y \wedge z) = (x \vee y) \wedge (x \vee z)$ for all $x, y, z \in L$.

Definition 3 [3]. Given a bounded lattice $(L, \leq, 0, 1)$ and $a, b \in L$, if a and b are incomparable, in this case, we use the notation $a \parallel b$. We denote the set of elements which are incomparable with a by I_a. So, $I_a = \{x \in L \mid x \parallel a\}$.

Definition 4 [3]. Given a bounded lattice $(L, \leq, 0, 1)$ and $a, b \in L$, $a \leq b$, the subinterval $[a, b]$ of L is defined as

$$[a, b] = \{x \in L \mid a \leq x \leq b\}.$$

Similarly, we define $(a, b] = \{x \in L \mid a < x \leq b\}, [a, b) = \{x \in L \mid a \leq x < b\}$ and $(a, b) = \{x \in L \mid a < x < b\}$.

Definition 5 [5,14,16]. An operation $T : L^2 \to L$ $(S : L^2 \to L)$ is called a t-norm (t-conorm) if it is commutative, associative, increasing with respect to both variables and has as neutral element $e = 1$ $(e = 0)$.

Definition 6 [15]. Let $(L, \leq, 0, 1)$ be a bounded lattice. A commutative, associative, non-decreasing in each variable function $V : L^2 \to L$ is called a nullnorm if there is an element $a \in L$ such that $V(x, 0) = x$ for all $x \leq a$ and $V(x, 1) = x$ for all $x \geq a$.

It can be easily obtained that $V(x, a) = a$ for all $x \in L$. So $a \in L$ is the zero element for V.

Consider the set $\mathscr{V}$ of all nullnorms on L with the following order: For $V_1, V_2 \in \mathscr{V}$,
$$V_1 \leq V_2 \Leftrightarrow V_1(x, y) \leq V_2(x, y) \text{ for all } (x, y) \in L^2.$$
It can be easily shown that $\mathscr{V}$ is a partially ordered set. If we denote the set of all nullnorms on L with the zero element $a \in L$ by $\mathscr{V}(a)$, then each $\mathscr{V}(a)$ is also a partially ordered set.

We use D_a to represent the following set:
$$D_a = [0, a] \times [a, 1] \cup [a, 1] \times [0, a] \text{ for } a \in L \backslash \{0, 1\}.$$

Definition 7 [13]. Let $(L, \leq, 0, 1)$ be a bounded lattice. An element $x \in L$ is called an idempotent element of a function $V : L \times L \to L$ if $V(x, x) = x$. The function V is called idempotent on L if all elements of L are idempotent.

Proposition 1 [9,11,15]. *Let $(L, \leq, 0, 1)$ be a bounded lattice, $a \in L \backslash \{0, 1\}$ and V be a nullnorm on L with the zero element e. Then*

(i) $V|_{[0,a]^2} : [0, a]^2 \to [0, a]$ is a t-conorm on $[0, a]$.
(ii) $V|_{[a,1]^2} : [a, 1]^2 \to [a, 1]$ is a t-norm on $[a, 1]$.

The next results characterizing general properties of nullnorms on a bounded lattice L are immediate from the definition of nullnorms.

Proposition 2 [9,15]. *Let $(L, \leq, 0, 1)$ be a bounded lattice, $a \in L \backslash \{0, 1\}$ and V be a nullnorm on L with the zero element a. The following properties hold:*

(i) $V(x, y) = a$ for all $(x, y) \in D_a$.
(ii) $a \leq V(x, y)$ for all $(x, y) \in [a, 1]^2 \cup [a, 1] \times I_a \cup I_a \times [a, 1]$.
(iii) $V(x, y) \leq a$ for all $(x, y) \in [0, a]^2 \cup [0, a] \times I_a \cup I_a \times [0, a]$.
(iv) $V(x, y) \leq y$ for all $(x, y) \in L \times [a, 1]$.
(v) $V(x, y) \leq x$ for all $(x, y) \in [a, 1] \times L$.
(vi) $x \leq V(x, y)$ for all $(x, y) \in [0, a] \times L$.
(vii) $y \leq V(x, y)$ for all $(x, y) \in L \times [0, a]$.
(viii) $x \vee y \leq V(x, y)$ for all $(x, y) \in [0, a]^2$.
(ix) $V(x, y) \leq x \wedge y$ for all $(x, y) \in [a, 1]^2$.
(x) $(x \wedge a) \vee (y \wedge a) \leq V(x, y)$ for all $(x, y) \in [0, a] \times I_a \cup I_a \times [0, a] \cup I_a \times I_a$.
(xi) $V(x, y) \leq (x \vee a) \wedge (y \vee a)$ for all $(x, y) \in [a, 1] \times I_a \cup I_a \times [a, 1] \cup I_a \times I_a$.

Corollary 1 [13]. *Let $(L, \leq, 0, 1)$ be a distributive bounded lattice and $a \in L \backslash \{0, 1\}$. Then*

$$V_*(x, y) = (x \wedge y) \vee (x \wedge a) \vee (y \wedge a) = (x \vee y) \wedge (x \vee a) \wedge (y \vee a) = V^*(x, y) \quad (1)$$

is an idempotent nullnorm with zero element $a \in L \backslash \{0, 1\}$.

3 Characterization of Idempotent Nullnorms on Bounded Lattices

In this section, we investigate the presence of idempotent nullnorms on bounded lattices. From Corollary 1, we know that there exists at least one idempotent nullnorm on distributive bounded lattices for given zero element. In the following theorem, we research whether there exists any idempotent nullnorm on distributive bounded lattices different from defined in Corollary 1.

Theorem 1. *Consider an arbitrary distributive bounded lattice* $(L, \leq, 0, 1)$ *and* $a \in L\backslash\{0, 1\}$. *In that case, there is no idempotent nullnorm* V *on* L *with the zero element* $a \in L\backslash\{0, 1\}$ *different from given by the formula* (1).

Note that if L is non-distributive bounded lattice, it does not need to satisfy the equality $(x \wedge y) \vee (x \wedge a) \vee (y \wedge a) = (x \vee y) \wedge (x \vee a) \wedge (y \vee a)$ (that is $V_*(x, y)$ does not need to equal to $V^*(x, y)$) in the formula (1) for all $x, y \in L$.

A natural question arises: for the bounded lattice L such that the operation $V : L^2 \to L$ defined as the formula (2) is an idempotent nullnorm with the zero element $a \in L\backslash\{0, 1\}$, does L need to be distributive?

$$V(x, y) = (x \vee y) \wedge (x \vee a) \wedge (y \vee a) \tag{2}$$

In the following, we give a negative example violating the above hypothesis.

Example 1. Given the non-distributive bounded lattice $L = \{0, a, t, z, 1\}$ with the order given in Fig. 1, define a mapping $V : L^2 \to L$ by Table 1. Then V is an idempotent nullnorm on L with the zero element a such that V is constructed by the formula (2). But the bounded lattice L is non-distributive.

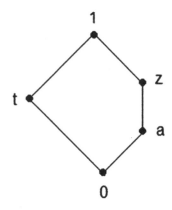

Fig. 1. The lattice L

Table 1. The idempotent nullnorm V on L

V	0	a	t	z	1
0	0	a	0	a	a
a	a	a	a	a	a
t	0	a	t	z	1
z	a	a	z	z	z
1	a	a	1	z	1

Consider an arbitrary bounded lattice L and $a \in L\backslash\{0,1\}$. Another genuine question arises: is there always an idempotent nullnorm V on L with the zero element a?

Let $(L, \leq, 0, 1)$ be a bounded lattice and $a \in L$. We know that there exists unique idempotent nullnorm V on L for $a = 0$, based on the fact that V is a t-norm for $a = 0$ and the only idempotent t-norm (inf) $T^{\wedge} : L^2 \to L$, $T^{\wedge}(x,y) = x \wedge y$ and there exists unique idempotent nullnorm V for $a = 1$, based on the fact that V is a t-conorm for $a = 1$ and the only idempotent t-conorm (sup) $S_{\vee} : L^2 \to L$, $S_{\vee}(x,y) = x \vee y$. One can wonder whether there exists an idempotent nullnorm on every bounded lattice L for the zero element $a \in L\backslash\{0,1\}$. In the following theorem, we show that there may not be existence an idempotent nullnorm on every bounded lattice L with the zero element $a \in L\backslash\{0,1\}$.

Theorem 2. *Suppose that a bounded lattice L contains a sublattice which is isomorphic to the sublattice characterized by Hasse diagram in Fig. 2. Then there is no idempotent nullnorm V on L for indicated zero element a.*

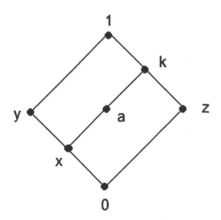

Fig. 2. The lattice L

Remark 1. Let $(L, \leq, 0, 1)$ be a bounded lattice, $a \in L \backslash \{0, 1\}$, S be a t-conorm on $[0, a]$ and T be a t-norm on $[a, 1]$. If all elements of L is comparable with a, then the following function $V : L^2 \rightarrow L$ is a nullnorm with the zero element a.

$$V(x, y) = \begin{cases} S(x, y) \ if \ (x, y) \in [0, a]^2, \\ T(x, y) \ if \ (x, y) \in [a, 1]^2, \\ a \qquad if \ (x, y) \in D_a. \end{cases} \quad (3)$$

If we put $S = S_\vee$ on $[0, a]$ and $T = T^\wedge$ on $[a, 1]$ in the formula (3), the following nullnorm is an idempotent nullnorm on L with the zero element a. Moreover, this idempotent nullnorm is unique, since the only idempotent t-conorm is $S_\vee$ and the only idempotent t-norm is $T^\wedge$.

$$V(x, y) = \begin{cases} x \vee y \ if \ (x, y) \in [0, a]^2, \\ x \wedge y \ if \ (x, y) \in [a, 1]^2, \\ a \qquad if \ (x, y) \in D_a. \end{cases} \quad (4)$$

By Theorem 2, we know that there is no nullnorm the bounded lattice L contains a sublattice which is isomorphic to the sublattice characterized in Fig. 2. So, we inverstigate that there always be in existence an idempotent nullnorm on which bounded lattice L with the zero element $a \in L \backslash \{0, 1\}$. In order that we propose the following theorem to characterize idempotent nullnorms on the bounded lattice L such that there is only one element in L incomparable with the zero element $a \in L \backslash \{0, 1\}$.

Theorem 3. *Let $(L, \leq, 0, 1)$ be a bounded lattice, $a \in L \backslash \{0, 1\}$ and there be only one element in L incomparable with a. If this element incomparable with a denotes by k, then the following function $V : L^2 \rightarrow L$ is an idempotent nullnorm with the zero element a.*

$$V(x, y) = \begin{cases} x \vee y & if \ (x, y) \in [0, a]^2, \\ x \wedge y & if \ (x, y) \in [a, 1]^2, \\ a & if \ (x, y) \in D_a, \\ x \vee (k \wedge a) \ if \ x \in [0, a] \ and \ y = k, \\ y \vee (k \wedge a) \ if \ x = k \ and \ y \in [0, a], \\ x \wedge (k \vee a) \ if \ x \in [a, 1] \ and \ y = k, \\ y \wedge (k \vee a) \ if \ x = k \ and \ y \in [a, 1], \\ k & if \ x = y = k. \end{cases} \quad (5)$$

Remark 2. Let $(L, \leq, 0, 1)$ be a bounded lattice and $a \in L \backslash \{0, 1\}$.

(i) If we take the bounded lattice L is distributive in Theorem 3, then the idempotent nullnorm obtained in the formula (5) coincides with the idempotent nullnorm given in the formula (1).

(ii) In Theorem 3, if the bounded lattice L is non-distributive and there are at least two elements incomparable with a, the formula (5) may not give an idempotent nullnorm on L due to Theorem 2.

Example 2. Given the bounded lattice $L = \{0, x, y, a, z, t, 1\}$, with the order given in Fig. 3, define a mapping $V : L^2 \to L$ by Table 2 such that V is constructed by using the formula (5). Then V is an idempotent nullnorm on L with the zero element a from Theorem 3.

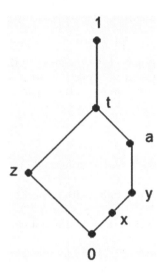

Fig. 3. The lattice L

Table 2. The idempotent nullnorm V on L

V	0	x	y	a	z	t	1
0	0	x	y	a	0	a	a
x	x	x	y	a	x	a	a
y	y	y	y	a	y	a	a
a	a	a	a	a	a	a	a
z	0	x	y	a	z	t	t
t	a	a	a	a	t	t	t
1	a	a	a	a	t	t	1

4   Concluding Remarks

In this paper, we study idempotent nullnorms on the bounded lattices. In the case of standard real unit interval $L = [0, 1]$, each nullnorm V with a zero element $a \in (0, 1)$ forms obtained in [15, Corollary 6]. And, here nullnorm V is the unique idempotent nullnorm with zero element $a \in (0, 1)$ based on the fact that the only idempotent t-norm on $[0,1]$ is $T^\wedge$ and the only idempotent t-conorm on $[0,1]$ is $S_\vee$. It can be found more details about idempotent nullnorms on unit interval in [8,12]. If we consider an arbitrary bounded lattice L, we show that there may not always be in existence idempotent nullnorms on L and in the case of L is a distributive, there is unique idempotent nullnorm on L. Moreover, we introduce a method to characterize idempotent nullnorms on an arbitrary bounded lattice L with zero element $a \in L\backslash\{0,1\}$, if there is only one element in L incomparable with a.

Acknowledgment. The full proofs of the theorems in this paper are contained in the paper [6]. We are grateful to the anonymous reviewers and editors for their valuable comments which have enabled us to improve the original version of our paper.

References

1. Aşıcı, E., Karaçal, F.: On the T-partial order and properties. Inf. Sci. **267**, 323–333 (2014)
2. Aşıcı, E., Karaçal, F.: Incomparability with respect to the triangular order. Kybernetika **52**, 15–27 (2016)
3. Birkhoff, G.: Lattice theory. American Mathematical Society Colloquium Publishers, Providence (1967)
4. Calvo, T., De Baets, B., Fodor, J.: The functional equations of Frank and Alsina for uninorms and nullnorm. Fuzzy Sets Syst. **120**, 385–394 (2001)
5. Çaylı, G.D., Karaçal, F., Mesiar, R.: On a new class of uninorms on bounded lattices. Inf. Sci. **367–368**, 221–231 (2016)
6. Çaylı, G.D., Karaçal, F.: Idempotent nullnorms on bounded lattices. Working paper
7. Drewniak, J., Drygaś, P., Rak, E.: Distributivity between uninorms and nullnorms. Fuzzy Sets Syst. **159**, 1646–1657 (2008)
8. Drygaś, P.: A characterization of idempotent nullnorms. Fuzzy Sets Syst. **145**, 455–461 (2004)
9. Drygaś, P.: Isotonic operations with zero element in bounded lattices. In: Atanassov, K., Hryniewicz, O., Kacprzyk, J. (eds.) Soft Computing Foundations and Theoretical Aspect, pp. 181–190. EXIT, Warszawa (2004)
10. Drygaś, P.: Distributivity between semi t-operators and semi nullnorms. Fuzzy Sets Syst. **264**, 100–109 (2015)
11. Ertuğrul, Ü., Kesicioğlu, M.N., Karaçal, F.: Ordering based on uninorms. Inf. Sci. **330**, 315–327 (2016)
12. Grabisch, M., Marichal, J.L., Mesiar, R., Pap, E.: Aggregation functions. Cambridge University Press, Cambridge (2009)
13. İnce, M.A., Karaçal, F., Mesiar, R.: Medians and nullnorms on bounded lattices. Fuzzy Sets Syst. **289**, 74–81 (2016)
14. Karaçal, F., Mesiar, R.: Uninorms on bounded lattices. Fuzzy Sets Syst. **261**, 33–43 (2015)
15. Karaçal, F., İnce, M.A., Mesiar, R.: Nullnorms on bounded lattices. Inf. Sci. **325**, 227–236 (2015)
16. Kesicioğlu, M.N., Mesiar, R.: Ordering based on implications. Fuzzy Sets Syst. **276**, 377–386 (2014)

17. Klement, E.P., Mesiar, R., Pap, E.: On the relationship of associative compensatory operation to triangular norms and conorms. Int. J. Uncertain Fuzziness Knowl. Based Syst. **4**, 129–144 (1996)
18. Klement, E.P., Mesiar, R., Pap, E.: Triangular norms. Kluwer Academic Publishers, Dordrecht (2000)
19. Mas, M., Mayor, G., Torrens, J.: t-operators. Int. J. Uncertain Fuzziness Knowl. Based Syst. **7**, 31–50 (1999)
20. Mas, M., Mayor, G., Torrens, J.: The distributivity condition for uninorms and t-operators. Fuzzy Sets Syst. **128**, 209–225 (2002)
21. Mas, M., Mayor, G., Torrens, J.: The modularity condition for uninorms and t-operators. Fuzzy Sets Syst. **126**, 207–218 (2002)
22. Mesiar, R., Pap, E.: Different interpretations of triangular norms and related operations. Fuzzy Sets Syst. **96**, 183–189 (1998)
23. Mesiarová-Zemánková, A.: Multi-polar t-conorms and uninorms. Inf. Sci. **301**, 227–240 (2015)
24. Schweizer, B., Sklar, A.: Statistical metric spaces. Pac. J. Math. **10**, 313–334 (1960)
25. Xie, A., Liu, H.: On the distributivity of uninorms over nullnorms. Fuzzy Sets Syst. **211**, 62–72 (2013)

Aggregating Fuzzy Subgroups
and T-vague Groups

D. Boixader[1], G. Mayor[2], and J. Recasens[1($\boxtimes$)]

[1] Universitat Politècnica de Catalunya, Sant Cugat del Vallès, Barcelona, Spain
{dionis.boixader,j.recasens}@upc.edu
[2] Universitat de les Illes Balears, Palma (Mallorca), Spain
gmayor@uib.es

Abstract. Fuzzy subgroups and T-vague groups are interesting fuzzy algebraic structures that have been widely studied. While fuzzy subgroups fuzzify the concept of crisp subgroup, T-vague groups can be identified with quotient groups of a group by a normal fuzzy subgroup and there is a close relation between both structures and T-indistinguishability operators (fuzzy equivalence relations).

In this paper the functions that aggregate fuzzy subgroups and T-vague groups will be studied. The functions aggregating T-indistinguishability operators have been characterized [9] and the main result of this paper is that the functions aggregating T-indistinguishability operators coincide with the ones that aggregate fuzzy subgroups and T-vague groups. In particular, quasi-arithmetic means and some OWA operators aggregate them if the t-norm is continuous Archimedean.

1 Introduction

T-indistinguishability operators, also called fuzzy equivalence relations or fuzzy equalities, fuzzify the concepts of crisp equivalence relation and crisp equality. They appear naturally when studying fuzzy systems and have an extensive amount of literature since its first definition by Zadeh in [18]. One important example is the study of fuzzy algebras (see [10] for example) and the works by Demirci [2]. In many situations more than one T-indistinguishability operator is defined on a system and it is necessary to aggregate them into a unique of such fuzzy relations. There are many works dealing with this problem [6,12,13]. The last one is [9] where the aggregation operators that aggregate T-indistinguishability operators have been completely characterized regardless of the nature of the t-norm T (continuity, archimedianity,...). This result will be recalled and used in this work.

Fuzzy subgroups of a group $(G, \circ)$ were introduced by Rosenfeld [16] as a natural generalization of subgroup of G and have been widely studied [11]. They are defined as fuzzy subsets of $(G, \circ)$ satisfying some properties fuzzifying the definition of crisp subgroup.

© Springer International Publishing AG 2018
V. Torra et al. (eds.), *Aggregation Functions in Theory and in Practice*,
Advances in Intelligent Systems and Computing 581, DOI 10.1007/978-3-319-59306-7_5

In the crisp case, if $(G, \circ)$ is a set with an operation $\circ : G \times G \to G$ and $\sim$ is an equivalence relation on G, then $\circ$ is compatible with $\sim$ if and only if

$$a \sim a' \text{ and } b \sim b' \text{ implies } a \circ b \sim a' \circ b'.$$

In this case, an operation $\tilde{\circ}$ can be defined on $\overline{G} = G/\sim$ by

$$\overline{a}\tilde{\circ}\overline{b} = \overline{a \circ b}$$

where $\overline{a}$ and $\overline{b}$ are the equivalence classes of a and b with respect to $\sim$.

If the equivalence relation is replaced by a T-indistinguishability operator we obtain fuzzy algebras [4,5,10]. Demirci generalized this idea by introducing the concept of vague algebra, which basically consists of a fuzzy operation on a set G (i.e.: a mapping $\tilde{\circ} : G \times G \times G \to [0,1]$ where $\tilde{\circ}(a,b,c)$ is interpreted as the degree in which $a \circ b$ is equivalent or indistinguishable from c) compatible with a given indistinguishability operator [2].

To every fuzzy subgroup μ of a group $(G, \circ)$ two T-indistinguishability operators E_μ and $_\mu E$ can be associated, that are left and a right invariant under translations. If the fuzzy subgroup μ is normal, then the left and right T-indistinguishability operators E_μ and $_\mu E$ coincide. The operation of the group is compatible with the T-indistinguishability operator E_μ if and only if μ is a normal fuzzy subgroup of G.

A group $(G, \circ)$ with the vague operation $\tilde{\circ}(a,b,c) = E(a \circ b, c)$ is a T-vague group and, reciprocally, for every T-vague group of G there exists a T-indistinguishability operator E that is invariant under translations such that $\tilde{\circ}(a,b,c) = E(a \circ b, c)$. Moreover, there is a bijection between the fuzzy normal subgroups of G and the T-vague groups of G. In particular, T-vague groups can be thought of as the fuzzy counterparts of crisp quotient groups (if μ is the normal fuzzy subgroup associated to the vague group $(G, \tilde{\circ})$, then $(G, \tilde{\circ})$ can be identified with G/μ).

Due to the importance of fuzzy groups and T-vague groups, it seems interesting to study how can they be fused or aggregated. More concrete, finding aggregation functions that aggregate them. This will be done in the present work, where the functions aggregating them will be characterized. Due to the close relation between fuzzy subgroups, T-vague groups and T-indistinguishability operator, it is not surprising that the aggregation functions aggregating them coincide. This is the main result of the work. As a consequence, quasi arithmetic means and some OWA operators aggregate fuzzy subgroups and T-vague groups if T is a continuous Archimedean t-norm.

2   Preliminaries

In this section, the basic definitions and properties of T-indistinguishability operators, fuzzy subgroups and T-vague groups are recalled.

T-indistinguishability operators fuzzify the concepts of crisp equivalence and crisp equality. An overview of these operators can be found in [14].

Definition 2.1. *Let T be a t-norm. A fuzzy relation E on a set X is a T-indistinguishability operator if and only if for all $x, y, z \in X$*

1. $E(x, x) = 1$ *(Reflexivity)*
2. $E(x, y) = E(y, x)$ *(Symmetry)*
3. $T(E(x, y), E(y, z)) \leq E(x, z)$ *(T-transitivity).*

E separates points if and only if

(c) $E(x, y) = 1$ implies $x = y$.

Fuzzy subgroups were introduced by Rosenfeld [16] by fuzzifying the definition of crisp subgroup. The use of the minimum to model the conjunction has been generalized to any t-norm later on. The reader interested in the study of fuzzy subgroups is referred to [11].

Definition 2.2. *Let $(G, \circ)$ be a group, e its identity element and μ a fuzzy subset of G. μ is a fuzzy subgroup of G if and only if for all $x, y \in G$*

- $\mu(e) = 1$
- $T(\mu(x), \mu(y)) \leq \mu(x \circ y^{-1}) \ \forall x, y \in X$.

T-vague algebras were introduced by Demirci considering fuzzy operations compatible with given T-indistinguishability operators and an extensive study of vague operations and T-vague groups can be found in [2].

Definition 2.3. *A fuzzy binary operation on a set G is a map $\tilde{\circ} : G \times G \times G \to [0, 1]$.*

$\tilde{\circ}(x, y, z)$ is interpreted as the degree in which z is $x \circ y$.

Definition 2.4. *Let E be a T-indistinguishability operator on G. A vague binary operation on G is a fuzzy binary operation $\tilde{\circ}$ satisfying for all $x, x', y, y' \in G$*

(a) $T(\tilde{\circ}(x, y, z), E(x, x'), E(y, y'), E(z, z')) \leq \tilde{\circ}(x', y', z')$.
(b) $T(\tilde{\circ}(x, y, z), \tilde{\circ}(x, y, z')) \leq E(z, z')$.
(c) For all $x, y \in G$ there exists a unique $z \in G$ such that $\tilde{\circ}(x, y, z) = 1$.

N.B. The usual definition of vague binary operation does not require uniqueness in the third property and the vague binary operations with this property are called perfect.

Definition 2.5. *Let $\tilde{\circ}$ be a T-vague binary operation on G with respect to a T-indistinguishability operator E on G. Then $(G, \tilde{\circ})$ is a T-vague group if and only if it satisfies the following properties.*

1. *Associativity.* $\forall x, y, z, t, m, q, w, \in G$

$$T(\tilde{\circ}(y, z, t), \tilde{\circ}(x, t, m), \tilde{\circ}(x, y, q), \tilde{\circ}(q, z, w)) \leq E(m, w)).$$

2. *Identity. There exists a (two sided) identity element $e \in G$ such that*

$$T(\tilde{o}(e, x, x), \tilde{o}(x, e, x)) = 1$$

for each $a \in G$.

3. *Inverse. For each $x \in G$ there exists a (two-sided) inverse element $x^{-1} \in G$ such that*

$$T(\tilde{o}(x^{-1}, x, e), \tilde{o}(x, x^{-1}, e)) = 1.$$

A T-vague group is Abelian or commutative if and only if

$$\forall x, y, m, w \in G, T((\tilde{o}(x, y, m), \tilde{o}(y, x, w))) \leq E(m, w)).$$

Proposition 2.6. *Let $(G, \tilde{o})$ be a T-vague group. Then $(G, \circ)$ with $\circ$ defined for all $x, y \in G$ by $x \circ y = z$, where z is the unique element of G with $\tilde{o}(x, y, z) = 1$, is a group.*

3 Relationship Between Indistinguishability Operators, Fuzzy Subgroups and Vague Groups

3.1 Fuzzy Subgroups

To every fuzzy subset μ of a group $(G, \circ)$ a pair of fuzzy relations can be associated that are indistinguishability operators if and only if μ is a fuzzy subgroup of G. This two indistinguishability operators coincide when μ is a fuzzy normal subgroup and there is a compatibility between it and the operation $\circ$ of the group. These properties and their relation with the invariance under translations will be analyzed in this section.

In the crisp case, given a subgroup H of a group $(G, \circ)$, the relations $\sim_r$ and $\sim_l$ on G defined by $x \sim_r y$ if and only if $x \circ y^{-1} \in H$ and $x \sim_l y$ if and only if $y^{-1} \circ x \in H$ respectively are equivalence relations. The operation $\circ$ of G is compatible with $\sim_r$ and $\sim_l$ if and only if H is a normal subgroup of G.

These results can be generalized to fuzzy subgroups and T-indistinguishability operators.

Definition 3.1. *Let $\circ$ be a binary operation on G and E a fuzzy relation on G. E is invariant under translations with respect to $\circ$ if and only if*

(a)

$$E(x, y) = E(z \circ x, z \circ y) \quad (left\ invariant)$$

and

(b)

$$E(x, y) = E(x \circ z, y \circ z) \quad (right\ invariant),$$

$\forall x, y, z \in G.$

To every fuzzy subset μ of a group $(G, \circ)$ two fuzzy relations E_μ and $_\mu E$ can be assigned that are right and left invariant T-indistinguishability operators respectively if and only if μ is a fuzzy subgroup of G.

Definition 3.2. *Let μ be a fuzzy subset of $(G, \circ)$. The fuzzy relations E_μ and $_\mu E$ on G defined by*

$$E_\mu(x, y) = \mu(x \circ y^{-1}) \ \forall x, y \in G$$

and

$$_\mu E(x, y) = \mu(y^{-1} \circ x) \ \forall x, y \in G$$

are the right and left fuzzy relations associated to μ respectively.

Proposition 3.3. *Let μ be a fuzzy subgroup of a group $(G, \circ)$. Then E_μ and $_\mu E$ are right and left invariant T-indistinguishability operators on G respectively.*

Lemma 3.4. *If μ is a fuzzy subgroup of $(G, \circ)$ and e is the identity element of G, then $E_\mu(x, y) = E_\mu(e, x \circ y^{-1})$ and $_\mu E(x, y) =_\mu E(e, y \circ x^{-1}) \ \forall x, y \in G$.*

Proof. Trivial.

Reciprocally, to every right (left) T-indistinguishability operator on $(G, \circ)$ a fuzzy subgroup of G can be assigned.

Proposition 3.5. *Let E be a T-indistinguishability operator on a group $(G, \circ)$ with identity element e such that E is right invariant. Then the column μ_e of E (i.e., the fuzzy subset μ_e of G defined by $\mu_e(x) = E(e, x) \ \forall x \in G$) is a fuzzy subgroup of G and $E = E_{\mu_e}$.*

Similarly,

Proposition 3.6. *Let E be a T-indistinguishability operator on a group $(G, \circ)$ with identity element e such that E is left invariant. Then the column μ_e of E is a fuzzy subgroup of G and $E =_{\mu_e} E$.*

Corollary 3.7. *Let $(G, \circ)$ be a group. There exist bijections between the set **FSG** of fuzzy subgroups of G, the set **RIG** of right invariant indistinguishability operators on G and the set **LIG** of left invariant indistinguishability operators on G mapping every fuzzy subgroup μ of G into its associated T-indistinguishability operators E_μ and $_\mu E$.*

The following definition fuzzifies the concept of normal subgroup.

Definition 3.8. *A fuzzy subgroup μ of a group $(G, \circ)$ is called a normal fuzzy subgroup if and only if $\mu(x \circ y) = \mu(y \circ x) \ \forall x, y \in G$.*

Proposition 3.9. *Let $(G, \circ)$ be a group and μ a normal fuzzy subgroup of G. The associated T-indistinguishability operators E_μ and $_\mu E$ to μ coincide and are invariant under translations.*

Reciprocally,

Proposition 3.10. *Let $(G, \circ)$ be a group, μ a fuzzy subgroup of G and E_μ and $_\mu E$ its associated T-indistinguishability operators. If E_μ and $_\mu E$ are invariant under translations, then they coincide and μ is a normal fuzzy subgroup of G.*

Corollary 3.11. *Let $(G, \circ)$ be a group. There is a bijection between the set of normal fuzzy subgroups of G and the set of T-indistinguishability operators on G invariant under translations with respect to $\circ$.*

The following proposition links normality of a fuzzy subgroup μ with compatibility with respect to its associated T-indistinguishability operator E_μ.

Proposition 3.12. *Let $(G, \circ)$ be a group, μ a fuzzy normal subgroup of G and E_μ its associated T-indistinguishability operator. Then $\circ$ is extensional with respect to E_μ (i.e., $T(E_\mu(x, x'), E_\mu(y, y')) \leq E_\mu(x \circ y, x' \circ y'))$.*

3.2 Vague Groups

Vague groups were introduced in [2] as structures compatible with given indistinguishability operators. They are also closely related to fuzzy normal subgroups [3].

The next two propositions relate normal fuzzy subgroups and indistinguishability operators invariant under translations with vague groups.

Proposition 3.13 [3]. *Let $(G, \circ)$ be a group, μ a normal fuzzy subgroup of $(G, \circ)$ and E_μ its associated T-indistinguishability operator on G. If $\tilde{\circ} : G \times G \times G \to [0, 1]$ is defined for all $x, y, z \in G$ by $\tilde{\circ}(x, y, z) = \mu(x \circ y \circ z^{-1}) = E_\mu(x \circ y, z)$, then $(G, \tilde{\circ})$ is a T-vague group.*

Proposition 3.14 [3]. *Let $(G, \tilde{\circ})$ be a T-vague group with respect to the T-indistinguishability operator E. Then,*

(a) $\tilde{\circ}(x, y, z) = E(x \circ y, z) \; \forall x, y, z \in G$.
(b) $\circ$ is extensional with respect to E.
(c) E is invariant under translations with respect to $\circ$.

Proposition 3.15. *Let $(G, \circ)$ be a group. There exist bijective maps between its T-vague groups, its fuzzy normal subgroups and its T-indistinguishability operators invariant under translations.*

Proof. The bijections are given by

$$\tilde{\circ}(x, y, z) = E(x \circ y, z) = \mu(x \circ y \circ z^{-1}).$$

4 Aggregating Fuzzy Subgroups and Vague Groups

In [9] the functions preserving indistinguishability operators have been characterized by means of the so called T-triangular triplets. In this section we first recall these results and we will use them in the next two subsections to obtain the functions preserving fuzzy subgroups and vague groups thanks to the results of the previous Sect. 3.

Definition 4.1. *We say that a triplet $(a, b, c) \in [0, \infty]^3$ is triangular if and only if*

$$a \leq b + c, \quad b \leq a + c, \quad c \leq a + b.$$

Being $\boldsymbol{a}, \boldsymbol{b}, \boldsymbol{c} \in [0, \infty]^m$, $m \geq 1$, we say that $(\boldsymbol{a}, \boldsymbol{b}, \boldsymbol{c})$ is a (m-dimensional) triangular triplet if (a_i, b_i, c_i) is triangular for all $i = 1, \ldots, m$, where $\boldsymbol{a} = (a_1, \ldots, a_m), \boldsymbol{b} = (b_1, \ldots, b_m), \boldsymbol{c} = (c_1, \ldots, c_m)$.

Definition 4.2. *Let T be a t-norm. We say that $(a, b, c) \in [0, 1]^3$ is T-triangular if and only if*

$$a \geq T(b, c), \quad b \geq T(a, c), \quad c \geq T(a, b).$$

Being $\boldsymbol{a}, \boldsymbol{b}, \boldsymbol{c} \in [0, 1]^m$, $m \geq 1$, we say that $(\boldsymbol{a}, \boldsymbol{b}, \boldsymbol{c})$ is a (m-dimensional) T-triangular triplet if (a_i, b_i, c_i) is T-triangular for all $i = 1, \ldots, m$, where $\boldsymbol{a} = (a_1, \ldots, a_m), \boldsymbol{b} = (b_1, \ldots, b_m), \boldsymbol{c} = (c_1, \ldots, c_m)$.

Proposition 4.3. *Let T be a left continuous t-norm and $\overleftrightarrow{T}$ its bi-residuation. A triplet $(a, b, c) \in [0, 1]^3$ is T-triangular if and only if $T(a, b) \leq c \leq \overleftrightarrow{T}(a, b)$.*

Example 4.4.

- *A triplet is T-triangular with respect to the minimum t-norm if and only if there exists a reordering (a, b, c) such that $a = b$ and $c \geq a$.*
- *A triplet is T-triangular with respect to the Łukasiewicz t-norm if and only if there exists a reordering (a, b, c) such that $max(a + b - 1, 0) \leq c \leq 1 - |a - b|$.*
- *A triplet is T-triangular with respect to the product t-norm if and only if it is $(0, 0, 0)$ or there exists a reordering (a, b, c) with $a, b, c > 0$, such that $ab \leq c \leq min(\frac{a}{b}, \frac{b}{a})$.*

Definition 4.5. *A function $F : [0, 1]^m \to [0, 1], m \geq 1$, aggregates T-indistinguishability operators if for any set X and any collection of T-indistinguishability operators on X, $(E_1, \ldots, E_m)$, then $F(E_1, \ldots, E_m)$ is also a T-indistinguishability operators on X, where $F(E_1, \ldots, E_m)$ is the fuzzy binary relation $F(E_1, \ldots, E_m)(x, y) = F(E_1(x, y), \ldots, E_m(x, y))$.*

The next result characterizes the functions that aggregate T-indistinguishability operators.

Proposition 4.6 [9]. *A function $F : [0, 1]^m \to [0, 1], m \geq 1$, aggregates T-indistinguishability operators if and only if the following conditions hold:*

(i) $F(\overbrace{1, \ldots, 1}^{m}) = 1$.

(ii) F transforms m-dimensional T-triangular triplets into 1-dimensional T-triangular triplets.

Proposition 4.7. *A function $F : [0, 1]^m \to [0, 1]$, aggregates min-indistinguishability operators if and only if it is increasing in each variable and $F(1, \ldots, 1) = 1$.*

When T is a continuous Archimedean t-norm, a characterization of those functions that aggregate T-equivalence relations can be formulated in terms of an additive generator of T as follows.

Proposition 4.8. *If T is a continuous Archimedean t-norm with additive generator g, then $F : [0,1]^m \longrightarrow [0,1]$ aggregates T-indistinguishability operators if and only if the function $G = gF(g^{(-1)} \times \ldots \times g^{(-1)})$ transforms (ordinary) triangular triplets of $[0,\infty]^m$ (with elements in $[0,g(0)]^m$) into (ordinary) triangle triplets of $[0,\infty]$ (with elements in $[0,g(0)]$) and $G(0,\ldots,0) = 0$.*

Example 4.9. *A function $F : [0,1]^m \rightarrow [0,1], m \geq 1$, aggregates T-indistinguishability operators with T the Łukasiewicz t-norm if and only if $G(a_1,\ldots,a_m) = 1 - F(max(1 - a_1,0),\ldots,max(1 - a_m,0))$ transforms triangular triplets of $[0,\infty]^m$ (with elements in $[0,1]^m$) into triangle triplets of $[0,\infty]$ (with elements in $[0,1]$) and $G(0,\ldots,0) = 0$.*

Under increasingness, subadditivity ($G(\mathbf{a} + \mathbf{b}) \leq G(\mathbf{a}) + G(\mathbf{b})$) is equivalent to the property of transforming triangular triplets into triangle triplets.

Proposition 4.10. *Consider $G : [0,\infty]^m \longrightarrow [0,\infty]$. Then:*

(i) If G transforms triangular triplets of $[0,\infty]^m$ into triangular triplets of $[0,\infty]$ then it is subadditive.

(ii) If G is increasing and subadditive then it transforms triangular triplets of $[0,\infty]^m$ into triangular triplets of $[0,\infty]$.

Thus, from the two previous propositions, we can enunciate the following

Proposition 4.11. *Let T be a continuous Archimedean t-norm with additive generator g. An increasing function $F : [0,1]^m \rightarrow [0,1]$, with $F(1,\ldots,1) = 1$, aggregates T-indistinguishability operators if and only if the function $G = gFg^{(-1)}$ is subadditive.*

Consequences of the previous propositions are two known results concerning the role of weighted arithmetic means and ordered weighted arithmetic means (OWA operators) in this approach.

Proposition 4.12. *Let T be a continuous Archimedean t-norm with additive generator g. Any weighted quasi-arithmetic mean $F(a_1,\ldots,a_m) = g^{-1}(\Sigma w_i g(a_i))$ where $(w_1,\ldots,w_m)$ are non-negative real numbers satisfying $\Sigma w_i = 1$ aggregates T-indistinguishability operators.*

Proposition 4.13. *Let T be a continuous Archimedean t-norm with additive generator g. An ordered weighted quasi-arithmetic mean $F(a_1,\ldots,a_m) = g^{-1}(\Sigma w_i g(a_{(m-i)}))$ where $a_{(k)}$ denotes the k-largest input in the list $(a_1,\ldots,a_m)$ aggregates T-indistinguishability operators if $w_i \geq w_j$ for $i \leq j$.*

4.1   Aggregating Fuzzy Subgroups

The relationship between fuzzy subgroups and indistinguishability operators will be used in this subsection to characterize the functions aggregating fuzzy subgroups.

Definition 4.14. *A function $S : [0,1]^m \to [0,1], m \geq 1$, aggregates T-fuzzy subgroups if for any group $(G, \circ)$ and any collection of T-fuzzy subgroups of G, $\mu_1, \ldots, \mu_m$, then $S(\mu_1, \ldots, \mu_m)$ is also a T-fuzzy subgroup of G, where $S(\mu_1, \ldots, \mu_m)$ is the fuzzy subset $S(\mu_1, \ldots, \mu_m)(x) = S(\mu_1(x), \ldots, \mu_m(x))$.*

Proposition 4.15. *Let $(G, \circ)$ be a group, $E_1, E_2, \ldots, E_m$ left (right) invariant T-indistinguishability operators on G and $F : [0,1]^m \to [0,1]$ a function aggregating T-indistinguishability operators. Then $F(E_1, E_2, \ldots, E_m)$ is a left (right) invariant T-indistinguishability operator.*

Proof.

$$F(E_1, E_2, \ldots, E_m)(z \circ x, z \circ y)$$
$$= F(E_1(z \circ x, z \circ y), E_2(z \circ x, z \circ y), \ldots, E_m(z \circ x, z \circ y))$$
$$= F(E_1(x, y), E_2(x, y), \ldots, E_m(x, y))$$
$$= F(E_1, E_2, \ldots, E_m)(x, y).$$

Proposition 4.16. *Let $(G, \circ)$ be a group, $F : [0,1]^m \to [0,1]$ a function, $\mu_1, \ldots \mu_m$ fuzzy subsets of G and $_{\mu_1}E, \ldots, _{\mu_m}E$ $(E_{\mu_1}, \ldots, E_{\mu_m})$ their respective left (right) invariant fuzzy relations. Then*

(a) $F(_{\mu_1}E, \ldots, _{\mu_m}E) = {}_{F(\mu_1, \ldots, \mu_m)}E$
(b) $F(E_{\mu_1}, \ldots, E_{\mu_m}) = E_{F(\mu_1, \ldots, \mu_m)}$.

Proof. We will prove (a):

$$F(_{\mu_1}E, \ldots, _{\mu_m}E)(x, y) = F(_{\mu_1}E(x, y), \ldots, _{\mu_m}E(x, y))$$
$$= F(\mu_1(y^{-1} \circ x), \ldots, \mu_m(y^{-1} \circ x))$$
$$= F(\mu_1, \ldots, \mu_m)(y^{-1} \circ x)$$
$$= {}_{F(\mu_1, \ldots, \mu_m)}E(x, y).$$

As a corollary we obtain:

Corollary 4.17. *With the same notations as in the preceding Proposition 4.16,*

- *if F aggregates T-indistinguishability operators and $\mu_1, \ldots, \mu_m$ are fuzzy subgroups of G, then $F(\mu_1, \ldots, \mu_m)$ is a fuzzy subgroup of G.*
- *if F aggregates T-fuzzy subgroups and $E_1, \ldots, E_m$ are left (right) invariant T-indistinguishability operators, then $F(E_1, \ldots, E_m)$ is a left (right) invariant T-indistinguishability operator.*

Corollary 4.18. *If $F([0,1]^m \to [0,1]$ aggregates T-indistinguishability operators, then F aggregates T-fuzzy subgroups.*

Proposition 4.19. *Let* $F : [0,1]^m \to [0,1]$ *be a function that aggregates fuzzy subgroups. Then* F *transforms* m-*dimensional* T-*triangular triplets into* 1-*dimensional triplets.*

Proof. Consider the group $\mathbb{Z}/(2) \times \mathbb{Z}/(2)$ and put $e = (\overline{0}, \overline{0})$, $a = (\overline{0}, \overline{1})$, $b = (\overline{1}, \overline{0})$, $c = (\overline{1}, \overline{1})$.

Let $\mathbf{a} = (a_1, \ldots, a_m), \mathbf{b} = (b_1, \ldots, b_m), \mathbf{c} = (c_1, \ldots, c_m)$ be m-dimensional T-triplets. For every $i = 1., m$ consider the fuzzy subgroup μ_i defined by

$$\mu_i(e) = 1, \quad \mu_i(a) = a_i, \quad \mu_i(b) = b_i, \quad \mu_i(c) = c_i.$$

Corollary 4.20. *Let* $F : [0,1] \to [0,1]$ *be a function that aggregates left (right) invariant* T-*indistinguishability operators. Then* F *transforms* m-*dimensional* T-*triangular triplets into* 1-*dimensional triplets.*

From Propositions 4.6 and 4.19 we obtain the following important result.

Proposition 4.21. *A function* $F : [0,1]^m \to [0,1], m \geq 1$, *aggregates fuzzy subgroups if and only if it aggregates* T-*indistinguishability operators.*

Proposition 4.22. *A function* $F : [0,1]^m \to [0,1], m \geq 1$, *aggregates fuzzy subgroups if and only if the following conditions hold:*

(i) $F(\overbrace{1, \ldots, 1}^{m}) = 1.$

(ii) F *transforms* m-*dimensional* T-*triangular triplets into* 1-*dimensional* T-*triangular triplets.*

4.2 Aggregating Vague Groups

In this subsection we will obtain results on the aggregation of T-vague groups similar to the ones in the previous subsection.

Definition 4.23. *A function* $V : [0,1]^m \to [0,1], m \geq 1$, *aggregates* T-*vague groups if for any group* $(G, \circ)$ *and any collection of* T-*vague groups of* G, $(G, \tilde{\circ}_1), \ldots, (G, \tilde{\circ}_m)$, *then* $V((G, \tilde{\circ}_1), \ldots, (G, \tilde{\circ}_m))$ *is also a* T-*vague group* G, *where* $V((G, \tilde{\circ}_1), \ldots, (G, \tilde{\circ}_m))$ *is the* T-*vague group of* G *with the vague operation defined for all* $x, y, z \in G$ *by* $V(\tilde{\circ}_1, \ldots, \tilde{\circ}_m)(x, y, z) = V(\tilde{\circ}_1(x, y, z), \ldots, \tilde{\circ}_m(x, y, z)).$

Proposition 4.24. *Let* $(G, \circ)$ *be a group,* $(G, \tilde{\circ}_1), \ldots, (G, \tilde{\circ}_m)$ T-*vague groups of* G, $E_1, \ldots, E_m$ *their respective associated* T-*indistinguishability operators (i.e.,* $\tilde{\circ}_i(x, y, z) = E_i(x \circ y, z)$ *for* $i = 1, \ldots, m)$ *and* $F : [0,1]^m \to [0,1]$ *a function. Then*

$$F(E_1, \ldots, E_m) = E_{F((G, \tilde{\circ}_1), \ldots, (G, \tilde{\circ}_m))}.$$

Corollary 4.25. *With the same notations as in the preceding Proposition 4.24,*

- if F aggregates T-indistinguishability operators and $(G, \tilde{o}_1), \ldots, (G, \tilde{o}_m)$ are T-vague groups of G, then $F((G, \tilde{o}_1), \ldots, (G, \tilde{o}_m))$ is a T-vague group of G.
- if F aggregates T-vague groups $E_1, \ldots, E_m$ are invariant T-indistinguishability operators, then $F(E_1, \ldots, E_m)$ is an invariant T-indistinguishability operator.

Similarly to fuzzy subgroups we obtain the following important result.

Proposition 4.26. *A function $F : [0,1]^m \to [0,1], m \geq 1$, aggregates T-indistinguishability operators if and only if it aggregates T-vague groups.*

Proposition 4.27. *A function $F : [0,1]^m \to [0,1], m \geq 1$, aggregates T-vague groups if and only if the following conditions hold:*

(i) $F(\overbrace{1, \ldots, 1}^{m}) = 1$.

(ii) F *transforms m-dimensional T-triangular triplets into 1-dimensional T-triangular triplets.*

5   Concluding Remarks

In this work we have dealt with the problem of aggregating fuzzy subgroups and T-vague groups. Thanks to the close relation between these objects and T-indistinguishability operators, it results that the functions that aggregate them are the functions preserving T-triplets. In other words, the functions preserving T-indistinguishability operators, fuzzy subgroups and T-vague groups coincide. Interesting examples of these functions are

- The t-norm
- The minimum
- If the t-norm T is continuous Archimedean and g and additive of T, then
  - The weighted quasi-arithmetic means m_g generated by g
  - The ordered quasi-arithmetic means generated by g with decreasing weights.

We have studied the aggregation of fuzzy subgroups and T-vague groups from a functional point of view. Nevertheless, there are other ways to fusion or aggregate them that are not functional. Given a collection of T-indistinguishability operators, fuzzy subgroups or T-vague groups, a very natural way to aggregate them is calculating the transitive closure of their union, which is not a functional procedure. We have pointed out in this paper that there are a bijections between the sets of left and of right invariant Tindistinguishability operators and the set of fuzzy subgroups, and the set of T-vague groups, normal fuzzy subgroups and invariant T-indistinguishability operators. In fact these sets are isomorphic lattices [15] and the bijections preserve transitive closures (if $E_1, \ldots, E_m$ are the left (right) T-indistinguishability operators associated to the fuzzy subgroups $\mu_1, \ldots, \mu_m$ respectively and E is the transitive closure of $E_1 \cup \ldots \cup E_m$, then

E is the left (right) T-indistinguishability operator associated to the transitive closure of $\mu_1 \cup \ldots \cup \mu_m$ and similarly with T-vague groups).

As an interesting consequence of the results of this work, crisp equivalence relations subgroups and quotient groups can be aggregated obtaining fuzzy objects.

Example 5.1. *Let T a continuous Archimedean t-norm with additive generator g. Then the weighted quasi-arithmetic means with generator g of crisp equivalence relations on a set X are T-indistinguishability operators on X.*

Example 5.2. *Let* (6) *and* (10) *be the subgroups of* $(\mathbb{Z}, +)$ *of multiples of* 6 *and* 10, *respectively,* $t(x) = 1 - x$ *an additive generator of the Łukasiewicz t-norm and* $p, q \geq 0$ *with* $p + q = 1$. *Then* $m_t^{p,q}((6), (10)) = p(6) + q(10)$ *is the fuzzy subgroup*

$$(p(6) + q(10))(x) = \begin{cases} 1 \text{ if } x \text{ is a multiple of the least common divisor lcd}(6, 10) = 30 \\ p \text{ if } x \text{ is a multiple of 6 and not a multiple of 10} \\ q \text{ if } x \text{ is a multiple of 10 and not a multiple of 6} \\ 0 \text{ if } x \text{ is neither a multiple of 6 nor a multiple of 10.} \end{cases}$$

References

1. Chon, I.: On T-fuzzy groups. Kangweon-Kyungki Math. J. **9**, 149–156 (2001)
2. Demirci, M.: Vague groups. J. Math. Anal. Appl. **230**, 142–156 (1999)
3. Demirci, M., Recasens, J.: Fuzzy groups, fuzzy functions and fuzzy equivalence relations. Fuzzy Sets and Syst. **144**, 441–458 (2004)
4. Gottwald, S.: Fuzzy Sets, Fuzzy Logic: The Foundations of Application from a Mathematical Point of View. Springer, Heidelberg (1993). Friedr. Vieweg & Sohn Verlagsgesellschaft mbH, Wiesbaden
5. Hájek, P.: Metamathematics of Fuzzy Logic. Kluwer Academic Publishers, New York (1998)
6. Jacas, J., Recasens, J.: Aggregation of T-transitive relations. Int. J. Intell. Syst. **18**, 1193–1214 (2003)
7. Klement, E.P., Mesiar, R., Pap, E.: Triangular Norms. Kluwer, Dordrecht (2000)
8. Mashour, A.S., Ghanim, M.H., Sidky, F.I.: Normal fuzzy subgroups. Univ. u Novom Sadu Zb. Rad. Prirod.-Mat. Fak. Ser. Mat **20**(2), 53–59 (1990)
9. Mayor, G., Recasens, J.: Preserving T-transitivity. In: CCIA 2016, Barcelona, pp. 79–87 (2016)
10. Mesiar, R., Novak, V.: Operations fitting triangular-norm-based biresiduation. Fuzzy Sets Syst. **104**, 77–84 (1999)
11. Mordeson, J., Bhutani, K., Rosenfeld, A.: Fuzzy Group Theory. Studies in Fuzziness and Soft Computing, vol. 182. Springer, Heidelberg (2005)
12. Pradera, A., Trillas, E., Castiñeira, E.: On the aggregation of some classes of fuzzy relations. In: Bouchon-Meunier, B., Gutiérrez, J., Magdalena, L., Yager, R. (eds.) Technologies for Constructing Intelligent Systems, pp. 125–147. Springer, Heidelberg (2002)
13. Pradera, A., Trillas, E.: A note on pseudometrics aggregation. Int. J. Gen. Syst. 41–51 (2002)

14. Recasens, J.: Indistinguishability Operators. Modelling Fuzzy Equalities and Fuzzy Equivalence Relations. Studies in Fuzziness and Soft Computing, vol. 260. Springer, Heidelberg (2010)
15. Recasens, J.: Permutable indistinguishability operators, perfect vague groups and fuzzy subgroups. Inf. Sci. **196**, 129–142 (2012)
16. Rosenfeld, A.: Fuzzy groups. J. Math. Anal. Appl. **35**, 512–517 (1971)
17. Schweizer, B., Sklar, A.: Probabilistic Metric Spaces. North-Holland, Amsterdam (1983)
18. Zadeh, L.A.: Similarity relations and fuzzy orderings Information. Science **3**, 177–200 (1971)

Families of Perturbation Copulas Generalizing the FGM Family and Their Relations to Dependence Measures

Jozef Komorník[1], Magdaléna Komorníková[2(✉)], and Jana Kalická[2]

[1] Faculty of Management, Comenius University, Odbojárov 10, P.O. BOX 95,
820 05 Bratislava, Slovakia
`jozef.komornik@fm.uniba.sk`
[2] Faculty of Civil Engineering, Slovak University of Technology, Radlinského 11,
810 05 Bratislava, Slovakia
{`magdalena.komornikova,jana.kalicka`}`@stuba.sk`

Abstract. In this paper we provide an extension of special parametric class of perturbations of an arbitrary copula (given in [3]) that represent a partial generalization of the FGM family of copulas for parameters from the unit interval. However the FGM family is defined for parameters from the interval $[-1, 1]$. We present a construction of perturbations of an arbitrary copula also for parameters from the interval $[-1, 0]$ so that together with the former family of perturbations of copulas we get a generalization of the FGM family for the whole interval $[-1, 1]$. We also investigated the influence of the parameters of the introduced class of perturbations of copulas on several measures of dependence (Spearman's rho, Blomqvist's beta, Gini's gamma, Kendall's tau).

Keywords: Copula · Perturbation of copulas · Measures of dependence

1 Introduction

A special parametric class of perturbations of an arbitrary copula C was introduced in [3] by the formula

$$C_\alpha(u, v) = C(u, v) + \alpha \left(u - C(u, v)\right)\left(v - C(u, v)\right) \text{ for } \alpha \in [0, 1]. \qquad (1)$$

It was shown that (1) provides a generalization of the Farlie–Gumbel–Morgenstern (FGM) family

$$FGM_\alpha(u, v) = u\,v + \alpha\,u\,(1 - u)\,v\,(1 - v), \alpha \in [-1, 1] \qquad (2)$$

(that represent perturbations of the product copula) for parameters from the interval $[0, 1]$. However the formula (2) for the FGM family is applicable for parameters from the interval $[-1, 1]$. We extend the class of copulas given by (1) by the formula

$$C_\alpha(u, v) = C(u, v) + \alpha C(u, v)\left[C(u, v) - (u + v - 1)\right], \ \alpha \in [-1, 0] \qquad (3)$$

© Springer International Publishing AG 2018
V. Torra et al. (eds.), *Aggregation Functions in Theory and in Practice*,
Advances in Intelligent Systems and Computing 581, DOI 10.1007/978-3-319-59306-7_6

and we obtain a generalization of the FGM family on the whole interval $[-1,1]$. We also investigate the influence of the parameters of the introduced class of perturbations of copulas on the values of several measures of dependence (Spearman's rho, Blomqvist's beta, Gini's gamma, Kendall's tau). We show that this influence is linear on both subintervals $[-1,0]$ and $[0,1]$ in case of the first three of the above coefficients, while it is quadratic on the both mentioned subintervals in case of Kendall's tau.

The paper is organized as follows. The second section presents a brief overview of the theory of copulas, their reflections and perturbations. In the third section, some selected dependence measures are reviewed. In the fourth section, some results concerning the values of selected dependence coefficients for reflections and considered perturbations of copulas are presented. Finally, some concluding remarks are added.

2   Copulas

Recall that for a 2–dimensional random vector (X,Y) with a joint distribution function F_{XY} and continuous marginal distribution functions F_X, F_Y a copula C satisfying the relations $F_{XY}(x,y) = C(F_X(x), F_Y(y))$ is the distribution function of the random vector (U,V), where $U = F_X(X)$ and $V = F_Y(Y)$ have uniform distributions on $[0,1]$. For more details we recommend monographs Joe [2] and Nelsen [4].

We follow the approach of Patton [5] and consider a so–called *survival copula* derived from a given copula C corresponding to the couple $(-X, -Y)$ by

$$\widehat{C}(u,v) = u + v - 1 + C(1-u, 1-v) \tag{4}$$

which is the copula corresponding to the couple $(-X, -Y)$.

Another natural transformations of the copula C are copulas LC and RC corresponding to the couples $(-X, Y)$ and $(X, -Y)$, respectively.

They have the form

$$LC(u,v) = v - C(1-u, v) \tag{5}$$

and

$$RC(u,v) = u - C(u, 1-v). \tag{6}$$

We will call the copulas LC and RC the *left* and the *right reflections* of the copula C, respectively (see e.g. [1]).

In [3], the following family of perturbations given by (1) of any copula C was introduced as a partial generalization of the FGM class of perturbations derived from the product copula

$$\Pi(u,v) = u\,v$$

by

$$\Pi_\alpha(u,v) = u\,v + \alpha\,(u - u\,v)\,(v - u\,v) = u\,v + \alpha\,u\,v(1-u)\,(1-v) \text{ for } \alpha \in [-1,1]. \quad (7)$$

However, the formula (1) can not be directly extended for $\alpha \in [-1,0]$ preserving the resulting functions in the copula class. For example, for the minimal copula

$$W(u,v) = \max(0, u + v - 1),$$

we get

$$W_{-0.5}(0.5, 0.5) = 0 - 0.5\,(0.5)^2 = -0.5^3 < 0.$$

We find a suitable extension of the class C_α for $\alpha \in [-1,0]$ in the form

$$C_\alpha = L\left((LC)_{(-\alpha)}\right) \text{ for } \alpha \in [-1,0]. \quad (8)$$

(Note that $L\left((LC)_0\right) = L(LC) = C$.)

We can obtain a more explicit form of C_α for $\alpha \in [-1,0)$ using (5) and (1). We have

$$(LC)_{(-\alpha)}(u,v) = [(v - C(1-u,v))] - \alpha\,[(u - (v - C(1-u,v)))\,(v - (v - C(1-u,v)))]$$

$$= [v - C(1-u,v)] - \alpha\,[u - v + C(1-u,v)]\,C(1-u,v),$$

hence

$$L((LC)_{(-\alpha)})(u,v) = v - [v - C(u,v)] - \alpha\,[(1-u) - v + C(u,v)]\,C(u,v)$$

$$= C(u,v) + \alpha\,[C(u,v) - (u+v-1)]\,C(u,v) \text{ for } \alpha \in [-1,0]$$

and thus the copulas given by (8) and (3) are identical.

Note that, defining $\tilde{C}(u,v) = C(v,u)$, it is not difficult to check that $\tilde{C}_\alpha(u,v) = C_\alpha(v,u)$. Hence an alternative expression

$$C_\alpha(u,v) = R((RC)_{(-\alpha)})(u,v)$$

also holds for $\alpha \in [-1,0]$. Put

$$D_1(u,v) = (u - C(u,v))\,(v - C(u,v))$$
$$D_2(u,v) = C(u,v)\,[C(u,v) - (u+v-1)]. \quad (9)$$

We can combine (1), (3) and (9) in the form

$$C_\alpha(u,v) = \begin{cases} C(u,v) + \alpha\,D_1(u,v) & \text{for } \alpha \in [0,1], \\ C(u,v) + \alpha\,D_2(u,v) & \text{for } \alpha \in [-1,0]. \end{cases} \quad (10)$$

We have

$$D_1(u,v) = u\,v + C(u,v)\,[C(u,v) - (u+v-1)] - C(u,v) = \Pi(u,v) + D_2(u,v) - C(u,v),$$

hence

$$C_\alpha(u,v) = C(u,v) - \alpha C(u,v)\left(u + v - C(u,v)\right) + \frac{\alpha + |\alpha|}{2} uv + \frac{\alpha - |\alpha|}{2} C(u,v)$$

for $\alpha \in [-1,1]$. Note that $C_\alpha(u,v)$ is a nondecreasing function of α for any $(u,v) \in [0,1]^2$. Moreover for $C = W$ we have $D_2(u,v) = 0$, $D_1(u,v) = \Pi(u,v) - W(u,v)$, thus for $(u,v) \in [0,1]^2$

$$W_\alpha(u,v) = \begin{cases} (1-\alpha)W(u,v) + \alpha\, \Pi(u,v) & \text{for } \alpha \in [0,1], \\ W(u,v) & \text{for } \alpha \in [-1,0]. \end{cases} \tag{11}$$

yielding $W_1(u,v) = \Pi(u,v)$. Furthermore, for any copula $C \neq W$ and $(u,v) \in [0,1]^2$ such that $C(u,v) > W(u,v)$ we have $D_2(u,v) > 0$ and thus $C_\alpha(u,v)$ is an increasing function of α for $\alpha \in [-1,0]$ and its minimum is

$$C_{-1}(u,v) = C(u,v) - D_2(u,v) = C(u,v)\left[u + v - C(u,v)\right] = uv - D_1(u,v) \leq \Pi(u,v).$$

Similarly for $M(u,v) = \min(u,v)$ we have $D_1(u,v) = 0$, thus $D_2(u,v) = M(u,v) - \Pi(u,v)$ and

$$M_\alpha(u,v) = \begin{cases} M(u,v) & \text{for } \alpha \in [0,1], \\ (1+\alpha)\, M(u,v) - \alpha\, \Pi(u,v) & \text{for } \alpha \in [-1,0]. \end{cases} \tag{12}$$

yielding $M_{-1}(u,v) = \Pi(u,v)$ for $(u,v) \in [0,1]^2$.

For any $C \neq M$ and $(u,v) \in [0,1]^2$ such that $C(u,v) < M(u,v)$ we have $D_1(u,v) > 0$ and thus $C_\alpha(u,v)$ is an increasing function of α for $\alpha \in [0,1]$ and its maximum

$$C_1(u,v) = C(u,v) + D_1(u,v) = uv + D_2(u,v) \geq \Pi(u,v).$$

Furthermore, for $C = \Pi$, we get from (3) that for $\alpha \in [-1,0]$

$$\Pi_\alpha(u,v) = uv + \alpha\left[uv - (u+v-1)\right] uv = uv + \alpha uv(1-u)(1-v). \tag{13}$$

Hence the formulas (13) and (7) give the same results for $\alpha \in [-1,1]$ and thus (10) provide a generalization of the FGM class.

Theorem 1. *Let $C : [0,1]^2 \to [0,1]$ be a copula. The perturbation C_α, $\alpha \in [-1,1]$ of copula C given by (9) and (10) is order preserving, i.e. for any two copulas $C^{(1)} \leq C^{(2)}$ and $\alpha \in [-1,1]$ we have $C_\alpha^{(1)} \leq C_\alpha^{(2)}$.*

Proof. For $\alpha \in [0,1]$ we have

$$C_\alpha^{(2)}(u,v) - C_\alpha^{(1)}(u,v) = \left(C^{(2)}(u,v) - C^{(1)}(u,v)\right)$$

$$+ \alpha\left[\left(uv - C^{(2)}(u,v)(u+v) + (C^{(2)})^2(u,v)\right) - \left(uv - C^{(1)}(u,v)(u+v) + (C^{(1)})^2(u,v)\right)\right]$$

$$= \left(C^{(2)}(u,v) - C^{(1)}(u,v)\right) + \alpha\left[\left((C^{(2)})^2(u,v) - (C^{(1)})^2(u,v)\right) - \left(C^{(2)}(u,v) - C^{(1)}(u,v)\right)(u+v)\right]$$

$$= (1-\alpha)\left(C^{(2)}(u,v) - C^{(1)}(u,v)\right) + \alpha\left(C^{(2)}(u,v) - C^{(1)}(u,v)\right)$$

$$\cdot\left[\left(C^{(2)}(u,v) + C^{(1)}(u,v)\right) - (u+v-1)\right] \geq 0.$$

Obviously

$$L(C^{(2)})(u,v) = v - C^{(2)}(1-u,v) \leq v - C^{(1)}(1-u,v) = L(C^{(1)})(u,v).$$

For $\alpha \in [-1, 0]$ we have

$$\left(L(C^{(2)})\right)_{-\alpha} \leq \left(L(C^{(1)})\right)_{-\alpha}$$

and thus

$$C_\alpha^{(2)} = L\left(\left(L(C^{(2)})\right)_{-\alpha}\right) \geq L\left(\left(L(C^{(1)})\right)_{-\alpha}\right) = C_\alpha^{(1)}.$$

3   Dependence measures for copulas

The dependence structure between random variables is completely described by their joint distribution function. Apart from linear correlation, there exist several other measures of association.

Linear correlation measures how well two random variables cluster around a linear function. A major shortcoming is that linear correlation is not invariant under non-linear monotonic transformations of random variables. The concordance and dependence measures (e.g. Kendall's tau, Spearman's rho) reflect the degree to which random variables cluster around a monotone function. This is a consequence of these measures being defined as only dependent on the copula and copulas are invariant under monotone transformations of the random variables (i.e., we deal with rank dependence measures only).

Two observations (x_1, y_1) and (x_2, y_2) from a pair of continuous random variables are *concordant* if $(x_l - x_2)(y_l - y_2) > 0$; and they are *discordant* if $(x_l - x_2)(y_l - y_2) < 0$ (see, e.g. [6]).

Spearman's rho

Let us denote $I^2 = [0,1] \times [0,1]$ (the unit square). Let (X, Y) is a random vector characterized by a copula C. It was shown in [4] that the Spearman's rho satisfies

$$\rho(C) = 12 \int\int_{I^2} C(u,v)dudv - 3. \tag{14}$$

Besides the notation $C_\alpha(u, v)$ for the perturbation of copulas given by (10) and (9) for parameters $\alpha \in [0, 1]$, we will also use notations $C_u(u, v) = \frac{\partial}{\partial u} C(u, v)$, $C_v(u, v) = \frac{\partial}{\partial v} C(u, v)$ for the indicated partial derivatives. Recall that the functions $C_u(u, v)$, $C_v(u, v)$ are almost everywhere defined and attain values in $[0, 1]$ (see Nelsen [4]).

Kendall's tau

Nelsen in [4] shows that if X, Y is a pair of continuous random variables with copula C, then the population version of Kendall's tau is

$$\tau(C) = 1 - 4 \int\int_{I^2} C_u(u,v)\, C_v(u,v)\, du\, dv, \tag{15}$$

(see Nelsen [4] for details).

Blomqvist's beta

   If X and Y are continuous random variables with copula C, then the population version of the median correlation coefficient (also known as Blomqvist's beta) is given by (see Nelsen [4])

$$\beta(C) = 4C\left(\frac{1}{2}, \frac{1}{2}\right) - 1. \tag{16}$$

Gini's gamma

   Denote $\delta_C(u) = C(u, u)$ and $\omega_C(u) = C(u, 1 - u)$.

   It was shown in [4] that Gini's gamma

$$\gamma(C) = 4\left[\int_0^1 C(u, 1 - u)du - \int_0^1 (u - C(u, u))\, du\right] = 4\left[\int_0^1 (\omega_C(u) + \delta_C(u) - u)\, du\right]. \tag{17}$$

Remark 1. It is a direct consequence of (14), (16) and (17) that the above dependence measures Spearman's rho, Blomqvist's beta and Gini's gamma are monotone, which means that for two copulas $C^{(1)}$, $C^{(2)}$ such that $C^{(1)}(u, v) \leq C^{(2)}(u, v)$ for $(u, v) \in I^2$ we have

$$\rho(C^{(1)}) \leq \rho(C^{(2)}), \quad \beta(C^{(1)}) \leq \beta(C^{(2)}), \quad \gamma(C^{(1)}) \leq \gamma(C^{(2)}).$$

4   Dependence Measures for Reflections and Perturbations of Copulas

Remark 2. 1. Note that for any continuous function $f : I \to R$ the equality

$$\int_0^1 f(x)\, dx = \int_0^1 f(1 - x)\, dx \tag{18}$$

holds.

2. The product copula $\Pi(u, v)$ is invariant under the reflections, i.e.

$$L\Pi = R\Pi = S\Pi = \Pi. \tag{19}$$

3. For any binary copulas C_1, C_2 we have

$$\int\int_{I^2} |LC_1(u, v) - LC_2(u, v)|\, du\, dv = \int\int_{I^2} |RC_1(u, v) - RC_2(u, v)|\, du\, dv =$$

$$\int\int_{I^2} |C_1(u, v) - C_2(u, v)|\, du\, dv.$$

Hint

$$|LC_1(u, v) - LC_2(u, v)| = |[u - C_1(1 - u, v)] - [u - C_2(1 - u, v)]|$$

$$= |C_2(1 - u, v) - C_1(1 - u, v)|.$$

Theorem 2. *The following equality holds:*

$$\gamma(LC) = \gamma(RC) = \gamma(SC) = \gamma(C).\tag{20}$$

Proof. According to (17) we have

$$\gamma(C) = 4\left|\int_0^1 [\delta_C(u) + \omega_C(u) - u]\, du\right|.$$

$$\delta_{RC}(u) = u - C(u, 1 - u) = u - \omega_C(u),\quad \omega_{RC}(u) = u - C(u, u) = u - \delta_C(u),$$

$$\delta_{RC}(u) + \omega_{RC}(u) - u = u - \omega_C(u) - \delta_C(u),$$

$$\delta_{LC}(u) = u - C(1 - u, u) = u - \omega_C(1 - u),$$

$$\omega_{LC}(u) = (1 - u) - C(1 - u, 1 - u) = (1 - u) - \delta_C(1 - u),$$

$$\delta_{LC}(u) + \omega_{LC}(u) - u = u - \omega_C(1 - u) + (1 - u) - \delta_C(1 - u) - u$$

$$= (1 - u) - \delta_C(1 - u) - \omega_C(1 - u).$$

$$\int_0^1 [\delta_{LC}(u) + \omega_{LC}(u) - u]\, du = -\int_0^1 [\delta_C(1 - u) + \omega_C(1 - u) - (1 - u)]\, du$$

$$= -\int_0^1 [\delta_C(u) + \omega_C(u) - u]\, du.$$

Hence $\gamma(LC) = \gamma(C)$. Similarly $\gamma(RC) = \gamma(C)$.

Theorem 3. *The following equalities hold:*

$$\rho(LC) = \rho(RC) = -\rho(C),\quad \rho(SC) = \rho(C))\tag{21}$$

for Spearman's rho,

$$\tau(LC) = \tau(RC) = -\tau(C),\quad \tau(SC) = \tau(C))\tag{22}$$

for Kendall's tau, and

$$\beta(LC) = \beta(RC) = -\beta(C),\quad \beta(SC) = \beta(C))\tag{23}$$

for Blomqvist's beta.

Proof. The validity of (21) and (22) follow directly from the fact that the couples of observations (x_1, y_1) and (x_2, y_2) are concordant when the couples $(-x_1, y_1)$ and $(-x_2, y_2)$ as well as the couples $(x_1, -y_1)$ and $(x_2, -y_2)$ are discordant (and vice versa). Moreover, if C corresponds to the couple (X, Y) of random variables then LC, RC and SC correspond to the couples $(-X, Y)$, $(X, -Y)$ and $(-X, -Y)$.

The above arguments yield validity of (23).

Theorem 4. *Let $C : [0, 1]^2 \rightarrow [0, 1]$ be a copula with Spearman's rho given by (14). For the perturbation C_α of copula C given by (9) and (10) is the Spearman's rho a function of α that is linear on both intervals $[-1, 0], [0, 1]$ and can be expressed in the form*

$$\rho(C_\alpha) = \begin{cases} \rho(C) + 12\alpha\, a_1 & \text{for } \alpha \in [0, 1], \\ \rho(C) + 12\alpha\, a_2 & \text{for } \alpha \in [-1, 0], \end{cases} \tag{24}$$

where the coefficients a_1, a_2 are given by

$$a_i = \int\int_{I^2} D_i(u, v)\, du\, dv \tag{25}$$

for $i = 1, 2$ and D_1, D_2 given by (9).

Proof. The relation (24) follows directly from (14), (10) and (9).

Theorem 5. *Let $C : [0, 1]^2 \rightarrow [0, 1]$ be a copula with Kendall's tau given by (15). For the perturbation C_α of copula C given by (9) and (10) is the Kendall's tau given by*

$$\tau(C_\alpha) = \begin{cases} \tau(C) + 4\alpha \int\int_{I^2} a_{1,1}(u, v)\, du\, dv + 4\alpha^2 \int\int_{I^2} a_{2,1}(u, v)\, du\, dv & \text{for } \alpha \in [0, 1], \\ \tau(C) + 4\alpha \int\int_{I^2} a_{1,2}(u, v)\, du\, dv + 4\alpha^2 \int\int_{I^2} a_{2,2}(u, v)\, du\, dv & \text{for } \alpha \in [-1, 0], \end{cases} \tag{26}$$

where the coefficients $a_{1,i}$, $a_{2,i}$ are given by

$$a_{1,i}(u, v) = - \left[C_u(u, v)\, (D_i)_v(u, v) + (D_i)_u(u, v)\, C_v(u, v) \right],$$

$$a_{2,i}(u, v) = -(D_i)_u(u, v)\, (D_i)_v(u, v)$$

for $i = 1, 2$ and D_1, D_2 given by (9).

Proof. The relation (26) follows directly from (15). Note that the partial derivative

$$(D_1)_u(u, v) = (1 - C_u(u, v))\, (v - C(u, v)) - (u - C(u, v))\, C_u(u, v)$$

exists almost everywhere and is a difference of products of bounded integrable functions, thus it is bounded and integrable. The same holds for $(D_1)_v(u, v)$, $(D_2)_u(u, v)$, $(D_2)_v(u, v)$.

Theorem 6. *Let $C : [0, 1]^2 \rightarrow [0, 1]$ be a copula with Blomqvist's beta given by (16). For the perturbation C_α of copula C given by (9) and (10) is the Blomqvist's beta given by*

$$\beta(C_\alpha) = \begin{cases} \beta(C) + 4\alpha D_1(\tfrac{1}{2}, \tfrac{1}{2}) & \text{for } \alpha \in [0, 1], \\ \beta(C) + 4\alpha D_2(\tfrac{1}{2}, \tfrac{1}{2}) & \text{for } \alpha \in [-1, 0]. \end{cases} \tag{27}$$

Proof. The relation (27) follows directly from (16), (10) and (9).

Theorem 7. *Let $C : [0,1]^2 \rightarrow [0,1]$ be a copula with Gini's gamma given by (17). For the perturbation C_α of copula C given by (9) and (10) is Gini's gamma given by*

$$\gamma(C_\alpha) = \begin{cases} \gamma(C) + 4\alpha \int_0^1 [D_1(u,u) + D_1(u,1-u)]\, du & \text{for } \alpha \in [0,1], \\ \gamma(C) + 4\alpha \int_0^1 [D_2(u,u) + D_2(u,1-u)]\, du, & \text{for } \alpha \in [-1,0]. \end{cases} \tag{28}$$

Proof. The relation (28) follows directly from (17), (10) and (9).

Remark 3. The fact that the functions $D_1(u,v)$ and $D_2(u,v)$ are nonnegative, together with (24), (25), (27) and (28) imply that the following functions of α: $\rho(C_\alpha)$, $\beta(C_\alpha)$ and $\gamma(C_\alpha)$ are linear and nondecreasing on both interval $[-1,0]$ and $[0,1]$.

Remark 4. Computationally more feasible alternative to (27) is

$$\beta(C_\alpha) = \begin{cases} \beta(C) + \alpha \left(\frac{1-\beta(C)}{2}\right)^2 & \text{for } \alpha \in [0,1], \\ \beta(C) + \alpha \left(\frac{1+\beta(C)}{2}\right)^2 & \text{for } \alpha \in [-1,0] \end{cases} \tag{29}$$

or

$$\beta(C_\alpha) = \beta(C) + \alpha \left(\frac{(1+\text{sign}(\alpha))\beta(C)}{2}\right)^2.$$

The relation (29) follows directly from (16) and (27).

Remark 5. **(a)** From (16) and (29) we get the lower and upper bounds for $\beta(C_\alpha)$.

$$\beta(C_{-1}) = -\left(\frac{1-\beta(C)}{2}\right)^2, \qquad \beta(C_1) = \left(\frac{1+\beta(C)}{2}\right)^2.$$

(b) For $\alpha \in [-1,1]$ we have

$$\beta(\Pi_\alpha) = \frac{\alpha}{4}, \quad \rho(\Pi_\alpha) = \frac{\alpha}{3}, \quad \gamma(\Pi_\alpha) = \frac{4\alpha}{15}.$$

For $\alpha \in [-1,0]$ we get from (11) and (12)

$$W_\alpha = W, \qquad M_\alpha = (1+\alpha)M - \alpha\,\Pi.$$

(c) Since

$$\rho(W) = \gamma(W) = \beta(W) = -1 \quad \text{and} \quad \rho(M) = \gamma(M) = \beta(M) = 1,$$

we have

$$\rho(W_\alpha) = \gamma(W_\alpha) = \beta(W_\alpha) = 0 \quad \text{and} \quad \rho(M_\alpha) = \gamma(M_\alpha) = \beta(M_\alpha) = 1 + \alpha.$$

Therefore, for any copula C the inequalities

$$\rho(C_\alpha) \leq 1 + \alpha, \qquad \gamma(C_\alpha) \leq 1 + \alpha, \qquad \beta(C_\alpha) \leq 1 + \alpha, \text{ for } \alpha \in [-1, 0]$$

hold. Hence

$$\rho(C_{-1}) \leq 0, \qquad \gamma(C_{-1}) \leq 0, \qquad \beta(C_{-1}) \leq 0.$$

For $\alpha \in [0, 1]$ we have

$$M_\alpha = M, \qquad W_\alpha = (1 - \alpha) W + \alpha \, \Pi.$$

Hence

$$\rho(W_\alpha) = \gamma(W_\alpha) = \beta(W_\alpha) = \alpha - 1 \quad \text{and} \quad \rho(M_\alpha) = \gamma(M_\alpha) = \beta(M_\alpha) = 1.$$

Therefore, for any copula C the inequalities

$$\rho(C_\alpha) \geq \alpha - 1, \qquad \gamma(C_\alpha) \geq \alpha - 1, \qquad \beta(C_\alpha) \geq \alpha - 1, \text{ hold for } \alpha \in [0, 1].$$

Hence

$$\rho(C_1) \geq 0, \qquad \gamma(C_1) \geq 0, \qquad \beta(C_1) \geq 0.$$

If we put

$$w_\alpha = \rho(W_\alpha) = \gamma(W_\alpha) = \beta(W_\alpha) = \max(-1, \alpha - 1)$$

and

$$m_\alpha = \rho(M_\alpha) = \gamma(M_\alpha) = \beta(M_\alpha) = \min(1, 1 + \alpha)$$

for $\alpha \in [-1, 1]$, we can synthetize the above inequalities in the form

$$w_\alpha \leq \rho(C_\alpha) \leq m_\alpha, \quad w_\alpha \leq \gamma(C_\alpha) \leq m_\alpha, \quad w_\alpha \leq \beta(C_\alpha) \leq m_\alpha$$

for $\alpha \in [-1, 1]$ and any copula C.

5 Concluding Remarks

We extended the definition of a special class of perturbations of copulas generalizing the FGM family of copulas for parameters from the unit interval to the
whole interval $[-1, 1]$. We proved that the class of perturbations of copulas is
order preserving. We showed that the values of Spearman's rho, Blomqvist's beta
and Gini's gamma are nondecreasing piecewise linear functions of perturbation
parameters (while Kendall's tau is a piecewise quadratic function of them). We
also found lower and upper bands for first three of the above parameters. We
expact applications of our results especially when fitting copulas to real data.
There are several methods for fitting copulas from particular classes, such as
Archimedean or Extreme Value copulas. When obtaining a copula C as the best
fitting one from the considered class of copulas, our approach allows to improve
this result, by considering copulas $(C_\alpha)_{\alpha \in [-1,1]}$ for a secondary fitting.

Acknowledgement. This work was supported by Slovak Research and Development
Agency under contracts No. APVV–14–0013 and by VEGA 1/0420/15.

References

1. De Baets, B., De Meyer, H., Kalická, J., Mesiar, R.: Flipping and cyclic shifting of binary aggregation functions. Fuzzy Sets Syst. **160**(6), 752–765 (2009)
2. Joe, H.: Multivariate Model and Dependence Concept. Monographs on Statistics and Applied Probability, vol. 73. Chapman & Hall, New York (1997)
3. Mesiar, R., Komorníková, M., Komorník, J.: Perturbation of bivariate copula. Fuzzy Sets and System **268**, 127–140 (2015)
4. Nelsen, R.B.: An Introduction to Copulas. Springer Series in Statistics, 2nd edn. Springer, New York (2006)
5. Patton, A.J.: Modelling asymmetric exchange rate dependence. Int. Econ. Rev. **47**(2), 527–556 (2006)
6. Úbeda-Flores, M.: Multivariate versions of Blomqvist's beta and Spearman's footrule. Ann. Inst. Stat. Math. **57**(4), 781–788 (2005)

k-maxitivity of Order-Preserving Homomorphisms of Lattices

Radko Mesiar[1(✉)] and Anna Kolesárová[2]

[1] Department of Mathematics and Descriptive Geometry,
Faculty of Civil Engineering, Slovak University of Technology, Radlinského 11,
810 05 Bratislava, Slovakia
radko.mesiar@stuba.sk
[2] Faculty of Chemical and Food Technology, Institute of Information Engineering,
Automation and Mathematics, Slovak University of Technology, Radlinského 9,
812 37 Bratislava, Slovakia
anna.kolesarova@stuba.sk

Abstract. The concept of k-maxitivity for order-preserving homomorphisms between bounded lattices is introduced and discussed. As particular cases, k-maxitive capacities and aggregation functions are studied and exemplified.

1 Introduction

Given any two lattices L_1, L_2, we can consider a maxitive mapping between them. Recall that a mapping $\varphi\colon L_1 \to L_2$ is maxitive if for any $x, y \in L_1$ we have $\varphi(x \vee_1 y) = \varphi(x) \vee_2 \varphi(y)$, where $\vee_i$ is the corresponding join in L_i, $i = 1, 2$. As a particular case, consider a finite space $X = \{1, \dots, n\}$ and its power set $2^X = L_1$ with $\vee_1 = \cup$ (union of sets), and the unit interval $[0, 1] = L_2$ with $\vee_2 = \vee$ (the standard maximum of reals). Then an order-preserving homomorphism (for the definition see Sect. 2) $m\colon 2^X \to [0, 1]$ is called a capacity [8]. The maxitivity of m means that $m(A \cup B) = m(A) \vee m(B)$ for all $A, B \subset X$, and in that case m is called a possibility measure [18]. Obviously, $m(A) = \bigvee_{i \in A} m(\{i\})$. For more details see, e.g., [6,17,18]. From now on, for simplicity of notation we will use the same symbol $\vee$ for the join in both lattices L_1 and L_2, if it is clear from the context on which one the operation $\vee$ is applied. Another distinguished example we obtain if, for any $n \in \mathbb{N}$, we consider $L_1 = [0, 1]^n$ and $L_2 = [0, 1]$. In this case, an order-preserving homomorphism $H\colon [0, 1]^n \to [0, 1]$ is called an (n-ary) aggregation function on $[0, 1]$, see, e.g., [1,3,8]. The maxitivity of H means that $H(\mathbf{x} \vee \mathbf{y}) = H(\mathbf{x}) \vee H(\mathbf{y})$ for all $\mathbf{x}, \mathbf{y} \in [0, 1]^n$ ($\vee$ stands for the standard maximum of reals or maximum of real n-tuples). For maxitive aggregation functions we have

$$H(\mathbf{x}) = \bigvee_{i=1}^{n} f_i(x_i),$$

© Springer International Publishing AG 2018
V. Torra et al. (eds.), *Aggregation Functions in Theory and in Practice*,
Advances in Intelligent Systems and Computing 581, DOI 10.1007/978-3-319-59306-7_7

where $f_i\colon [0,1] \to [0,1]$, $f_i(x_i) = H(0,\ldots,x_i,\ldots,0)$, $i \in \{1,\ldots,n\}$. Clearly, all functions f_i are monotone non-decreasing and satisfy the properties $f_1(0) = \cdots = f_n(0) = 0$, $\bigvee_{i=1}^{n} f_i(1) = 1$ (i.e., $f_{i_0}(1) = 1$ for some $i_0 \in \{1,\ldots,n\}$).

Twenty years ago, the concept of k-maxitive capacities on $X = \{1,\ldots,n\}$, $k \le n$, was introduced [10–12]. The k-maxitivity of a capacity $m\colon 2^X \to [0,1]$ was characterized by the condition $m(A) = m(B)$ satisfied for any subset $A \subset X$, where B is some subset of A with card(B) $\le$ k. Clearly, each capacity m on X is n-maxitive and the 1-maxitivity of capacities is just the standard maxitivity. Inspired by this fact, we have proposed the concept of k-maxitivity for aggregation functions [13]. The aim of this paper is to generalize the concept of k-maxitivity for all order-preserving homomorphisms between (bounded) lattices L_1, L_2, and also to study some basic properties and construction methods of k-maxitive order-preserving homomorphisms. The paper is organized as follows. In the next section, k-maxitivity of order-preserving homomorphisms is introduced, exemplified and studied. In Sect. 3, we discuss k-maxitive capacities as well as k-maxitive aggregation functions, especially in connection with some integrals when particular cases $L_1 = L^n$, $L_2 = L$ are considered.

2   *k*-maxitive Order-Preserving Homomorphisms

In what follows, we will consider any two bounded lattices $(L_1, \le_1, 0_1, 1_1)$ and $(L_2, \le_2, 0_2, 1_2)$.

Definition 1. A mapping $\varphi\colon L_1 \to L_2$ is called an order-preserving homomorphism whenever it preserves the order and bounds, i.e., if $\varphi(x) \le_2 \varphi(y)$ for all $x, y \in L_1$ satisfying $x \le_1 y$, and $\varphi(0_1) = 0_2$, $\varphi(1_1) = 1_2$.

Remark 1. Recall that a lattice homomorphism from L_1 to L_2 is a mapping preserving join and meet of any two elements, and also the top and bottom elements. Observe that an order-preserving homomorphism $\varphi\colon L_1 \to L_2$ need not be a lattice homomorphism; it need not preserve the join and meet. Neither a maxitive order-preserving homomorphism need not preserve the meet, and thus need not be a lattice homomorphism. On the other hand, each lattice homomorphism is a maxitive order-preserving homomorphism.

As already mentioned, an order-preserving homomorphism is maxitive whenever $\varphi(x \vee y) = \varphi(x) \vee \varphi(y)$ for all $x, y \in L_1$. Clearly, if L_1 is a bounded chain then each order-preserving homomorphism φ is maxitive. If $L_1 = L^n$ (Cartesian product lattice) and $L_2 = L$ for some bounded chain L, then an order-preserving homomorphism $\varphi\colon L^n \to L$ is an n-ary aggregation function on L [5] and its maxitivity implies the following structure of φ.

Proposition 1. *Let $(L, \le_L, 0_L, 1_L)$ be a bounded chain and $\varphi\colon L^n \to L$ be an aggregation function on L. Then φ is maxitive if and only if there are monotone non-decreasing functions $f_1,\ldots,f_n\colon L^n \to L$ satisfying $f_1(0_L) = \cdots = f_n(0_L) = 0_L$ and $f_i(1_L) = 1_L$ for some $i \in \{1,\ldots,n\}$, such that*
$$\varphi(x_1,\ldots,x_n) = \bigvee_{i=1}^{n} f_i(x_i).$$

Inspired by the k-maxitivity of capacities [10,11] we extend the concept of k-maxitivity to order-preserving homomorphisms of bounded lattices.

Definition 2. Let $(L_1, \leq_1, \mathbf{0}_1, \mathbf{1}_1)$ and $(L_2, \leq_2, \mathbf{0}_2, \mathbf{1}_2)$ be bounded lattices and let $k \in \mathbb{N}$. An order-preserving homomorphism $\varphi \colon L_1 \to L_2$ is called k-maxitive whenever for any $r \in \mathbb{N}$ and $x_1, \ldots, x_r \in L_1$ there is a subset $I \subset \{1, \ldots, r\}$ with $\mathrm{card}(I) \leq k$ such that

$$\varphi\left(\bigvee_{i=1}^{r} x_i\right) = \varphi\left(\bigvee_{i \in I} x_i\right). \tag{1}$$

For $k \geq 2$, φ is called proper k-maxitive if it is k-maxitive but not $(k-1)$-maxitive.

Observe that, in general, maxitivity and 1-maxitivity differ. Consider, for example, the diamond lattice $L = \{0, a, b, 1\}$ with incomparable elements a, b and $0 < a < 1$, $0 < b < 1$. Then $\varphi \colon L \to L$, $\varphi(x) = x$, is a maxitive order-preserving homomorphism that is not 1-maxitive. Note that in this case, φ is 2-maxitive.

Proposition 2. *Let $\varphi \colon L_1 \to L_2$ be an order-preserving homomorphism. The maxitivity and 1-maxitivity of φ coincide whenever L_2 is a bounded chain.*

In general, for any pair (L_1, L_2) of bounded lattices, the greatest order-preserving homomorphism $\varphi^* \colon L_1 \to L_2$ given by $\varphi^*(x) = \begin{cases} \mathbf{0}_2 & \text{if } x = \mathbf{0}_1, \\ \mathbf{1}_2 & \text{otherwise,} \end{cases}$ is simultaneously maxitive and 1-maxitive.

Depending on the structure of lattices L_1 and/or L_2, one can determine possible values k for the proper k-maxitivity of order-preserving homomorphisms $\varphi \colon L_1 \to L_2$. So, for example, if L_2 is a chain with $\mathrm{card}(L_2) = n$ then each order-preserving homomorphism φ is n-maxitive (i.e., proper k-maxitive for some $k \in \{1, \ldots, n\}$). Similarly, if L_1 satisfies the property (P): for any $x_1, \ldots, x_r \in L_1$ there is an $I \subset \{1, \ldots, r\}$, $\mathrm{card}(I) \leq n$, such that $\bigvee_{i=1}^{r} x_i = \bigvee_{i \in I} x_i$, then φ is proper k-maxitive for some $k \in \{1, \ldots, n\}$. So, for example, the diamond lattice $L = \{0, a, b, 1\}$ satisfies the property (P) with $n = 2$. For any fixed $n \in N$, if $X = \{1, \ldots, n\}$ and $L_1 = 2^X$, then L_1 satisfies the property (P) with the considered n, and, consequently, each capacity $m \colon 2^X \to [0, 1]$ is n-maxitive and hence proper k-maxitive for some $k \in \{1, \ldots, n\}$. Similarly, for a bounded chain L, if $L_1 = L^n$ then L_1 satisfies the property (P) with the considered n, and any order-preserving homomorphism $\varphi \colon L^n \to L_2$ is proper k-maxitive for some $k \in \{1, \ldots, n\}$. Putting $L_2 = L$, we see that this claim is valid for any n-ary aggregation function H on a bounded chain L (i.e., if an aggregation function $H \colon L^n \to L$ is considered).

The degree of maxitivity of compositions can be estimated as follows.

Proposition 3. *Let L_1, L_2, L_3 be bounded lattices and $\varphi_1 \colon L_1 \to L_2$, $\varphi_2 \colon L_2 \to L_3$ order-preserving homomorphisms. Then the composed mapping $\varphi \colon L_1 \to L_3$, $\varphi(x) = \varphi_2(\varphi_1(x))$, is also an order-preserving homomorphism. Moreover, if φ_1 is k-maxitive, then so is φ.*

We can also consider more general compositions.

Proposition 4. *Let $n \in \mathbb{N}$. If L_1, L_2, L_3 are bounded lattices, $\varphi_1, \ldots, \varphi_n \colon L_1 \to L_2$ and $\varphi \colon L_2^n \to L_3$ order-preserving homomorphisms, then the composed mapping $\eta \colon L_1 \to L_3$ given by*

$$\eta(x_1, \ldots, x_n) = \varphi(\varphi_1(x_1), \ldots, \varphi_n(x_n))$$

is also an order-preserving homomorphism. Moreover, if φ_i, $i = 1, \ldots, n$, are k_i-maxitive and φ is r-maxitive then η is k-maxitive, where $k = k_{(1)} + \cdots + k_{(\min(n,r))}$ and $(\cdot) \colon \{1, \ldots, n\} \to \{1, \ldots, n\}$ is any permutation such that $k_{(1)} \geq \cdots \geq k_{(n)}$.

Note that the previous proposition determines the upper bound of the order k of maxitivity only.

Example 1. Let $L_1 = [0,1]^3$, $L_2 = L_3 = [0,1]$ and $n = 3$. Consider $\varphi_1, \varphi_2, \varphi_3 \colon [0,1]^3 \to [0,1]$, $\varphi_1 = Med$, $\varphi_2 = Max$ and $\varphi_3 = Min$, where Med, Max and Min are the standard ternary median, maximum and minimum. Let $\varphi \colon [0,1]^3 \to [0,1]$, $\varphi = Med$. Then φ_1 and φ are proper 2-maxitive, φ_2 is 1-maxitive and φ_3 is proper 3-maxitive, i.e., $k_1 = r = 2$, $k_2 = 1$ and $k_3 = 3$. Based on Proposition 4, the composed mapping $\eta \colon [0,1]^3 \to [0,1]$ given by

$$\eta(x_1, x_2, x_2) = Med(Med(x_1, x_2, x_2), Max(x_1, x_2, x_2), Min(x_1, x_2, x_2))$$

is k-maxitive where $k = k_{(1)} + k_{(2)} = k_3 + k_1 = 5$. However η is (proper) 2-maxitive.

As already mentioned, 1-maxitivity and maxitivity of order-preserving homomorphisms $\varphi \colon L_1 \to L_2$ coincide whenever L_2 is a chain. We have the following characterization of the 1-maxitivity of order-preserving homomorphisms.

Proposition 5. *Let L_1, L_2 be bounded lattices and $\varphi \colon L_1 \to L_2$ an order-preserving homomorphism. Then φ is 1-maxitive if and only if the range $Ran\, \varphi$ is a chain in L_2 and φ is maxitive.*

Note that a 1-maxitive order-preserving homomorphism $\varphi \colon L_1 \to L_2$ allows to introduce k-maxitive order-preserving homomorphisms $\varphi_k \colon L_1^k \to L_2^k$ given by

$$\varphi_k(x_1, \ldots, x_k) = (\varphi(x_1), \ldots, \varphi(x_k)).$$

For some particular lattices we also have other methods for construction of k-maxitive order-preserving homomorphisms. Some of them are discussed in the next section.

3 k-maxitive Capacities and k-maxitive Aggregation Functions

The notion of k-maxitive capacities $m \colon 2^X \to [0,1]$, where $X = \{1, \ldots, n\}$, was introduced in [11] (and independently in [10]) and further discussed in [12]. The k-maxitivity of m was characterized by the following axiom:

for any $A \subset X$ there exists $B \subset A$, $\mathrm{card}(B) \leq k$ such that $m(A) = m(B)$. (2)

It is not difficult to check that for capacities on a finite universe, k-maxitivity as introduced in Definition 2 is equivalent to (2). Note that Definition 2 allows to introduce the k-maxitivity of capacities on any measurable space $(X, \mathscr{A})$. Based on Proposition 4, we have the following result.

Proposition 6. *Let $(X, \mathscr{A})$ be a measurable space and let $m_1, \ldots, m_n \colon \mathscr{A} \to [0, 1]$ be capacities which are k_1-, $\ldots$, k_n-maxitive, respectively. Then for any aggregation function $H \colon [0, 1]^n \to [0, 1]$, the mapping $m \colon \mathscr{A} \to [0, 1]$ given by*

$$m(A) = H(m_1(A), \ldots, m_n(A))$$

is a k-maxitive capacity with $k = k_1 + \cdots + k_n$.

So, for example, the product of two 1-maxitive capacities (i.e., possibility measures) is a 2-maxitive capacity.

Example 2. Let $X = \mathbb{N}$, $\mathscr{A} = 2^{\mathbb{N}}$, and let $m_1, m_2 \colon 2^{\mathbb{N}} \to [0, 1]$ be given by

$$m_1(A) = \begin{cases} 0 & \text{if } A \text{ is finite,} \\ 1 & \text{if } A \text{ is infinite,} \end{cases} \qquad m_2(A) = \frac{1}{\min\{n \in A\}}, \quad A \neq \emptyset.$$

Then both m_1, m_2 are 1-maxitive (maxitive) capacities and their product $m = m_1 m_2$ is a 2-maxitive capacity. Indeed, consider r subsets $A_1, \ldots, A_r$ of $\mathbb{N}$, $r > 2$. If all of them are finite then $A = \bigcup_{i=1}^{r} A_i$ is also finite and thus $m(A) = m(A_1)$. If at least one of the sets $A_1, \ldots, A_r$ is infinite, say, e.g., A_1, then A is also infinite. Let $n_0 = \min\{n \in A\}$. Then necessarily $n_0 \in A_j$ for some $j \in \{1, \ldots, r\}$. Clearly, $m(A) = \frac{1}{n_0} = m(A_1 \cup A_j)$, which proves the 2-maxitivity of m.

Our approach to k-maxitivity introduced in Definition 2 also covers the k-maxitivity of aggregation functions proposed in [13]. There is a close connection between k-maxitive capacities and k-maxitive aggregation functions, compare also [2].

Proposition 7. *Let $H \colon [0, 1]^n \to [0, 1]$ be a k-maxitive aggregation function. Then the mapping $m \colon 2^{\{1, \ldots, n\}} \to [0, 1]$ given by $m(A) = H(\mathbf{1}_A)$ is a k-maxitive capacity.*

Note that $\mathbf{1}_A \colon \{1, \ldots, n\} \to \{0, 1\}$, $\mathbf{1}_A(i) = 1$ if $i \in A$, and $\mathbf{1}_A(i) = 0$ otherwise, is the characteristic function of A.

Due to the above result, when the sets are represented by the related characteristic functions, each k-maxitive aggregation function H can be seen as a monotone extension of some k-maxitive capacity. Some of such possible extensions are based on universal integrals [9] linked to semicopulas [7]. Recall that a semicopula $\otimes \colon [0, 1]^2 \to [0, 1]$ is a binary aggregation function with neutral element $e = 1$ ($x \otimes 1 = 1 \otimes x = x$ for each $x \in [0, 1]$).

Recall that for any fixed semicopula $\otimes$ and capacity $m \colon 2^{\{1, \ldots, n\}} \to [0, 1]$, the mapping $\mathfrak{I}_{\otimes, m} \colon [0, 1]^n \to [0, 1]$ given by

$$\mathfrak{I}_{\otimes, m}(\mathbf{x}) = \bigvee_{i=1}^{n} x_i \otimes m(\{j \in \{1, \ldots, n\} \mid x_j \geq x_i\}), \tag{3}$$

is the smallest universal integral based on $\otimes$, see [9].

Proposition 8. *Let $n \geq 2$ and let $m\colon 2^{\{1,\ldots,n\}} \to [0,1]$ be a k-maxitive capacity. Then for any semicopula $\otimes$ the smallest universal integral $\mathfrak{I}_{\otimes,m}\colon [0,1]^n \to [0,1]$ defined by (3) is a k-maxitive aggregation function.*

Observe that if $\otimes = \wedge$ (minimum), $\mathfrak{I}_{\wedge,m}$ is just the Sugeno integral [16]. The Sugeno integral with respect to k-maxitive capacities was studied, e.g., in [4]. If $\otimes = \cdot$ (the standard product), $\mathfrak{I}_{\cdot,m}$ is the Shilkret integral [15].

There are also other methods for constructing k-maxitive aggregation functions, see [13]. We introduce here a representation of k-maxitive symmetric aggregation functions only.

Proposition 9. *Let $H\colon [0,1]^n \to [0,1]$ be an aggregation function. Then H is symmetric and k-maxitive for some $k \in \{1,\ldots,n\}$ if and only if there is an aggregation function $G\colon [0,1]^k \to [0,1]$ such that*

$$H(x_1,\ldots,x_n) = G(x_{(1)},\ldots,x_{(k)}), \tag{4}$$

where $(\cdot)\colon \{1,\ldots,n\} \to \{1,\ldots,n\}$ is a permutation satisfying $x_{(1)} \geq \cdots \geq x_{(n)}$.

Observe that H given by (4) is proper k-maxitive if and only if the values of the aggregation function G depend on the last coordinate.

Example 3. Let $H\colon [0,1]^{2k-1} \to [0,1]$ be the median,

$$H(\mathbf{x}) = Med(x_1,\ldots,x_{2k-1}).$$

Then $H(\mathbf{x}) = x_{(k)}$, i.e., $H(\mathbf{x}) = G(x_{(1)},\ldots,x_{(k)})$, where G is the projection into the last coordinate. Thus H is a proper k-maxitive symmetric aggregation function.

4   Concluding Remarks

We have introduced and discussed the concept of k-maxitivity of order-preserving homomorphisms $\varphi\colon L_1 \to L_2$ between bounded lattices. Our approach generalizes the k-maxitivity of capacities as well as the k-maxitivity of aggregation functions. The main reason for introducing k-maxitivity is the reduction of computational complexity. This issue in the case of maxitive capacities was discussed, e.g., in [14]. Similarly, for any lattice L_1 generated by atoms $a_1,\ldots,a_r$, i.e., $L_1 = \bigvee\limits_{I \subset \{1,\ldots,r\}}^{i \in I} a_i$, an order-preserving homomorphism $\varphi\colon L_1 \to L_2$ requires the knowledge of $2^r - 2$ values (as two values are known from the boundary conditions) while if φ is 1-maxitive then it is enough to know r values, namely $\varphi(a_1),\ldots,\varphi(a_r)$. If φ is 2-maxitive, it is enough to know $r \cdot (r+1)/2$ values of φ, etc. We expect several interesting results concerning k-maxitivity, in particular for product lattices and some other special lattices.

Acknowledgement. Both authors kindly acknowledge the support of the project of Science and Technology Assistance Agency under the contract No. APVV–14–0013. Moreover, the work of R. Mesiar on this paper was supported by the VEGA grant 1/0420/15.

References

1. Beliakov, G., Pradera, A., Calvo, T.: Aggregation Functions: A Guide for Practitioners. Springer, Heidelberg (2007)
2. Calvo, T., De Baets, B.: Aggregation operators defined by k-order additive/maxitive fuzzy measures. Int. J. Uncertain. Fuzzyness Knowl. Based Syst. **6**, 533–550 (1998)
3. Calvo, T., Kolesárová, A., Komorníková, M., Mesiar, R.: Aggregation operators: properties, classes and construction methods. In: Calvo, T., Mayor, G., Mesiar, R. (eds.) Aggregation Operators. New Trends and Applications, pp. 3–107. Physica-Verlag, Heidelberg (2002)
4. Cao-Van, K., De Baets, B.: A decomposition of k-additive Choquet and k-maxitive Sugeno integrals. Int. J. Uncertain. Fuzzyness Knowl. Based Syst. **9**, 127–144 (2001)
5. De Cooman, G., Kerre, E.: Order norms on bounded partially ordered sets. J. Fuzzy Math. **2**(2), 281–310 (1994)
6. Dubois, D., Prade, H.: Possibility Theory. Plenum Press, New York (1988)
7. Durante, F., Sempi, C.: Semicopulae. Kybernetika **41**, 281–310 (2005)
8. Grabisch, M., Marichal, J.-L., Mesiar, R., Pap, E.: Aggregation Functions. Cambridge University Press, Cambridge (2009)
9. Klement, E.P., Mesiar, R., Pap, E.: A universal integral as common frame for Choquet and Sugeno integral. IEEE Trans. Fuzzy Syst. **18**, 178–187 (2010)
10. Marichal, J.-L.: Aggregation operations for multicriteria decision aid. Ph.D. thesis. Institute of Mathematics University of Liege, Liege, Belgium (1998)
11. Mesiar, R.: k-order pan-additive discrete fuzzy measures. In: Proceedings of the 7th IFSA World Congress, pp. 488–490. Prague (1997)
12. Mesiar, R.: k-order additivity and maxitivity. Atti del Seminario Matematico e Fisico dell'Universita di Modena **23**(51), 179–189 (2003)
13. Mesiar, R., Kolesárová, A.: k-maxitive aggregation functions. Fuzzy Sets Syst. (submitted)
14. Murillo, J., Guillaume, S., Bulacio, P.: k-maxitive fuzzy measures: a scalable approach to model interactions. Fuzzy Sets Syst. (submitted)
15. Shilkret, N.: Maxitive measure and integration. Indag. Math. **33**, 109–116 (1971)
16. Sugeno, M.: Theory of fuzzy integrals and its applications. Ph.D. thesis. Tokyo Institute of Technology (1974)
17. Wang, Z., Klir, G.J.: Generalized Measure Theory. Springer, New York (2008)
18. Zadeh, L.A.: Fuzzy sets as a basis for a theory of possibility. Fuzzy Sets Syst. **1**, 3–28 (1978)

On Some Classes of RU-Implications Satisfying U-Modus Ponens

Margarita Mas, Daniel Ruiz-Aguilera$^{(\boxtimes)}$, and Joan Torrens

SCOPIA Research Group, Department of Mathematics and Computer Science,
University of the Balearic Islands, Crta. Valldemossa, Km. 7.5, 07122 Palma, Spain
{mmg448,daniel.ruiz,jts224}@uib.es

Abstract. The Modus Ponens property for fuzzy implication functions is essential in the inference process in approximate reasoning. It is usually considered with respect to a continuous t-norm T but it can be generalized to any conjunctor and, in particular, to a conjunctive uninorm U. In this paper, it is investigated when RU-implications derived from uninorms satisfy the Modus Ponens with respect to a conjunctive uninorm U. The new property, called here U-Modus Ponens, is studied in detail for RU-implications derived from uninorms lying in the classes of representable uninorms and uninorms continuous in the open unit square.

Keywords: Fuzzy implication · Uninorm · RU-implication · Modus Ponens

1 Introduction

Fuzzy implication functions are used in fuzzy logic not only to modelize fuzzy conditionals, but also to make inferences. Thus, when the Zadeh's compositional rule of inference is used to manage forward inferences, the Modus Ponens becomes essential in the process and this rule of inference is guaranteed when the fuzzy operators used, that is, the conjunction and the fuzzy conditional, satisfy the following inequality:

$$T(x, I(x, y)) \leq y \quad \text{for all} \quad x, y \in [0, 1], \tag{1}$$

where T is usually considered a (continuous) t-norm and I a fuzzy implication function. The previous inequality is also known as the Modus Ponens property or T-conditionality.

Due to its importance in the inference process, those t-norms T and fuzzy implication functions I that satisfy Eq. (1) have been investigated by many researchers for decades (see for instance, [2,3,18–20,27–30]). The main studies are related to implications derived from t-norms and t-conorms, like residual implications and (S, N)-implications investigated in detail in [2,27,28], and QL and D-implications in [29]. Moreover, these results were collected and completed later in [3] (see Sect. 7.4).

© Springer International Publishing AG 2018
V. Torra et al. (eds.), *Aggregation Functions in Theory and in Practice,*
Advances in Intelligent Systems and Computing 581, DOI 10.1007/978-3-319-59306-7_8

However, there are other studies involving different kinds of implication functions, specially those derived from more general aggregation functions than t-norms and t-conorms (see for instance [22]). In particular, implication functions derived from uninorms have been extensively investigated (see [1,4,6,17,23–25]) and recently, the Modus Ponens with respect to a continuous t-norm T has been already studied for these kinds of implication functions, that is, for the so-called RU-implications and (U, N)-implications (see [14,15]).

Recently, the authors have proposed a generalization of the Modus Ponens based on the idea of considering a conjunctive uninorm U as the conjunction, instead of a continuous t-norm, that will be called the U-Modus Ponens or U-conditionality (see [16]). In the mentioned paper, it is proved that this new property is never satisfied by the usual kinds of implications, that only implications derived from uninorms are available in this framework and the investigation involving RU-implications derived from some kinds of uninorms was initialized. In particular, RU-implications derived from uninorms in $\mathscr{U}_{\min}$ and from idempotent uninorms were considered leading to many new solutions of the U-Modus Ponens.

The idea of this paper is to extend such study to RU-implications derived from uninorms lying in other well known families of uninorms, that is, to representable uninorms and to uninorms continuous in the open unit square. We will see that many new solutions appear when these classes of uninorms are considered.

The paper is organized as follows. After this introduction, Sect. 2 is devoted to some preliminaries in order to make the paper as self-contained as possible. Section 3 deals with the Modus Ponens with respect to a uninorm U, including some general results for any kind of implication functions as well as some particular ones for the case of RU-implications. This last part is divided then into two subsections, one for each class of uninorms considered. Finally, the paper ends with Sect. 4 devoted to some conclusions and future work.

2   Preliminaries

We will suppose the reader to be familiar with the theory of t-norms, t-conorms and fuzzy negations (all necessary results and notations can be found in [11]). We also suppose that some basic facts on uninorms are known (see for instance [9] and the recent survey [13]). We recall here only some facts on implication functions and uninorms in order to stablish the necessary notation that we will use along the paper.

Definition 1. *A binary operator* $I : [0, 1] \times [0, 1] \to [0, 1]$ *is said to be a fuzzy implication function, or an implication, if it satisfies:*

(I1) $I(x, z) \geq I(y, z)$ *when* $x \leq y$, *for all* $z \in [0, 1]$.
(I2) $I(x, y) \leq I(x, z)$ *when* $y \leq z$, *for all* $x \in [0, 1]$.
(I3) $I(0, 0) = I(1, 1) = 1$ *and* $I(1, 0) = 0$.

Note that, from the definition, it follows that $I(0, x) = 1$ and $I(x, 1) = 1$ for all $x \in [0, 1]$ whereas the symmetrical values $I(x, 0)$ and $I(1, x)$ are not derived from the definition.

Lemma 1. *Given a fuzzy implication function I, the function $N_I(x) = I(x, 0)$ for all $x \in [0, 1]$ is always a fuzzy negation, known as the* natural negation *of I.*

Definition 2. *A* uninorm *is a two-place function $U : [0, 1]^2 \rightarrow [0, 1]$ which is associative, commutative, increasing in each place and such that there exists some element $e \in [0, 1]$, called* neutral element, *such that $U(e, x) = x$ for all $x \in [0, 1]$.*

Evidently, a uninorm with neutral element $e = 1$ is a t-norm and a uninorm with neutral element $e = 0$ is a t-conorm. For any other value $e \in]0, 1[$ the operation works as a t-norm in the $[0, e]^2$ square, as a t-conorm in $[e, 1]^2$ and its values are between minimum and maximum in the set of points $A(e)$ given by

$$A(e) = [0, e[\times]e, 1] \cup]e, 1] \times [0, e[.$$

We will usually denote a uninorm with neutral element e and underlying t-norm and t-conorm, T_U and S_U, by $U \equiv \langle T_U, e, S_U \rangle$. For any uninorm it is satisfied that $U(0, 1) \in \{0, 1\}$ and a uninorm U is called *conjunctive* if $U(1, 0) = 0$ and *disjunctive* when $U(1, 0) = 1$. On the other hand, let us recall two of the most studied classes of uninorms in the literature (more details can be found for instance in the recent survey [13]).

Definition 3. *A* uninorm *U, with neutral element $e \in]0, 1[$, is called* representable *if there exists a strictly increasing function $h : [0, 1] \rightarrow [-\infty, +\infty]$ (called an* additive generator *of U, which is unique up to a multiplicative constant $k > 0$), with $h(0) = -\infty$, $h(e) = 0$ and $h(1) = +\infty$, such that U is given by*

$$U(x, y) = h^{-1}(h(x) + h(y))$$

for all $(x, y) \in [0, 1]^2 \setminus \{(0, 1), (1, 0)\}$. We have either $U(0, 1) = U(1, 0) = 0$ or $U(0, 1) = U(1, 0) = 1$.

A representable uninorm with neutral element $e \in]0, 1[$ and additive generator h will be denoted by $U \equiv \langle e, h \rangle_{\text{rep}}$ and the class of all representable uninorms by $\mathscr{U}_{\text{rep}}$.

This class is clearly contained in the class of uninorms continuous in $]0, 1[^2$ which was characterized in [10] as follows (see again [13] for more details):

Theorem 1. *Suppose U is a uninorm continuous in $]0, 1[^2$ with neutral element $e \in]0, 1[$. Then either one of the following cases is satisfied:*

(a) *There exist $u \in [0, e[$, $\lambda \in [0, u]$, a continuous t-norm T and a representable uninorm R such that U can be represented as*

$$U(x, y) = \begin{cases} uT\left(\frac{x}{u}, \frac{y}{u}\right) & \text{if } x, y \in [0, u], \\ u + (1 - u)R\left(\frac{x-u}{1-u}, \frac{y-u}{1-u}\right) & \text{if } x, y \in]u, 1[, \\ 1 & \text{if } \min(x, y) \in]\lambda, 1] \text{ and } \max(x, y) = 1, \\ \lambda \text{ or } 1 & \text{if } (x, y) = (\lambda, 1) \text{ or } (x, y) = (1, \lambda), \\ \min(x, y) & \text{elsewhere.} \end{cases} \quad (2)$$

(b) *There exist $v \in]e, 1]$, $\omega \in [v, 1]$, a continuous t-conorm S and a representable uninorm R such that U can be represented as*

$$U(x,y) = \begin{cases} vR\left(\frac{x}{v}, \frac{y}{v}\right) & if\ x, y \in]0, v[, \\ v + (1-v)S\left(\frac{x-v}{1-v}, \frac{y-v}{1-v}\right) & if\ x, y \in [v, 1], \\ 0 & if\ \max(x,y) \in [0, \omega[\ and\ \min(x,y) = 0, \\ \omega\ or\ 0 & if\ (x,y) = (0, \omega)\ or\ (x,y) = (\omega, 0), \\ \max(x,y) & elsewhere. \end{cases} \quad (3)$$

The class of uninorms continuous in $]0, 1[^2$ will be denoted by $\mathscr{U}_{\cos}$. A uninorm as in (2) will be denoted by $U \equiv \langle \lambda, T, u, (R, e) \rangle_{\cos,\min}$ and the class of all uninorms continuous in the open unit square of this form will be denoted by $\mathscr{U}_{\cos,\min}$. Analogously, a uninorm as in (3) will be denoted by $U \equiv \langle (R, e), v, S, \omega \rangle_{\cos,\max}$ and the class of all uninorms of this form will be denoted by $\mathscr{U}_{\cos,\max}$.

We do not recall here other classes of uninorms, like uninorms in $\mathscr{U}_{\min}$ and $\mathscr{U}_{\max}$ [9] or idempotent uninorms [5, 12, 26], because they will be scarcely used in this work. In any case, the structure and characterization of all the classes mentioned here can be found in [13]. On the other hand, different classes of implications derived from uninorms have been studied. We recall here RU-implications.

Definition 4. *Let U be a uninorm. The* residual operation *derived from U is the binary operation given by*

$$I_U(x,y) = \sup\{z \in [0, 1] \mid U(x, z) \le y\}\ for\ all\ x, y \in [0, 1].$$

The residual operator derived from U is a fuzzy implication in many cases.

Proposition 1. *Let U be a uninorm and I_U its residual operation. hen I_U is an implication if and only if the following condition holds*

$$U(x, 0) = 0 \quad for\ all \quad x < 1. \quad (4)$$

In this case I_U is called an RU-implication.

Uninorms satisfying Eq. (4) include all conjunctive uninorms but also many disjunctive ones, like for instance disjunctive representable uninorms and some disjunctive uninorms in $\mathscr{U}_{\cos,\max}$. Some properties of RU-implications have been studied involving the main classes of uninorms including those previously stated. Recently, the Modus Ponens property with respect to a t-norm T has been studied in detail also for implications derived from uninorms (not only for RU, but also for (U, N)-implications) in [14, 15].

3 U-Modus Ponens

In this section we want to deal with the generalization of the Modus Ponens with respect to a t-norm T by substituting the t-norm T by a conjunctive uninorm U, leading to the so-called U-Modus Ponens or also U-conditionality:

Definition 5. *Let I be an implication function and U a uninorm. It is said that I satisfies the* Modus Ponens *property with respect to U, or that I is an U-conditional if*

$$U(x, I(x,y)) \leq y \quad for\,all\,x, y \in [0,1]. \tag{5}$$

The purpose of this paper is to study which conditions must satisfy a fuzzy implication I and a uninorm U in order to be I a U-conditional, when the implication I is an RU-implication derived from a uninorm lying in some of the three classes recalled in the preliminaries. The same study was already done for RU-implications derived from uninorms in $\mathscr{U}_{\min}$ and from idempotent uninorms in [16]. In that paper the following general result was already stated.

Proposition 2. *Let U be a conjunctive uninorm with neutral element $e \in]0,1[$ and let I be an implication satisfying U-Modus Ponens. The following properties hold:*

1. $I(e,y) \leq y$ *for all* $y \in [0,1]$.
2. *The natural negation N_I must satisfy*

$$N_I(x) = 0 \quad for\,all\,x \geq e, \quad and \quad N_I(x) < e \quad for\,all\,0 < x < e.$$

In particular, N_I can not be continuous.
3. *It must be $U(x, N_I(x)) = 0$ for all $x \in [0,1]$.*
4. $I(x,y) < e$ *for all* $x > y \geq e$. *In particular, $I(1,y) < e$ for all $y < 1$.*

The previous proposition gives some necessary conditions on the uninorm U as well as on the implication I in order they satisfy U-conditionality. From Property 4 in the previous proposition it is clear that the usual classes of fuzzy implication functions, that is, R, (S,N), QL and D-implications derived from t-norms and t-conorms, as well as f and g-generated Yager's implications, can not be U-conditionals (note that all of them satisfy $I(1,y) = y$ for all $y \in [0,1]$). However, this is not the case of RU and (U,N)-implications derived from uninorms.

For instance, RU-implications (see Definition 4) satisfy $I_U(e,y) = y$ for all $y \in [0,1]$ and so they are good candidates to be U-conditionals. In fact, the following two partial results were also presented in [16].

Proposition 3. *Let U, U_0 be two uninorms with neutral elements $e, e_0 \in]0,1[$ respectively, such that one of them is left-continuous and let I_{U_0} be the residual implication derived from U_0. If $U \leq U_0$ then I_{U_0} is an U-conditional.*

Proposition 4. *Let U, U_0 be two uninorms with neutral elements $e, e_0 \in]0,1[$ respectively and let I_{U_0} be the residual implication derived from U_0. If I_{U_0} is a U-conditional then it must be $e_0 \leq e$.*

In what follows we will deal with U-Modus Ponens for RU-implications derived from uninorms in one of the classes recalled in the preliminaries, that is, when U_0 is in $\mathscr{U}_{\mathrm{rep}}$, or in $\mathscr{U}_{\cos}$. We will divide our study in two subsections, one for each class of uninorms.

3.1   The Case When U_0 Is In $\mathscr{U}_{\text{rep}}$

In this section we want to deal with residual implications derived from representable uninorms. Let us recall first how are this kind of implications that can be found in [6] (see also [3]). Suppose that U_0 is a conjunctive representable uninorm with neutral element $e_0 \in]0,1[$ and additive generator h. Denote by U_0^* the disjunctive representable uninorm with the same additive generator h. Then both, the residuated RU-implication derived from U_0 and the one derived from U_0^*, coincide and they are given as follows.

Proposition 5. *Let $U_0 \in \mathscr{U}_{\text{rep}}$ be a representable uninorm with neutral element $e_0 \in]0,1[$ and additive generator h. Then the RU-implication derived from U_0 is given by:*

$$I_{U_0}(x,y) = \begin{cases} h^{-1}(h(y) - h(x)) & if\ (x,y) \notin \{(0,0),(1,1)\}, \\ 1 & if\ (x,y) \in \{(0,0),(1,1)\}. \end{cases}$$

For this kind of implications the characterization can be easily stated. In fact, the sufficient condition given in Proposition 3 is also necessary in this case as we can see in the next theorem.

Theorem 2. *Let U be a uninorm with neutral element $e \in]0,1[$ and let $U_0 \equiv \langle e_0, h \rangle_{\text{rep}}$ be a representable uninorm with neutral element $e_0 \leq e$ and additive generator h. If I_{U_0} is the residual implication derived from U_0 then*

$$I_{U_0}\ is\ a\ U - conditional \quad \Longleftrightarrow \quad U(x,y) \leq U_0(x,y)\ for\ all\ x,y \in]0,1[.$$

Example 1. Let $U_0 \equiv \langle e, h \rangle_{\text{rep}}$ be a representable uninorm with neutral element $e \in]0,1[$. It is well known that the underlying t-norm T_U and the underlying t-conorms S_U are then strict. Consider the uninorms in $\mathscr{U}_{\min}$ given by

$$U \equiv \langle T_U, e, S_U \rangle_{\min} \qquad \text{and} \qquad U' \equiv \langle \min, e, S_U \rangle_{\min}.$$

Then it is clear that $U \leq U_0$ but $U' \not\leq U_0$ and consequently in this case, from the theorem above, we have that I_{U_0} is a U-conditional but it is not and U'-conditional.

3.2   The Case When U_0 Is In $\mathscr{U}_{\cos}$

Let us deal in this section with residual implications derived from uninorms continuous in the open unit square $]0,1[^2$. However, since there are two different classes of these uninorms, we will divide our study into two subsections, one devoted to uninorms in $\mathscr{U}_{\cos,\min}$ and the other to uninorms in $\mathscr{U}_{\cos,\max}$.

Case in $\mathscr{U}_{\cos,\min}$ Recall first how are the residual implications derived from uninorms in this case, that can be found in [25].

Proposition 6. *Let $U_0 \equiv \langle \lambda, T, u, (R, e_0) \rangle_{\text{cos,min}}$ be a uninorm lying in $\mathscr{U}_{\text{cos,min}}$. Then the RU-implication derived from U_0 is given by:*

$$I_{U_0}(x,y) = \begin{cases} uI_{T_U}\left(\frac{x}{u}, \frac{y}{u}\right) & \text{if } x \in [0,u] \text{ and } y < x, \\ 1 & \text{if } x \in [0,u] \text{ and } y \geq x, \\ y & \text{if } (x,y) \in]u,1[\times [0,u], \\ u + (1-u)I_R\left(\frac{x-u}{1-u}, \frac{y-u}{1-u}\right) & \text{if } (x,y) \in]u,1[^2, \\ 1 & \text{if } y = 1, \\ y & \text{if } x = 1 \text{ and } y \leq \lambda, \\ \lambda & \text{if } x = 1 \text{ and } y > \lambda. \end{cases} \tag{6}$$

For this kind of implications we give first some partial results before to be able to give the characterization those that are U-conditionals.

Proposition 7. *Let U be a uninorm with neutral element $e \in]0,1[$ and let U_0 be a uninorm in $\mathscr{U}_{\text{cos,min}}$ given by $U_0 \equiv \langle \lambda, T, u, (R, e_0) \rangle_{\text{cos,min}}$, with $e_0 \leq e$. Let I_{U_0} the residual implication derived from U_0. If I_{U_0} is a U-conditional then*

$$U(x,y) = \min(x,y) \quad \text{for all} \quad \min(x,y) \leq u < e_0 \leq e < \max(x,y). \tag{7}$$

Now, we are able to characterize all residual implications from uninorms in $\mathscr{U}_{\text{cos,min}}$ that are U-conditionals, in the case when U has continuous underlying t-norm and $U(u,u) = u$. Note that in this case there exists a continuous t-norm T_1 and a uninorm U_1 with neutral element $\frac{e-u}{1-u}$ such that U can be written as

$$U(x,y) = \begin{cases} uT_1\left(\frac{x}{u}, \frac{y}{u}\right) & \text{if } x,y \in [0,u], \\ u + (1-u)U_1\left(\frac{x-u}{1-u}, \frac{y-u}{1-u}\right) & \text{if } x,y \in [u,1], \\ U(x,y) & \text{otherwise.} \end{cases} \tag{8}$$

Theorem 3. *Let U be a uninorm with neutral element $e \in]0,1[$ and let U_0 be a uninorm in $\mathscr{U}_{\text{cos,min}}$ given by $U_0 \equiv \langle \lambda, T, u, (R, e_0) \rangle_{\text{cos,min}}$, with $e_0 \leq e$, and I_{U_0} its residual implication. Suppose that $U(u,u) = u$ and T_U is continuous. Then there exist a continuous t-norm T_1 and a uninorm U_1 with neutral element $\frac{e-u}{1-u}$ such that U can be written as in (8) and I_{U_0} is a U-conditional if and only if the following conditions hold:*

(i) *U satisfies Eq. (7),*
(ii) *I_T is a T_1-conditional,*
(iii) *$U_1(x,y) \leq R(x,y)$ for all $x,y \in]0,1[$.*

For instance, we have the following example showing residual implications I_{U_0} derived from uninorms U_0 in $\mathscr{U}_{\text{cos,min}}$ that are U-conditionals.

Example 2. Let U be the conjunctive uninorm with neutral element $e \in]0,1[$ given by

$$U(x,y) = \begin{cases} uT\left(\frac{x}{u}, \frac{y}{u}\right) & \text{if } x, y \in [0, u], \\ u + (1-u)R\left(\frac{x-u}{1-u}, \frac{y-u}{1-u}\right) & \text{if } x, y \in [u, 1], \\ \min(x, y) & \text{otherwise,} \end{cases}$$

where T is any continuous t-norm and R any representable uninorm with neutral element $\frac{e-u}{1-u}$. Consider U_0 the uninorm in $\mathscr{U}_{\cos,\min}$ given by $U_0 \equiv \langle \lambda, \min, u, (R, e) \rangle_{\cos,\min}$. Then we have that I_{U_0} is given by

$$I_{U_0}(x,y) = \begin{cases} 1 & \text{if } x \in [0, u] \text{ and } y \geq x, \\ y & \text{if } y \in [0, u] \text{ and } y < x < 1, \\ u + (1-u)I_R\left(\frac{x-u}{1-u}, \frac{y-u}{1-u}\right) & \text{if } (x, y) \in]u, 1[^2, \\ 1 & \text{if } y = 1, \\ y & \text{if } x = 1 \text{ and } y \leq \lambda, \\ \lambda & \text{if } x = 1 \text{ and } y > \lambda. \end{cases} \tag{9}$$

and that I_{U_0} is a U-conditional by Theorem 3. The structures of U, U_0 and I_{U_0} are depicted in Fig. 1.

Case in $\mathscr{U}_{\cos,\max}$ In this case the residual operator derived from a uninorm $U_0 \equiv \langle (R, e_0), v, S, \omega \rangle_{\cos,\max}$ in $\mathscr{U}_{\cos,\max}$ is not always a fuzzy implication. Since it is necessary that $U_0(x, 0) = 0$ for all $x < 1$ by Proposition 1, it must be $\omega = 1$ and so this condition will be assumed from now on. Let us recall also how are the residual implications derived from this kind of uninorms.

Proposition 8. *Let $U_0 \equiv \langle (R, e_0), v, S, \omega \rangle_{\cos,\max}$ be a uninorm lying in $\mathscr{U}_{\cos,\max}$ with $\omega = 1$. Then the RU-implication derived from U_0 is given by:*

$$I_{U_0}(x,y) = \begin{cases} vI_R\left(\frac{x}{v}, \frac{y}{v}\right) & \text{if } (x, y) \in]0, v]^2, \\ v + (1-v)R_S\left(\frac{x-v}{1-v}, \frac{y-v}{1-v}\right) & \text{if } (x, y) \in]v, 1[^2 \text{ and } y \geq x, \\ 0 & \text{if } x \in]v, 1[\text{ and } y < x, \\ y & \text{if } (x, y) \in]0, v] \times [v, 1[, \\ 1 & \text{if } y = 1 \text{ or } x = 0, \\ 0 & \text{if } y = 0 \text{ and } x \neq 0. \end{cases} \tag{10}$$

For this kind of implications the characterization of those that are U-conditionals can be easily derived when $U(v, v) = v$ and has underlying t-conorm continuous. In this case there exist a continuous t-conorm S_1 and a uninorm U_1 with neutral element $\frac{e}{v}$ such that U can be written as

$$U(x,y) = \begin{cases} vU_1\left(\frac{x}{v}, \frac{y}{v}\right) & \text{if } x, y \in [0, v], \\ v + (1-v)S_1\left(\frac{x-v}{1-v}, \frac{y-v}{1-v}\right) & \text{if } x, y \in [v, 1], \\ U(x, y) & \text{otherwise.} \end{cases} \tag{11}$$

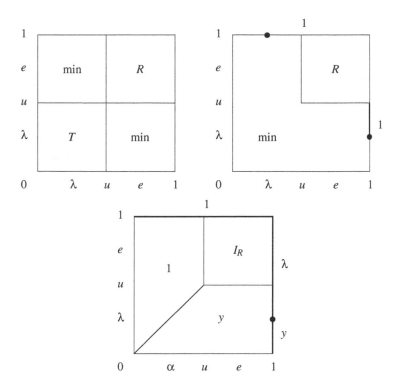

Fig. 1. Structure of U (top, left), U_0 (top, right) and I_{U_0} (bottom), with $U_0 \in \mathcal{U}_{\cos,\min}$ from Example 2.

Given a t-conorm S we denote by R_S the residual operator $R_S(x,y) = \sup\{z \in [0,1] \mid S(x,z) \leq y\}$. Then we have the following result.

Theorem 4. *Let U be a uninorm with neutral element $e \in]0,1[$ and let U_0 be a uninorm in $\mathcal{U}_{\cos,\max}$ given by $U_0 \equiv \langle (R,e_0), v, S, \omega \rangle_{\cos,\max}$ with $\omega = 1$ and $e_0 \leq e$. Let I_{U_0} be the residual implication derived from U_0 and suppose that $U(v,v) = v$ and its underlying t-conorm is continuous. Then there exist a continuous t-conorm S_1 and a uninorm U_1 with neutral element $\frac{e}{v}$ such that U can be written as in (11) and I_{U_0} is a U-conditional if and only if the following conditions hold:*

(i) $U_1(x,y) \leq R(x,y)$ *for all* $x,y \in]0,1[$.
(ii) $S_1(x, R_S(x,y)) \leq y$ *for all* $x \leq y$.

For instance, we have the following example showing residual implications I_{U_0} derived from uninorms U_0 in $\mathcal{U}_{\cos,\max}$ that are U-conditionals.

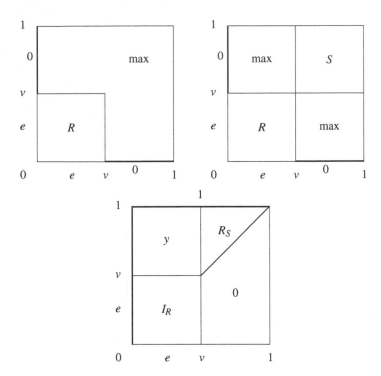

Fig. 2. Structure of U (top, left), U_0 (top, right) and I_{U_0} (bottom), with $U_0 \in \mathscr{U}_{\cos,\max}$ from Example 3.

Example 3. Let U be the conjunctive uninorm with neutral element $e \in]0,1[$ given by

$$U(x,y) = \begin{cases} vR\left(\frac{x}{v}, \frac{y}{v}\right) & \text{if } x, y \in [0,v], \\ \min(x,y) & \text{if } x = 0 \text{ and } y > v, \\ \max(x,y) & \text{otherwise,} \end{cases}$$

where R is any representable uninorm with neutral element $\frac{e}{v}$. Consider U_0 the uninorm in $\mathscr{U}_{\cos,\max}$ given by $U_0 \equiv \langle (R,e), v, S, \omega \rangle_{\cos,\max}$ with $\omega = 1$ and S any continuous t-conorm. Then we have that I_{U_0} is given by

$$I_{U_0}(x,y) = \begin{cases} vI_R\left(\frac{x}{v}, \frac{y}{v}\right) & \text{if } (x,y) \in]0,v]^2, \\ v + (1-v)R_S\left(\frac{x-v}{1-v}, \frac{y-v}{1-v}\right) & \text{if } (x,y) \in]v,1[^2 \text{ and } y \geq x, \\ y & \text{if } y \in]v,1[\text{ and } x < v, \\ 1 & \text{if } y = 1 \text{ or } x = 0, \\ 0 & \text{otherwise.} \end{cases} \tag{12}$$

and that I_{U_0} is a U-conditional by the previous theorem. The structures of U, U_0 and I_{U_0} are depicted in Fig. 2.

4 Conclusions and Future Work

Forward inferences schemes in approximate reasoning are based on the Modus Ponens property, also called T-conditionality and given by Eq. (1). In this paper we have enlarged such property to the so-called U-Modus Ponens or U-conditionality, just by substituting the t-norm T by a conjunctive uninorm U. Fixed a uninorm U we have investigated in this paper which RU-implications satisfy U-conditionality. We have given a detailed study in the cases when the uninorm used to derive the RU-implication lies in the class of representable uninorms and when it lies in the class of uninorms continuous in the open unit square, in a similar way as it was done in [16] for the class of uninorms in $\mathscr{U}_{\min}$ and the class of idempotent uninorms.

As a future work, we want to extend this study to RU-implications derived from uninorms in the class of locally internal uninorms (see [7,8]). In fact, we are currently working in this direction but we have not included the results because of the restriction of space. Moreover, we want to deal also in next future with other kind of implications like (U, N)-implications derived from disjunctive uninorms (see [3]) or h and (h, e)-implications recently introduced in [21]. Finally, a similar generalization through uninorms of the Modus Tollens property would be also worth of study.

Acknowledgements. This paper has been supported by the Spanish Grant TIN2016-75404-P AEI/FEDER, UE.

References

1. Aguiló, I., Suñer, J., Torrens, J.: A characterization of residual implications derived from left-continuous uninorms. Inf. Sci. **180**, 3992–4005 (2010)
2. Alsina, C., Trillas, E.: When (S, N)-implications are (T, T_1)-conditional functions? Fuzzy Sets Syst. **134**, 305–310 (2003)
3. Baczyński, M., Jayaram, B.: Fuzzy Implications. Studies in Fuzziness and Soft Computing, vol. 231. Springer, Heidelberg (2008)
4. Baczyński, M., Jayaram, B.: (U, N)-implications and their characterizations. Fuzzy Sets Syst. **160**, 2049–2062 (2009)
5. Baets, B.: Idempotent uninorms. Eur. J. Oper. Res. **118**, 631–642 (1999)
6. Baets, B., Fodor, J.C.: Residual operators of uninorms. Soft. Comput. **3**, 89–100 (1999)
7. Drygaś, P.: On properties of uninorms with underlying t-norm and t-conorm given as ordinal sums. Fuzzy Sets Syst. **161**, 149–157 (2010)
8. Drygaś, P., Ruiz-Aguilera, D., Torrens, J.: A characterization of a class of uninorms with continuous underlying operators. Fuzzy Sets Syst. **287**, 137–153 (2016)
9. Fodor, J.C., Yager, R.R., Rybalov, A.: Structure of uninorms. Int. J. Uncertain. Fuzziness Knowl.-Based Syst. **5**, 411–427 (1997)
10. Hu, S.K., Li, Z.F.: The structure of continuous uninorms. Fuzzy Sets Syst. **124**, 43–52 (2001)
11. Klement, E.P., Mesiar, R., Pap, E.: Triangular norms. Kluwer Academic Publishers, Dordrecht (2000)

12. Martín, J., Mayor, G., Torrens, J.: On locally internal monotonic operators. Fuzzy Sets Syst. **137**, 27–42 (2003)
13. Mas, M., Massanet, S., Ruiz-Aguilera, D., Torrens, J.: A survey on the existing classes of uninorms. J. Intell. Fuzzy Syst. **29**, 1021–1037 (2015)
14. Mas, M., Monserrat, M., Ruiz-Aguilera, D., Torrens, J.: Residual implications derived from uninorms satisfying the Modus Ponens. In: IFSA-EUSFLAT-2015, pp. 233–240. Atlantis Press, Gijón (2015)
15. Mas, M., Monserrat, M., Ruiz-Aguilera, D., Torrens, J.: RU and (U, N)-implications satisfying Modus Ponens. Int. J. Approx. Reason. **73**, 123–137 (2016)
16. Mas, M., Monserrat, M., Ruiz-Aguilera, D., Torrens, J.: On a generalization of the Modus Ponens: U-conditionality. In: Carvalho, J.P., et al. (eds.) Information Processing and Management of Uncertainty in Knowledge-Based Systems in the series Communications in Computer and Information Science, vol. 610, pp. 387–398. Springer, Cham (2016)
17. Mas, M., Monserrat, M., Torrens, J.: Two types of implications derived from uninorms. Fuzzy Sets Syst. **158**, 2612–2626 (2007)
18. Mas, M., Monserrat, M., Torrens, J.: Modus Ponens and Modus Tollens in discrete implications. Int. J. Approx. Reason. **49**, 422–435 (2008)
19. Mas, M., Monserrat, M., Torrens, J.: A characterization of (U, N), RU, QL and D-implications derived from uninorms satisfying the law of importation. Fuzzy Sets Syst. **161**, 1369–1387 (2010)
20. Mas, M., Monserrat, M., Torrens, J., Trillas, E.: A survey on fuzzy implication functions. IEEE Trans. Fuzzy Syst. **15**(6), 1107–1121 (2007)
21. Massanet, S., Torrens, J.: On a new class of fuzzy implications: h-implications and generalizations. Inf. Sci. **181**, 2111–2127 (2011)
22. Massanet, S., Torrens, J.: An overview of construction methods of fuzzy implications. In: Baczyński, M., Beliakov, G., Bustince Sola, H., Pradera, A. (eds.) Advances in Fuzzy Implication Functions. Studies in Fuzziness and Soft Computing, vol. 300, pp. 1–30. Springer, Berlin (2013)
23. Ruiz, D., Torrens, J.: Residual implications and co-implications from idempotent uninorms. Kybernetika **40**, 21–38 (2004)
24. Ruiz-Aguilera, D., Torrens, J.: Distributivity of residual implications over conjunctive and disjunctive uninorms. Fuzzy Sets Syst. **158**, 23–37 (2007)
25. Ruiz-Aguilera, D., Torrens, J.: S- and R-implications from uninorms continuous in $]0, 1[^2$ and their distributivity over uninorms. Fuzzy Sets Syst. **160**, 832–852 (2009)
26. Ruiz-Aguilera, D., Torrens, J., Baets, B., Fodor, J.: Some remarks on the characterization of idempotent uninorms. In: Hüllermeier, E., Kruse, R., Hoffmann, F. (eds.) IPMU 2010. LNCS, vol. 6178, pp. 425–434. Springer, Heidelberg (2010). doi:10.1007/978-3-642-14049-5_44
27. Trillas, E., Alsina, C., Pradera, A.: On MPT-implication functions for fuzzy logic. Revista de la Real Academia de Ciencias Serie A. Matemáticas (RACSAM) **98**(1), 259–271 (2004)
28. Trillas, E., Alsina, C., Renedo, E., Pradera, A.: On contra-symmetry and MPT-conditionality in fuzzy logic. Int. J. Intell. Syst. **20**, 313–326 (2005)
29. Trillas, E., Campo, C., Cubillo, S.: When QM-operators are implication functions and conditional fuzzy relations. Int. J. Intell. Syst. **15**, 647–655 (2000)
30. Trillas, E., Valverde, L.: On Modus Ponens in fuzzy logic. In: 15th International Symposium on Multiple-Valued Logic, pp. 294–301. Kingston, Canada (1985)

CMin-Integral: A Choquet-Like Aggregation Function Based on the Minimum t-Norm for Applications to Fuzzy Rule-Based Classification Systems

Graçaliz Pereira Dimuro[1,2(✉)], Giancarlo Lucca[3], José António Sanz[4], Humberto Bustince[4], and Benjamín Bedregal[5]

[1] Institute of Smart Cities, Universidad Pública de Navarra, 31006 Pamplona, Spain
gracaliz.pereiraa@unavarra.es
[2] Centro de Ciências Computacionais, Universidade Federal do Rio Grande, Rio Grande 96201-900, Brazil
[3] Departamento de Automática y Computación, Universidad Pública de Navarra, 31006 Pamplona, Spain
lucca.112793@e.unavarra.es
[4] Departamento de Automática y Computación and Institute of Smart Cities, Universidad Pública de Navarra, 31006 Pamplona, Spain
{bustince,joseantonio.sanz}@unavarra.es
[5] Departamento de Informática e Matemática Aplicada, Universidade Federal do Rio Grande do Norte, Natal, Brazil
bedregal@dimap.ufrn.br

Abstract. This paper studies the concept of Choquet-like copula-based aggregation function (CC-integral), introduced by Lucca et al. [1], when one considers the Minimum t-norm, showing an application in fuzzy rule-based classification systems. The CC-integral is built from the standard Choquet integral, which is expanded by distributing the product operation, and, then, the product operation is generalized by a copula. In this paper, we study the behavior of this aggregation function in fuzzy rule-based classification systems, when one considers the Minimum t-norm as de copula of the CC-integral, which we call the CMin-integral. We show that the CMin-integral obtains a performance that is, with a high level of confidence, better than the approach that adopts the winning rule (maximum). Moreover, its behaviour is similar to the best Choquet-like pre-aggregation functions, introduced by Lucca et al. [10], with excellent performance. Consequently, the CMin-integral enlarge the scope of the applications by offering new possibilities for defining fuzzy reasoning methods with a similar gain in performance.

1 Introduction

In [1], a new promising concept in the field of aggregation functions [2,3], namely, the concept of pre-aggregation function, was introduced by Lucca et al. This type of functions fulfills the basic propriety of boundary conditions of any aggregation

© Springer International Publishing AG 2018
V. Torra et al. (eds.), *Aggregation Functions in Theory and in Practice*,
Advances in Intelligent Systems and Computing 581, DOI 10.1007/978-3-319-59306-7_9

function, but, however, the monotonicity is considered just along some fixed direction (i.e., it is directionally increasing [4]). The authors proposed three different methods to build pre-aggregation functions, one of them consists in replace the product of the Choquet integral [5] by t-norms [6]. An application to fuzzy rule-based classification systems (FRBCS) [7,8] was presented, and it was shown that when the minimum or the Hamacher product t-norms are considered for such construction and applied in the fuzzy reasoning method (FRM), the obtained results were better than two classical averaging operators, namely, the Maximum and the standard Choquet integral, the latter in the approach proposed by Barrenechea et al. [9], which was the winner method at that time. Properties of pre-aggregation functions as well as other constructions methods were studied by Dimuro et al. [10] and Lucca et al. [11]

In a similar line of research, Lucca et al. [12] introduced the concept of CC-integrals, which are Choquet-like Copula-based aggregation functions obtaining by expanding the standard Choquet integral by distributing the product operation, and then replacing the product by a copula [13]. These functions were used in the FRM of FRBCS, presenting a behavior similar to the best Choquet-like pre-aggregation function.

The aim of this paper is to study the concept of CC-integral when one considers the Minimum t-norm, which we call CMin-integral, in order to compare the performance of the FRBCS adopting this aggregation function in its FRM with the approaches that adopts the winning rule (Maximum) and, also, the one that considers the best Choquet-like pre-aggregation functions. For this analysis, we have selected 30 datasets that are accessible in KEEL[1]database repository [14]. Our conclusions are supported by the well-known Wilcoxon signed-rank test [15].

The paper is organized as follows. Section 2 presents some preliminary concepts that are necessary to develop the paper. Section 3 contains the method for the construction of the CMin-integral. In Sect. 4, the FRM used in this study is presented. We describe the experimental framework, the results achieved in testing by the application of the CMin-integral in FRBCSs besides the analysis of these results in Sect. 5. The main conclusions are drawn in Sect. 6.

2   Preliminaries

This section aims at introducing the background necessary to understand the paper. One important class of fuzzy operators are the *aggregation operators* [2,3].

Definition 1. A function $A : [0,1]^n \to [0,1]$ is said to be an n-ary aggregation function whenever the following conditions are satisfied:

(A1) A is increasing[2] in each argument: for each $i \in \{1, \ldots, n\}$, if $x_i \leq y$, then
$A(x_1, \ldots, x_n) \leq A(x_1, \ldots, x_{i-1}, y, x_{i+1}, \ldots, x_n)$;

---

[1] http://www.keel.es.

[2] For an increasing (decreasing) function we do not mean a strictly increasing (decreasing) function.

(A2) A satisfies the boundary conditions: $A(0,\ldots,0) = 0$ and $A(1,\ldots,1) = 1$.

Definition 2. Let $\mathbf{r} = (r_1,\ldots,r_n)$ be a real n-dimensional vector, $\mathbf{r} \neq \mathbf{0}$. A function $F : [0,1]^n \to [0,1]$ is directionally increasing [4] with respect to $\mathbf{r}$ ($\mathbf{r}$-increasing, for short) if for all $(x_1,\ldots,x_n) \in [0,1]^n$ and $c > 0$ such that $(x_1 + cr_1,\ldots,x_n + cr_n) \in [0,1]^n$ it holds that

$$F(x_1 + cr_1,\ldots,x_n + cr_n) \geq F(x_1,\ldots,x_n). \tag{1}$$

Similarly, one defines an $\mathbf{r}$-decreasing function.

Definition 3 [1]. Let $\mathbf{r} = (r_1,\ldots,r_n)$ be a real n-dimensional vector, $\mathbf{r} \neq \mathbf{0}$. A function $F : [0,1]^n \to [0,1]$ is is said to be an n-ary $\mathbf{r}$-pre-aggregation function if the following conditions hold:

(PA1) F is $\mathbf{r}$-increasing;
(PA2) F satisfies the boundary conditions: $F(0,\ldots,0) = 0$ and $F(1,\ldots,1) = 1$.

Definition 4. An aggregation function $T : [0,1]^2 \to [0,1]$ is a t-norm if, for all $x,y,z \in [0,1]$, it satisfies the following properties:

(T1) Commutativity: $T(x,y) = T(y,x)$;
(T2) Associativity: $T(x,T(y,z)) = T(T(x,y),z)$;
(T3) Boundary condition: $T(x,1) = x$.

If T satisfies just **(T3)** (and also $T(1,x) = x$), then it is called a semi-copula.

Examples of t-norms are the minimum t-norm $T_M : [0,1]^2 \to [0,1]$, defined, for all $x,y \in [0,1]$, by
$$T_M(x,y) = \min\{x,y\}, \tag{2}$$
the Hamacher product $Ham : [0,1]^2 \to [0,1]$, defined, for all $x,y \in [0,1]$, by

$$Ham(x,y) = \begin{cases} 0 & \text{if } x = y = 0 \\ \frac{xy}{x+y-xy} & \text{otherwise} \end{cases} \tag{3}$$

and the Łukasiewicz t-norm $T_{\text{Ł}} : [0,1]^2 \to [0,1]$, defined, for all $x,y \in [0,1]$, by

$$T_{\text{Ł}}(x,y) = \max\{0, x + y - 1\}. \tag{4}$$

Definition 5. A bivariate function $C : [0,1]^2 \to [0,1]$ is a copula if it satisfies the following conditions, for all $x,x',y,y' \in [0,1]$ with $x \leq x'$ and $y \leq y'$:

(C1) $C(x,y) + C(x',y') \geq C(x,y') + C(x',y)$;
(C2) $C(x,0) = C(0,x) = 0$;
(C3) $C(x,1) = C(1,x) = x$.

Proposition 1 [6, Proposition 9.8] [16, Lemma 6.1.8, Lemma 6.3.1]. *Consider the Łukasiewicz and Minimum T-norms $T_L, T_M : [0,1]^2 \to [0,1]$, defined by Eqs. (4) and (2), respectively. For each copula $C : [0,1]^2 \to [0,1]$, it holds that:*

(i) $T_L \leq C \leq T_M$;
(ii) C is increasing;
(iii) C satisfies the Lipschitz property with constant 1, that is, for all $x_1, x_2, y_1, y_2 \in [0,1]$, *one has that:*

$$| C(x_1, y_1) - C(x_2, y_2) | \leq | x_1 - x_2 | + | y_1 - y_2 | .$$

An immediate consequence of Proposition 1 is that any copula is continuous. Then, each associative copula is a continuous t-norm [6, Corollary 9.9].

Observe that the minimum t-norm T_M and the Hamacher product t-norm Ham, given in Eqs. (2) and (3), are both copulas.

Now, we recall the concept of fuzzy measure [5,17], which is a central tool for defining the Choquet integral. In what follows, denote $N = \{1, \ldots, n\}$, for an arbitrary $n > 0$.

Definition 6. A function $\mathfrak{m} : 2^N \rightarrow [0,1]$ is said to be a fuzzy measure if, for all $X, Y \subseteq N$, it satisfies the following properties:

(m1) Increasing: if $X \subseteq Y$, then $\mathfrak{m}(X) \leq \mathfrak{m}(Y)$;
(m2) Boundary conditions: $\mathfrak{m}(\emptyset) = 0$ and $\mathfrak{m}(N) = 1$.

In this paper, we adopt the power measure $\mathfrak{m}_{PM} : 2^N \rightarrow [0,1]$, which is defined, for all $X \subseteq N$, by

$$\mathfrak{m}_{PM}(X) = \left(\frac{|X|}{n} \right)^q, \text{ with } q > 0. \tag{5}$$

The choice for this fuzzy measure was based on the results obtained by Barrenechea et al. [9], who introduced an evolutionary algorithm to define the most suitable q to be used in the definition of the measure for each class. We point out that we consider the same approach to learn the parameter q as adopted in [1,9,11,12,18].

Definition 7 [5]. Let $\mathfrak{m} : 2^N \rightarrow [0,1]$ be a fuzzy measure. The discrete Choquet integral is the function $\mathfrak{C}_\mathfrak{m} : [0,1]^n \rightarrow [0,1]$, defined, for all of $\mathbf{x} = (x_1, \ldots, x_n) \in [0,1]^n$, by:

$$\mathfrak{C}_\mathfrak{m}(\mathbf{x}) = \sum_{i=1}^{n} \left(x_{(i)} - x_{(i-1)} \right) \cdot \mathfrak{m} \left(A_{(i)} \right), \tag{6}$$

where $\left(x_{(1)}, \ldots, x_{(n)} \right)$ is an increasing permutation on the input x, that is, $0 \leq x_{(1)} \leq \ldots \leq x_{(n)}$, where $x_{(0)} = 0$ and $A_{(i)} = \{(i), \ldots, (n)\}$ is the subset of indices corresponding to the $n - i + 1$ largest components of $\mathbf{x}$.

Observe that the Eq. (6) can be also written as:

$$\mathfrak{C}_\mathfrak{m}(\mathbf{x}) = \sum_{i=1}^{n} \left(x_{(i)} \cdot \mathfrak{m} \left(A_{(i)} \right) - x_{(i-1)} \cdot \mathfrak{m} \left(A_{(i)} \right) \right), \tag{7}$$

which we call the Choquet Integral in its expanded form [18].

Definition 8 [12, Definition 7]. Let $\mathfrak{m} : 2^N \rightarrow [0, 1]$ be a fuzzy measure and $C : [0, 1]^2 \rightarrow [0, 1]$ be a bivariate copula. The Choquet-like copula-based integral (CC-integral) with respect to $\mathfrak{m}$ is defined as a function $\mathfrak{C}_\mathfrak{m}^C : [0, 1]^n \rightarrow [0, 1]$, given, for all $x \in [0, 1]^n$, by

$$\mathfrak{C}_\mathfrak{m}^C(\mathbf{x}) = \sum_{i=1}^{n} C\left(x_{(i)}, \mathfrak{m}\left(A_{(i)}\right)\right) - C\left(x_{(i-1)}, \mathfrak{m}\left(A_{(i)}\right)\right), \qquad (8)$$

where $(x_{(1)}, \ldots, x_{(n)})$ is an increasing permutation on the input $\mathbf{x}$, that is, $0 \leq x_{(1)} \leq \ldots \leq x_{(n)}$, with the convention that $x_{(0)} = 0$, and $A_{(i)} = \{(i), \ldots, (n)\}$ is the subset of indices of $n - i + 1$ largest components of $\mathbf{x}$.

3   Constructing the CMin-Integral

In this section, we construct the CMin-integral, the aggregation function obtained by considering the minimum t-norm as the copula C in Eq. (8).

In the following, consider $N = \{1, \ldots, n\}$.

Definition 9. Let $\mathfrak{m} : 2^N \rightarrow [0, 1]$ be a fuzzy measure and $T_M : [0, 1]^2 \rightarrow [0, 1]$ be the minimum t-norm given in Eq. (2). The Choquet-like Minimum-based integral with respect to $\mathfrak{m}$ (CMin-integral) is defined as a function $\mathfrak{C}_\mathfrak{m}^{\min} : [0, 1]^n \rightarrow [0, 1]$, given, for all $x \in [0, 1]^n$, by

$$\mathfrak{C}_\mathfrak{m}^{\min}(\mathbf{x}) = \sum_{i=1}^{n} \min\left\{x_{(i)}, \mathfrak{m}\left(A_{(i)}\right)\right\} - \min\left\{x_{(i-1)}, \mathfrak{m}\left(A_{(i)}\right)\right\}, \qquad (9)$$

where $(x_{(1)}, \ldots, x_{(n)})$ is an increasing permutation on the input x, that is, $0 \leq x_{(1)} \leq \ldots \leq x_{(n)}$, with the convention that $x_{(0)} = 0$, and $A_{(i)} = \{(i), \ldots, (n)\}$ is the subset of indices of $n - i + 1$ largest components of $\mathbf{x}$.

Proposition 2. *For any fuzzy measure* $\mathfrak{m} : 2^N \rightarrow [0, 1]$, $\mathfrak{C}_\mathfrak{m}^{\min}$ *is idempotent.*

Proof. Considering $\mathbf{x} = (x, \ldots, x) \in [0, 1]^n$, one has that:

$$\mathfrak{C}_\mathfrak{m}^{\min}(\mathbf{x}) = \min\{x, \mathfrak{m}(A_{(1)})\} - \min\{0, \mathfrak{m}(A_{(1)})\}$$
$$+ \sum_{i=2}^{n} \min\{x, \mathfrak{m}(A_{(i)})\} - \min\{x, \mathfrak{m}(A_{(i)})\}$$
$$= \min\{x, 1\} - \min\{0, 1\} + 0$$
$$= x.$$

Proposition 3. *For any copula fuzzy measure* $\mathfrak{m} : 2^N \rightarrow [0, 1]$, $\mathfrak{C}_\mathfrak{m}^{\min}$ *satisfies the boundary conditions (A2).*

Proof. Considering $\mathbf{0} = (0, \ldots, 0) \in [0, 1]^n$ and $\mathbf{1} = (1, \ldots, 1) \in [0, 1]^n$, by Proposition 2, one has that $\mathfrak{C}_\mathfrak{m}^{\min}(\mathbf{0}) = 0$ and $\mathfrak{C}_\mathfrak{m}^{\min}(\mathbf{1}) = 1$.

Proposition 4. *For any fuzzy measure* $\mathfrak{m} : 2^N \to [0,1]$, $\mathfrak{C}_{\mathfrak{m}}^{\min}$ *is increasing (A1).*

Proof. Since $\mathfrak{C}_{\mathfrak{m}}^{\min}$ is trivially commutative, then it is sufficient to consider the case when the input $\mathbf{x}$ is ordered, that is, $x_i = x_{(i)}$, for each $i = 1, \ldots, n$. Also, by the transitivity, is it is sufficient to consider the following cases:

(i) Consider $x_{(j)} \le y \le x_{(j+1)}$, for some $j = 1, \ldots, n-1$. Observe that:

$$\min\{x_{(j)}, \mathfrak{m}(A_{(j)})\} - \min\{x_{(j-1)}, \mathfrak{m}(A_{(j)})\} + \min\{x_{(j+1)}, \mathfrak{m}(A_{(j+1)})\}$$
$$- \min\{x_{(j)}, \mathfrak{m}(A_{(j+1)})\}$$
$$\le \min\{y, \mathfrak{m}(A_{(j)})\} - \min\{x_{(j-1)}, \mathfrak{m}(A_{(j)})\} + \min\{x_{(j+1)}, \mathfrak{m}(A_{(j+1)})\}$$
$$- \min\{y, \mathfrak{m}(A_{(j+1)})\},$$

and, then, it follows that

$$\mathfrak{C}_{\mathfrak{m}}^{\min}(x_1, \ldots, x_j, \ldots, x_n)$$
$$= \sum_{i=1}^{n} \min\{x_{(i)}, \mathfrak{m}(A_{(i)})\} - \min\{x_{(i-1)}, \mathfrak{m}(A_{(i)})\}$$
$$\le \left(\sum_{i=1}^{j-1} \min\{x_{(i)}, \mathfrak{m}(A_{(i)})\} - \min\{x_{(i-1)}, \mathfrak{m}(A_{(i)})\} \right)$$
$$+ \left(\min\{y, \mathfrak{m}(A_{(j)})\} - \min\{x_{(j-1)}, \mathfrak{m}(A_{(j)})\} \right)$$
$$+ \left(\min\{x_{(j+1)}, \mathfrak{m}(A_{(j+1)})\} - \min\{y, \mathfrak{m}(A_{(j+1)})\} \right)$$
$$+ \left(\sum_{i=j+2}^{n} \min\{x_{(i)}, \mathfrak{m}(A_{(i)})\} - \min\{x_{(i-1)}, \mathfrak{m}(A_{(i)})\} \right)$$
$$= \mathfrak{C}_{\mathfrak{m}}^{\min}(x_1, \ldots, x_{j-1}, y, x_{j+1}, \ldots, x_n).$$

(ii) Consider $x_{(n)} \le y$. Observe that:

$$\min\{x_{(n)}, \mathfrak{m}(A_{(n)})\} - \min\{x_{(n-1)}, \mathfrak{m}(A_{(n)})\}$$
$$\le \min\{y, \mathfrak{m}(A_{(n)})\} - \min\{x_{(n-1)}, \mathfrak{m}(A_{(n)})\},$$

and, then, it follows that:

$$\mathfrak{C}_{\mathfrak{m}}^{\min}(x_1, \ldots, x_n)$$
$$= \sum_{i=1}^{n} \min\{x_{(i)}, \mathfrak{m}(A_{(i)})\} - \min\{x_{(i-1)}, \mathfrak{m}(A_{(i)})\}$$
$$\le \left(\sum_{i=1}^{n-1} \min\{x_{(i)}, \mathfrak{m}(A_{(i)})\} - \min\{x_{(i-1)}, \mathfrak{m}(A_{(i)})\} \right)$$
$$+ \left(\min\{y, \mathfrak{m}(A_{(n)})\} - \min\{x_{(n-1)}, \mathfrak{m}(A_{(n)})\} \right)$$
$$= \mathfrak{C}_{\mathfrak{m}}^{\min}(x_1, \ldots, x_{n-1}, y).$$

Corollary 1. *For any fuzzy measure* $\mathfrak{m} : 2^N \to [0,1]$, *it holds that* $\min \leq \mathfrak{C}_{\mathfrak{m}}^{\min} \leq$ max.

Theorem 1. *For any fuzzy measure* $\mathfrak{m} : 2^N \to [0,1]$, $\mathfrak{C}_{\mathfrak{m}}^{\min}$ *is an average aggregation function.*

Proof. It follows from Propositions 3 and 4, and Corollary 1.

4   The Fuzzy Reasoning Method with the CMin-Integral

In this section, we present the new FRM generalized by the CMin-integral $\mathfrak{C}_{\mathfrak{m}}^{\min}$, as presented in Eq. (9).

In the following, consider that a classification problem, consists of m training examples $\mathbf{x}_p = (x_{p1}, \ldots, x_{pn}, y_p)$, with $p = 1, \ldots, m$, where x_{pi}, with $i = 1, \ldots, n$, is the value of the i-th attribute and $y_p \in \mathbb{C} = \{C_1, C_2, \ldots, C_M\}$ is the label of the class of the p-th training example.

In this work, we use a FRBCS to deal with classification problems. Specifically, we have selected FARC-HD [19] to accomplish the learning process and the form of the fuzzy rules used by this algorithm is:

$$\text{Rule } R_j : \text{ If } x_{p1} \text{ is } A_{j1} \text{ and } \ldots \text{ and } x_{pn} \text{ is } A_{jn} \text{ then Class is } C_j \text{ with } RW_j,$$
(10)

where $x_p = (x_{p1}, \ldots, x_{pn})$ is the n-dimensional vector of attribute values corresponding to an example $\mathbf{x}_p$, R_j is the label of the jth rule, A_{ji} is an antecedent fuzzy set modeling a linguistic term, C_j is the class of the j-th rule, and $RW_j \in [0,1]$ is the rule weight [20], which, in this case, is computed using the certainty factor.

Our proposal is a modification of the third step of the FRM in the FARC-HD fuzzy classifier [19]. More precisely, we propose the usage of the CMin-integral in order to obtain the information associated with each class of the problem. Specifically, the new classification soundness degree in the FRM is the following:

- **Example classification soundness degree for all classes.** In this step, we apply our CMin-integral to combine the positive association degrees (fired fuzzy rules) obtained in the previous steps of the FRM, $b_i^k(x_p) > 0$, as follows:

$$Y_k(x_p) = \mathfrak{C}_{\mathfrak{m}}^{\min}\left(b_1^k(x_p), \ldots, b_L^k(x_p)\right), \text{ with } k = 1, \ldots, M,$$
(11)

where $\mathfrak{C}_{\mathfrak{m}}^{\min}$ is the CMin-integral defined in Eq. (9), x_p is the example to be classified, M is the number of classes of the problem and L is the number of fuzzy rules in the system.

5   Experimental Results

In this section, we firstly describe the 30 real world classification problems selected from the KEEL dataset repository [14]. After that, we present the achieved results in testing, using the FRM generalized by our CMin-integral (denoted, for sake of simplicity, by CMin), along with an analysis of these obtained results (Sect. 5.1).

The properties of the datasets, containing for each dataset, the identifier (Id.), along with the name (Dataset), the number of instances (#*Inst*), the number of attributes (#*Att*) and the number of classes (#*Class*) are summarized in Table 1. The *magic, page-blocks, penbased, ring, shuttle, satimage* and *twonorm* datasets have been stratified sampled at 10% in order to reduce their size for training. Examples with missing values have been removed, e.g., in the *wisconsin* dataset.

Table 1. Summary of the datasets used in this study

Id.	Dataset	#Inst	#Att	#Class	Id.	Dataset	#Inst	#Att	#Class
App	Appendiciticis	106	7	2	Pho	Phoneme	5, 404	5	2
Bal	Balance	625	4	3	Pim	Pima	768	8	2
Ban	Banana	5300	2	2	Rin	Ring	740	20	2
Bnd	Bands	365	19	2	Sah	Saheart	462	9	2
Bup	Bupa	345	6	2	Sat	Satimage	6, 435	36	7
Cle	Cleveland	297	13	5	Seg	Segment	2, 310	19	7
Eco	Ecoli	336	7	8	Shu	Shuttle	5, 800	9	7
Gla	Glass	214	9	6	Spe	Spectfheart	267	44	2
Hab	Haberman	306	3	2	Tit	Titanic	2, 201	3	2
Hay	Hayes-Roth	160	4	3	Two	Twonorm	740	20	2
Iri	Iris	150	4	3	Veh	Vehicle	846	18	4
Mag	Magic	1, 902	10	2	Vow	Vowel	990	13	11
New	Newthyroid	215	5	3	Win	Wine	178	13	3
Pag	Pageblocks	5, 472	10	5	Wis	Wisconsin	683	11	2
Pen	Penbased	1, 099	16	10	Yea	Yeast	1, 484	8	10

As proposed in [1,9,21], we adopt the 5-fold cross-validation model, in other words, a dataset is splitted in five random partitions, where each partition have 20% of the examples, and a combination of four of them is used for training and the remainder one is used for testing. This process is repeated five times by using a different partition to test the created system each time. In order to measure the quality of each partition, the accuracy rate is calculated, that is, we divide the number of correctly classified examples divided by the total number of examples for each partition. Then, as the final result of the algorithm we consider the average of the achieved accuracy in this five partitions.

5.1 Experimental Results

This subsection present the results achieved in testing by the FRM considering the CMin-integral, with the power measure where the exponent q is learned genetically, as in [1,9].

Table 2. Accuracy results achieved in test by the CMin-Integral and the other considered operators

Dataset	CMin	Choquet	Ham	WR
App	**85.84**	80.13	82.99	83.03
Bal	81.60	82.40	**82.72**	81.92
Ban	84.30	**86.32**	85.96	83.94
Bnd	71.06	68.56	**72.13**	69.40
Bup	61.45	**66.96**	65.80	62.03
Cle	54.88	55.58	55.58	**56.91**
Eco	77.09	76.51	**80.07**	75.62
Gla	**69.17**	64.02	63.10	64.99
Hab	**74.17**	72.52	72.21	70.89
Hay	**81.74**	79.49	79.49	78.69
Iri	92.67	91.33	93.33	**94.00**
Mag	**79.81**	78.86	79.76	78.60
New	93.95	94.88	**95.35**	94.88
Pag	93.97	94.16	**94.34**	94.16
Pen	91.27	90.55	90.82	**91.45**
Pho	82.94	82.98	**83.83**	82.29
Pim	**75.78**	74.60	73.44	74.60
Rin	87.97	**90.95**	88.78	90.00
Sah	**70.78**	69.69	70.77	68.61
Sat	79.01	79.47	**80.40**	79.63
Seg	92.25	**93.46**	93.33	93.03
Shu	**98.16**	97.61	97.20	96.00
Spe	**78.99**	77.88	76.02	77.90
Tit	**78.87**	**78.87**	**78.87**	**78.87**
Two	85.14	84.46	85.27	**86.49**
Veh	**69.86**	68.44	68.20	66.67
Vow	**68.89**	67.58	68.18	67.98
Win	93.83	93.79	**96.63**	96.60
Wis	95.90	**97.22**	96.78	96.34
Yea	**57.01**	55.73	56.53	55.32
Mean	**80.28**	79.83	80.26	79.70

In order to determine the quality of the our proposal, considering the same approach for the fuzzy measure, we also present the results achieved by the best pre-aggregation functions proposed by Lucca et al. in [1] (the one that generalizes the standard Choquet integral using the Hamacher product t-norm, denoted here by Ham), the standard Choquet Integral (as it was studied in [9], denoted here by Choquet) and the classical FRM of the Winning Rule (WR) (the FRM that uses the Maximum as the aggregation function instead of the CMin-integral in Eq. (11)).

The results achieved in testing by these approaches are presented in Table 2 by columns, where the best result achieved among the different datasets if highlighted in **boldface**.

Looking at the results of these four approaches, it is noticeable that CMin-integral obtained the best global mean accuracy result, being similar with Ham and it is superior to those of WR and the standard Choquet integral. In a closer look we can observe that the CMin-integral obtains the best result in 13 datasets, whereas Ham achieves the best mean accuracy in 9 datasets, the standard Choquet integral achieves the best mean accuracy in 6 datasets, and, finally, the WR achieves the best mean accuracy in 5 datasets.

However, analyzing exclusively the achieved mean accuracy is not enough to draw any conclusion. For this reason, in order to support our previous results, we have carried out a set of pairwise statistical comparisons using the well-known Wilcoxon signed-rank test [15]. Specifically, we have compared the the CMin-integral, versus WR, Ham and Choquet integral. Table 3 shows the results of these comparisons, where R^+ indicates the ranks obtained by the CMin and R^- represents the ranks achieved by the method used in each comparison.

Table 3. Wilcoxon test to compare the best CC-integral versus the Ham_{PA}, the winning rule and the standard Choquet integral

Comparison	R^+	R^-	p-value
CMin vs. Ham_{PA}	215	250	0.72
CMin vs. WR	311.5	153.5	0.09
CMin vs. Choquet	303.5	161.5	0.14

According to the obtained statistical results presented in Table 3, we can affirm, with a high level of confidence, that the CMin-integral is better than WR. Regarding the standard Choquet integral, we can observe that, although there are not statistical differences, the obtained p-value is low. Furthermore, the CMin-integral improves the results of the Choquet integral in 18 out of the 30 datasets considered in this study. These two facts, show that the CMin-integral is enhancing the results provided by the standard Choquet integral. Finally, we point out that, when comparing the CMin-integral with Ham, the obtained p-value is high, which implies that the behavior of these two approaches is similar.

6   Conclusion

In this paper, we consider the notion of Choquet-like copula-based aggregation function (CC-integral), considering the minimum t-norm in the place of the copula, obtaining the CMin-Integral. We applied the CMin-integral in FRBCSs, showing that this function allows to enhance the results of the classical FRM of the winning rule as well as those of the standard Choquet integral, and provides results that are competitive with those obtained by the best pre-aggregation function presented in [1], offering new possibilities in defining FRMs with similar gain in performance.

In future works, we intend to study the properties satisfied by the CMin-integral. We will also consider the CMin-integral in a fuzzy interval approach [22–25], as, e.g., in [26,27].

Acknowledgment. This work is supported by Brazilian National Counsel of Technological and Scientific Development CNPq (under the Processes 233950/2014-1, 305882/2016-3, 307781/2016-0) and by the Spanish Ministry of Science and Technology (under project TIN2016-77356-P). G.P. Dimuro is also supported by Caixa and Fundación Caja Navarra of Spain.

References

1. Lucca, G., Sanz, J., Pereira Dimuro, G., Bedregal, B., Mesiar, R., Kolesárová, A., Sola, H.B.: Pre-aggregation functions: construction and an application. IEEE Trans. Fuzzy Syst. **24**(2), 260–272 (2016)
2. Beliakov, G., Pradera, A., Calvo, T.: Aggregation Functions: A Guide for Practitioners. Springer, Berlin (2007)
3. Mayor, G., Trillas, E.: On the representation of some aggregation functions. In: Proceedings of IEEE International Symposium on Multiple-Valued Logic, pp. 111–114. IEEE, Los Alamitos (1986)
4. Bustince, H., Fernandez, J., Kolesárová, A., Mesiar, R.: Directional monotonicity of fusion functions. Eur. J. Oper. Res. **244**(1), 300–308 (2015)
5. Choquet, G.: Theory of capacities. Annales de l'Institut Fourier **5**, 131–295 (1953–1954)
6. Klement, E.P., Mesiar, R., Pap, E.: Triangular Norms. Kluwer Academic Publisher, Dordrecht (2000)
7. Ishibuchi, H., Nakashima, T., Nii, M.: Classification and Modeling with Linguistic Information Granules: Advanced Approaches to Linguistic Data Mining. Advanced Information Processing. Springer, Berlin (2005)
8. Alpaydin, E.: Introduction to Machine Learning, 2nd edn. The MIT Press, Cambridge (2010)
9. Barrenechea, E., Bustince, H., Fernandez, J., Paternain, D., Sanz, J.A.: Using the Choquet integral in the fuzzy reasoning method of fuzzy rule-based classification systems. Axioms **2**(2), 208–223 (2013)

10. Dimuro, G.P., Bedregal, B., Bustince, H., Fernandez, J., Lucca, G., Mesiar, R.: New results on pre-aggregation functions. In: Uncertainty Modelling in Knowledge Engineering and Decision Making, Proceedings of the 12th International FLINS Conference (FLINS 2016) World Scientific Proceedings Series on Computer Engineering and Information Science, vol. 10, pp. 213–219. World Scientific, Singapura (2016)

11. Bustince, H., Sanz, J.A., Lucca, G., Dimuro, G.P., Bedregal, B., Mesiar, R., Kolesárová, A., Ochoa, G: Pre-aggregation functions: definition, properties and construction methods. In: 2016 IEEE International Conference on Fuzzy Systems (FUZZ-IEEE), pp. 294–300. IEEE, July 2016

12. Lucca, G., Sanz, J.A., Dimuro, G.P., Bedregal, B., Asiain, M.J., Elkano, M., Bustince, H.: CC-integrals: Choquet-like copula-based aggregation functions and its application in fuzzy rule-based classification systems. Knowl.-Based Syst. **119**, 32–43 (2017)

13. Alsina, C., Frank, M.J., Schweizer, B.: Associative Functions: Triangular Norms and Copulas. World Scientific Publishing Company, Singapore (2006)

14. Alcalá-Fdez, J., Sánchez, L., García, S., Jesus, M., Ventura, S., Garrell, J., Otero, J., Romero, C., Bacardit, J., Rivas, V., Fernández, J., Herrera, F.: KEEL: a software tool to assess evolutionary algorithms for data mining problems. Soft. Comput. **13**(3), 307–318 (2009)

15. Wilcoxon, F.: Individual comparisons by ranking methods. Biometrics **1**, 80–83 (1945)

16. Schweizer, B., Sklar, A.: Probabilistic Metric Spaces. North-Holland, New York (1983)

17. Murofushi, T., Sugeno, M., Machida, M.: Non-monotonic fuzzy measures and the Choquet integral. Fuzzy Sets Syst. **64**(1), 73–86 (1994)

18. Lucca, G., Dimuro, G.P., Mattos, V., Bedregal, B., Bustince, H., Sanz, J.A.: A family of Choquet-based non-associative aggregation functions for application in fuzzy rule-based classification systems. In: 2015 IEEE International Conference on Fuzzy Systems (FUZZ-IEEE), pp. 1–8. IEEE, Los Alamitos (2015)

19. Alcalá-Fdez, J., Alcalá, R., Herrera, F.: A fuzzy association rule-based classification model for high-dimensional problems with genetic rule selection and lateral tuning. IEEE Trans. Fuzzy Syst. **19**(5), 857–872 (2011)

20. Ishibuchi, H., Nakashima, T.: Effect of rule weights in fuzzy rule-based classification systems. IEEE Trans. Fuzzy Syst. **9**(4), 506–515 (2001)

21. Sanz, J.A., Galar, M., Jurio, A., Brugos, A., Pagola, M., Bustince, H.: Medical diagnosis of cardiovascular diseases using an interval-valued fuzzy rule-based classification system. Appl. Soft Comput. **20**, 103–111 (2014)

22. Bedregal, B.C., Dimuro, G.P., Reiser, R.H.S.: An approach to interval-valued R-implications and automorphisms. In: Carvalho, J.P., Dubois, D., Kaymak, U., da Costa Sousa, J.M. (eds.) Proceedings of the Joint International Fuzzy Systems Association World Congress and European Society of Fuzzy Logic and Technology Conference, IFSA/EUSFLAT, pp. 1–6 (2009)

23. Bedregal, B.C., Dimuro, G.P., Santiago, R.H.N., Reiser, R.H.S.: On interval fuzzy S-implications. Inf. Sci. **180**(8), 1373–1389 (2010)

24. Dimuro, G.P.: On interval fuzzy numbers. In: Workshop-School on Theoretical Computer Science, WEIT 2011, pp. 3–8. IEEE, Los Alamitos (2011)

25. Dimuro, G.P., Bedregal, B.C., Santiago, R.H.N., Reiser, R.H.S.: Interval additive generators of interval t-norms and interval t-conorms. Inf. Sci. **181**(18), 3898–3916 (2011)

26. Bustince, H., Galar, M., Bedregal, B., Kolesárová, A., Mesiar, R.: A new approach to interval-valued Choquet integrals and the problem of ordering in interval-valued fuzzy set applications. IEEE Trans. Fuzzy Syst. **21**(6), 1150–1162 (2013)
27. Sanz, J.A., Fernández, A., Bustince, H., Herrera, F.: Improving the performance of fuzzy rule-based classification systems with interval-valued fuzzy sets and genetic amplitude tuning. Inf. Sci. **180**(19), 3674–3685 (2010)

Directional and Ordered Directional Monotonicity of Mixture Functions

Jana Špirková[1]([✉]), Gleb Beliakov[2], Humberto Bustince[3,4], and Javier Fernández[3,4]

[1] Faculty of Economics, Matej Bel University, Tajovského 10,
975 90 Banská Bystrica, Slovakia
jana.spirkova@umb.sk
[2] School of Information Technology, Deakin University, 221 Burwood Hwy,
Burwood 3125, Australia
gleb@deakin.edu.au
[3] Dept. de Automática y Computación, Universidad Pública de Navarra,
Campus Arrosadia, s/n, 31.006 Pamplona, Spain
{bustince,fcojavier.fernandez}@unavarra.es
[4] Institute of Smart Cities, Universidad Pública de Navarra,
Campus Arrosadia, s/n, 31.006 Pamplona, Spain

Abstract. In this contribution, we discuss the concepts of so-called fusion functions, pre-aggregation functions and their directional and ordered directional monotonicity in the context of mixture functions. Mixture functions represent a special class of weighted averaging functions whose weights are determined by continuous weighting functions which depend on the input values. They need not be monotone, in general. If they are monotone increasing, they also belong to the important class of aggregation functions. If the are directionally monotone, they belong to the class of pre-aggregation functions.

Currently there is increased interest in studying generalized forms of monotonicity such as weak, directional or ordered directional monotonicity due to their possible application in fields such classification or image processing.

This paper discusses properties of selected mixture functions with special emphasis on their directional and ordered directional monotonicity. The concept of directional and ordered directional monotonicity of mixture functions is investigated with respect to linear and quadratic weighting functions.

1 Introduction

According to [1,8], aggregation functions play a very important role in many computational problems. Very interesting approaches can be found, for example, in [3,9].

Boundary conditions and standard monotonicity are the basic properties of aggregation functions. In general, mixture functions are not aggregation functions, but under certain (sufficient) conditions, mixture functions can be

V. Torra et al. (eds.), *Aggregation Functions in Theory and in Practice*,
Advances in Intelligent Systems and Computing 581, DOI 10.1007/978-3-319-59306-7_10

aggregation functions. In [1,11–16] and [17] the authors provided several sufficient conditions for standard monotonicity of mixture functions.

Wilkin and Beliakov [19] introduced the concept of so-called weak monotonicity. The property of weak monotonicity is very useful for calculating representative values of clusters of data in the presence of outliers. Recently, authors in [2] proposed weak monotonicity as a relaxation of the monotonicity condition for averaging functions and discussed the concept of directional and cone monotonicity, and monotonicity with respect to majority of inputs and coalitions of inputs. In [4], the authors investigated weak monotonicity of the Lehmer and Gini means.

Regarding the generalization of the concept of aggregation functions, in [5], the authors introduced and discussed so-called fusion functions and their directional monotonicity, considering monotonicity along arbitrary rays. Their results generalize the results of [19] concerning weak monotonicity.

In [10], the authors introduce the concept of so-called pre-aggregation functions, their construction and its applications. An investigation of directional and ordered directional monotonicity is in the focus of attention of the authors in [6].

Our paper builds on the research of the last two papers mentioned above.

The paper consists of five sections. Section 2 contains the basic definitions related to standard, weak, directional and ordered directional monotonicity. Moreover, this section contains basic definitions of aggregation, pre-aggregation, fusion and mixture functions. Section 3 presents the latest sufficient conditions of directional monotonicity of mixture functions and relevant results concerning this topic. The attention is mainly focused on directional monotonicity of the mixture functions generated by linear and quadratic weighting functions. Section 4 discusses ordered directional monotonicity of the mentioned mixture functions. Section 5 summarizes the presented results and brings some ideas for the future research.

2   Preliminaries

Throughout the paper, we investigate directional monotonicity of mixture functions on the interval $[0, 1]$. The choice of the unit interval is not restrictive. In general, we could study our functions on arbitrary any closed non-empty interval $[a, b] \subset [-\infty, \infty]$.

In this part, we offer basic definitions of different types of monotonicity recently investigated in an aggregation.

Definition 1 (Standard Monotonicity). A function $F : [0, 1]^n \to [0, 1]$ is monotone increasing if for every $(x_1, \ldots, x_n), (y_1, \ldots, y_n) \in [0, 1]^n$ such that $x_i \geq y_i$ for every $i = 1, \ldots, n$, the inequality $F(x_1, \ldots, x_n) \geq F(y_1, \ldots, y_n)$ holds.[1]

---

[1] The term "increasing" is understood in a non-strict sense.

With respect to application, it is not always necessary that processing functions are monotone but it is sufficient if functions are so-called *weakly monotone*, see [18,19].

Definition 2 (Weak Monotonicity). A function $F : [0,1]^n \rightarrow [0,1]$ is weakly increasing if $F(x_1 + k, x_2 + k, \ldots, x_n + k) \geq F(x_1, x_2, \ldots, x_n)$ for all $(x_1, x_2, \ldots, x_n)$, for any $k > 0$ such that $(x_1, x_2, \ldots, x_n)$, $(x_1+k, x_2+k, \ldots, x_n+k)$ $\in [0,1]^n$.

It is clear that each standard increasing function is also weakly increasing. Inspired by the notion of weak monotonicity the researchers have recently opened investigation of the so-called directional monotonicity which is defined as follows.

Definition 3 ([5] Directional Monotonicity). Let $\mathbf{r} = (r_1, r_2, \ldots, r_n)$ be a real n-dimensional vector, $\mathbf{r} \neq \mathbf{0}$. A function $F : [0,1]^n \rightarrow [0,1]$ is $\mathbf{r}$-increasing if for all points $(x_1, x_2, \ldots, x_n) \in [0,1]^n$ and all $k > 0$ such that $(x_1 + kr_1, x_2 + kr_2, \ldots, x_n + kr_n) \in [0,1]^n$, it holds that $F(x_1 + kr_1, x_2 + kr_2, \ldots, x_n + kr_n) \geq F(x_1, x_2, \ldots, x_n)$.

Vectors $\mathbf{r} \neq \mathbf{0}$ are called *directions*. It is clear, that weakly increasing functions are $\mathbf{r}$-increasing in the direction of vector $\mathbf{r} = (1, 1, \ldots, 1)$.

Teams of authors in [6,7,10] investigated not only directional monotonicity but also so-called ordered directional monotonicity which has significant application in image processing.

For ordered directionally monotone functions, on the contrary, the direction along which monotonicity is required varies depending on the ordinal size of the coordinates of the considered input.

Definition 4 ([7] Ordered Directional Monotonicity). Let $\mathbf{r} = (r_1, r_2, \ldots, r_n)$ be a real n-dimensional vector, $\mathbf{r} \neq \mathbf{0}$. A function $F : [0,1]^n \rightarrow [0,1]$ is $\mathbf{r}$-ordered increasing if for all $\mathbf{x} = (x_1, x_2, \ldots, x_n) \in [0,1]^n$ and for any permutation $\sigma : \{1, 2, \ldots, n\} \rightarrow \{1, 2, \ldots, n\}$ with $x_{\sigma(1)} \geq \ldots \geq x_{\sigma(n)}$ and any $k > 0$ such that $1 \geq x_{\sigma(1)} + kr_1 \geq \ldots \geq x_{\sigma(n)} + kr_n \geq 0 \in [0,1]^n$ we have $F(\mathbf{x} + k\mathbf{r}_{\sigma^{-1}}) \geq F(\mathbf{x})$; $\mathbf{r}_{\sigma^{-1}} = (r_{\sigma^{-1}(1)}, \ldots, r_{\sigma^{-1}(n)})$.

Standard monotonicity belongs to the basic properties of aggregation functions both in the real setting and when dealing with extensions of fuzzy sets, [6]. Apart from monotonicity, the definition of an aggregation function includes boundary conditions, too.

Definition 5. A function $F : [0,1]^n \rightarrow [0,1]$ is called an n-ary aggregation function if the following conditions hold:

(A1) F satisfies the boundary conditions $F(0, 0, \ldots, 0) = 0$ and $F(1, 1, \ldots, 1) = 1$;
(A2) F is (standard) increasing.

Pre-aggregation functions relax the monotonicity condition, as follows.

Definition 6 [10]. A function $F : [0,1]^n \rightarrow [0,1]$ is called an n-ary pre-aggregation function if the following conditions hold:

(PA1) There exists a real vector $\mathbf{r} \in [0,1]^n$, $\mathbf{r} \neq \mathbf{0}$ such that F is $\mathbf{r}$-increasing.
(PA2) F satisfies the boudary conditions $F(0,0,\ldots,0)=0$ and $F(1,1,\ldots,1)=1$.

Because in our paper we examine directional and ordered directional monotonicity of mixture functions, we present their definition as follows.

Definition 7 [11]. A function $M_g : [0,1]^n \rightarrow [0,1]$ given by

$$M_g(x_1,\ldots,x_n) = \frac{\sum\limits_{i=1}^{n} g(x_i) \cdot x_i}{\sum\limits_{i=1}^{n} g(x_i)}, \tag{1}$$

where $g : [0,1] \rightarrow [0,\infty[$ is a continuous weighting function, is called a mixture function.

Definition 8 [11]. A continuous differentiable function $g(x) : [0,1] \rightarrow [0,\infty[$ is called a weighting function.

3   Directional Monotonicity

In this section, we present our result regarding to directional monotonicity of mixture function (1) with linear and quadratic weighting functions.

In the case of determination of sufficient conditions for $\mathbf{r}$-increasingness for vector $\mathbf{r} = (r, 1 - r)$, $r \geq 0$, mixture functions represent a special class of pre-aggregation functions.

3.1   Mixture Function with Linear Weighting Function

In this part, we introduce two sufficient conditions for $\mathbf{r}$-increasingness of mixture function with linear weighting function. The first condition is stronger than the second one.

Proposition 1. Let $M_g : [0,1]^2 \rightarrow [0,1]$ be a mixture function defined by (1) with the weighting function $g(x) = x + l$, $l \geq 0$ and let $\mathbf{r} = (r, 1 - r)$, $r \geq 0$. If M_g is $\mathbf{r}$-increasing, then the coefficient l must satisfy the condition

$$l \geq 1. \tag{2}$$

Proof. Let $\mathbf{r} = (r, 1 - r)$, $r \geq 0$. Let $\mathbf{x} = (x, y) \in [0,1]^2$ and $k > 0$ such that $\mathbf{x} + k\mathbf{r} \in [0,1]^2$.

From Definition 3, we get

$$\frac{(x + kr)(x + kr + l) + (y + k(1 - r))(y + k(1 - r) + l)}{x + y + 2l + k} \geq \frac{x(x + l) + y(y + l)}{x + y + 2l}, \tag{3}$$

whence

$$2(rx + (1 - r)y) + k(r^2 + (1 - r)^2) + l \geq \frac{x(x + l) + y(y + l)}{(x + y + 2l)}. \tag{4}$$

Since the right-hand side of the previous inequality can be maximal 1, we get

$$2(rx + (1 - r)y) + k(r^2 + (1 - r)^2) + l \geq 1. \tag{5}$$

Without loss of generality, for $k \to 0$, we obtain inequality

$$2(rx + (1 - r)y) + l \geq 1. \tag{6}$$

For $x \to 0$, $y \to 0$, we obtain condition (2).   □

Now, we introduce weaker sufficient condition for **r**-increasingness of mixture function which gives us a greater range of the coefficients l.

Proposition 2. *Let $M_g : [0,1]^2 \to [0,1]$ be a mixture function defined by (1) with the weighting function $g(x) = x + l$, $l \geq 0$ and let $\mathbf{r} = (r, 1 - r)$, $r \geq 0$. If M_g is r-increasing, then the coefficient l must satisfy the conditions*

$$l \geq -r + \sqrt{\left(r - \frac{1}{2}\right)^2 + \frac{1}{4}} \quad for \quad 0 \leq r \leq \frac{1}{2}$$

or $\tag{7}$

$$l \geq r - 1 + \sqrt{\left(r - \frac{1}{2}\right)^2 + \frac{1}{4}} \quad for \quad \frac{1}{2} \leq r \leq 1.$$

Proof. Let $\mathbf{r} = (r, 1 - r)$, $r \geq 0$. Let $\mathbf{x} = (x, y) \in [0,1]^2$ and $k > 0$ such that $\mathbf{x} + k\mathbf{r} \in [0,1]^2$.

From Definition 3, we get

$$\frac{(x + kr)(x + kr + l) + (y + k(1 - r))(y + k(1 - r) + l)}{x + y + 2l + k} \geq \frac{x(x + l) + y(y + l)}{x + y + 2l}, \tag{8}$$

whence

$$(x + y + 2l)\left[2(rx + (1 - r)y) + k(r^2 + (1 - r)^2) + l\right] \geq x^2 + y^2 + xl + yl. \tag{9}$$

Without loss of generality, for $k \to 0$ and after some modification, we obtain inequality

$$2l^2 + 4l\left(r(x - y) + y\right) + 2xy + (x^2 - y^2)(2r - 1) \geq 0. \tag{10}$$

For $x \to 0$, $y \to 1$ or $x \to 1$, $y \to 0$, we obtain conditions (7). See Fig. 1.   □

Remark 1. From Proposition 1, we can state that the Lehmer mean $LeM(x, y) = \frac{x^2 + y^2}{x + y}$ is **r**-increasing only for vector $\mathbf{r} = (\frac{1}{2}, \frac{1}{2})$, that means it is only weakly monotone increasing.

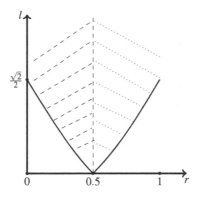

Fig. 1. The set of directional monotonicity of $M_g : [0,1]^2 \to [0,1]$ with $g(x) = x + l$.

Remark 2. Let $M_g : [0,1]^2 \to [0,1]$ be a mixture function defined by (1). If M_g is **r**-increasing, then $M_{B \cdot g}$, with $B > 0$ is also **r**-increasing.

On the basis of Remark 2, we present the next sufficient condition.

Proposition 3. *Let $M_g : [0,1]^2 \to [0,1]$ be a mixture function defined by (1) with the weighting function $g(x) = cx + 1 - c$, $c \in [0,1]$ and let $\mathbf{r} = (r, 1 - r)$, $r \geq 0$. If M_g is $\mathbf{r}$-increasing, then the coefficient c must satisfy the condition*

$$0 \leq c \leq \frac{1}{2}. \tag{11}$$

Proof. Using the same procedure as in proof of Proposition 1 or Remark 2 and substitution $l = \frac{1}{c} - 1$, we obtain condition (11). $\qquad\square$

Corollary 1. *Let $M_g : [0,1]^2 \to [0,1]$ be a mixture function defined by (1) with the weighting function $g(x) = cx + 1 - c$, $c \in [0,1]$ and let $\mathbf{r} = (r, 1 - r)$, $r \geq 0$. If M_g is $\mathbf{r}$-increasing, then the coefficient c must satisfy the conditions*

$$0 \leq c \leq \frac{1 - r - \sqrt{(r - \frac{1}{2})^2 + \frac{1}{4}}}{\frac{1}{2} - r} \quad for \quad 0 \leq r \leq \frac{1}{2}$$

or $\qquad\qquad\qquad\qquad\qquad\qquad\qquad\qquad\qquad\qquad\qquad\qquad\qquad\qquad\qquad$ (12)

$$0 \leq c \leq \frac{r - \sqrt{(r - \frac{1}{2})^2 + \frac{1}{4}}}{r - \frac{1}{2}} \quad for \quad 1 \geq r \geq \frac{1}{2}.$$

Proof. Using the same procedure as in proof of Proposition 2, Remark 2 and substitution $l = \frac{1}{c} - 1$, we obtain conditions (12) with the convention $\frac{0}{0} = 1$. See Fig. 2. $\qquad\square$

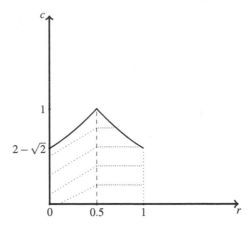

Fig. 2. The set of directional monotonicity of $M_g : [0,1]^2 \rightarrow [0,1]$ with $g(x) = cx+1-c$.

Proposition 4. *Let* $M_g : [0,1]^2 \rightarrow [0,1]$ *be a mixture function defined by (1) with the weighting function* $g(x) = kx + q$, $k \geq 0$, $q \geq 0$ *and let* $\mathbf{r} = (r, 1 - r)$, $r \geq 0$. *If* M_g *is* $\mathbf{r}$*-increasing, then the coefficients* q *and* k *must satisfy the condition*

$$q \geq k.$$

Proof. Direct on the basis of proofs of Proposition 1 and Remark 2.   □

Corollary 2. *Let* $M_g : [0,1]^2 \rightarrow [0,1]$ *be a mixture function defined by (1) with the weighting function* $g(x) = kx + q$, $k \geq 0$, $q \geq 0$ *and let* $\mathbf{r} = (r, 1 - r)$, $r \geq 0$. *If* M_g *is* $\mathbf{r}$*-increasing, then the coefficients* q *and* k *must satisfy the conditions*

$$q \geq k \cdot \left(-r + \sqrt{\left(r - \frac{1}{2} \right)^2 + \frac{1}{4}} \right) \quad for \quad 0 \leq r \leq \frac{1}{2}$$

or   (13)

$$q \geq k \cdot \left(r - 1 + \sqrt{\left(r - \frac{1}{2} \right)^2 + \frac{1}{4}} \right) \quad for \quad \frac{1}{2} \leq r \leq 1.$$

Proof. Direct on the basis of proofs of Proposition 2 and Remark 2.   □

3.2   Mixture Function with Quadratic Weighting Function

Proposition 5. *Let* $M_g : [0,1]^2 \rightarrow [0,1]$ *be a mixture function defined by (1) with the weighting function* $g(x) = x^2 + p$, $p \geq 0$ *and let* $\mathbf{r} = (r, 1 - r)$, $r \geq 0$. *Then* M_g *is* $\mathbf{r}$*-increasing for all coefficients* p.

Proof. Using Definition 3, we can put down

$$\frac{(x + kr)((x + kr)^2 + p) + (y + k(1 - r))((y + k(1 - r))^2 + p)}{(x + kr)^2 + (y + k(1 - r))^2 + 2p} \tag{14}$$

$$\geq \frac{x(x^2 + p) + y(y^2 + p)}{x^2 + y^2 + 2p},$$

whence

$$(x^2 + y^2 + 2p)\left[(x^2 + p)r + (y^2 + p)(1 - r) + 2x^2r + 2y^2(1 - r)\right] \tag{15}$$

$$\geq \left[(x^2 + p)x + (y^2 + p)y\right] \cdot \left[2xr + 2y(1 - r)\right].$$

Without loss of generality, for $x \to 0$ and $y \to 0$, we obtain condition $p \geq 0$.   □

4   Ordered Directional Monotonicity

Proposition 6. *Let $M_g : [0,1]^2 \to [0,1]$ be a mixture function defined by (1) with the weighting function $g(x) = x + l$, $l \geq 0$ and let $\mathbf{r} = (r, 1 - r)$, $r \geq 0$. If M_g is $\mathbf{r}$-ordered increasing, then the coefficient l must satisfy the condition*

$$l \geq -r + \sqrt{\left(r - \frac{1}{2}\right)^2 + \frac{1}{4}} \quad for \quad 0 \leq r \leq \frac{1}{2}. \tag{16}$$

Proof. Let $\mathbf{r} = (r, 1 - r)$, $r \geq 0$. Let $\mathbf{x} = (x, y) \in [0,1]^2$ and $k > 0$ such that $\mathbf{x} + k\mathbf{r} \in [0,1]^2$.

Using Definition 4, if $x > y$, we get gradually (8), (9), (10).

For $x \to 1$, $y \to 0$, we obtain condition (16).

If $x < y$, it is enough to replace x and y in inequality (10) and use boundary input vector $(0, 1)$, from where we obtain condition (16), again. The set of $\mathbf{r}$-ordered directional increasing M_g represents the left part of Fig. 1 which is highlighted by dashed lines. This result follows directly from a symmetry of the mixture function.   □

From a symmetry of mixture functions, it is obvious that if we consider vector $\mathbf{r} = (1 - r, r)$, we obtain the second part of condition (7), i.e.,

$$l \geq r - 1 + \sqrt{\left(r - \frac{1}{2}\right)^2 + \frac{1}{4}} \quad for \quad \frac{1}{2} \leq r \leq 1. \tag{17}$$

and graphically, the right part of Fig. 1.

5   Conclusion

In this paper, we introduced sufficient conditions for $\mathbf{r}$-increasingness and $\mathbf{r}$-ordered increasnigness of mixture functions with selected weighting functions.

If $\mathbf{r} = (r, 1 - r)$, $r \geq 0$, $\mathbf{r} \neq \mathbf{0}$, under our sufficient conditions, mixture functions create a special class of pre-aggregation functions. Because mixture functions are symmetric, regarding to its $\mathbf{r}$-ordered increasnigness, we obtained the same sufficient conditions.

Our future investigation will focus on statement of sufficient conditions of $\mathbf{r}$-increasingness and $\mathbf{r}$-ordered increasnigness of n-ary mixture functions with different types of weighting functions not only on the unit interval and $n = 2$ but on a general interval $[a, b] \subset [-\infty, \infty]$ and $n > 2$.

Acknowledgements. Jana Špirková has been supported by the Project VEGA no. 1/0093/17 Identification of risk factors and their impact on products of the insurance and savings schemes.

Humberto Bustince and Javier Fernández have been supported by the Spanish Government grant TIN2016-77356-P.

References

1. Beliakov, G., Pradera, A., Calvo, T.: A Guide for Practitioners. Springer, Berlin (2007)
2. Beliakov, G., Calvo, T., Wilkin, T.: Three types of monotonicity of averaging functions. Knowl.-Based Syst. **72**, 114–122 (2014)
3. Bustince, H., Barrenechea, E., Calvo, T., James, S., Beliakov, G.: Consensus in multi-expert decision making problems using penalty functions defined over a Cartesian product of lattices. Inf. Fusion **17**, 56–64 (2014)
4. Beliakov, G., Špirková, J.: Weak monotonicity of Lehmer and Gini means. Fuzzy Sets Syst. **299**, 26–40 (2016)
5. Bustince, H., Fernandez, J., Kolesárová, A., Mesiar, R.: Directional monotonicity of fusion functions. Eur. J. Oper. Res. **244**(1), 300–308 (2015)
6. Bustince, H., Fernandez, J., Mesiar, R., Kolesárová, Bedregal B., Dimuro, G. P., Barrenechea, E.: Directional monotonicity and ordered directional monotonicity. In: International Symposium on Aggregation and Structures, ISAS 2016, Book of Abstracts, vol. 18 (2016)
7. Bustince, H., Barrenechea, L. J., Mesiar, R., Kolesárová, A.: Ordered directionally monotone functions. Justification and application. IEEE Trans. Fuzzy Syst. (2016, submitted)
8. Grabisch, J., Marichal, J., Mesiar, R., Pap, E.: Aggregation Functions. Cambridge University Press, Cambridge (2009)
9. Jurio, A., Bustince, H., Pagola, M., Pradera, A., Yager, R.: Some properties of overlap and grouping functions and their application to image thresholding. Fuzzy Sets Syst. **229**, 69–90 (2013)
10. Lucca, G., Sanz, J.A., Dimuro, G.P., Bedregal, B., Mesiar, R., Kolesárová, A., Bustince, H.: Preaggregation functions: construction and an application. IEEE Trans. Fuzzy Syst. **24**(2), 260–272 (2016)
11. Marques Pereira, R.A., Pasi, G.: On non-monotonic aggregation: mixture operators. In: Proceedings of the 4th Meeting of the EURO Working Group on Fuzzy Sets (EUROFUSE 1999) and 2nd International Conference on Soft and Intelligent Computing (SIC 1999), Budapest, Hungary, pp. 513–517 (1999)

12. Ribeiro, R.A., Marques Pereira, R.A.: Generalized mixture operators using weighting functions: a comparative study with WA and OWA. Eur. J. Oper. Res. **145**, 329–342 (2003)
13. Ribeiro, R.A., Pereira, R.A.M.: Aggregation with generalized mixture operators using weighting functions. Fuzzy Sets Syst. **137**, 43–58 (2003)
14. Mesiar, R., Špirková, J.: Weighted means and weighting functions. Kybernetika **42**(2), 151–160 (2006)
15. Mesiar, R., Špirková, J., Vavríková, L.: Weighted aggregation operators based on minimization. Inf. Sci. **17**(4), 1133–1140 (2008)
16. Špirková, J.: Weighted operators based on dissimilarity function. Inf. Sci. **281**, 172–181 (2014)
17. Špirková, J.: Induced weighted operators based on dissimilarity functions. Inf. Sci. **294**, 530–539 (2015)
18. Wilkin, T., Beliakov, G., Calvo, T.: Weakly monotone averaging functions. In: Laurent, A., Strauss, O., Bouchon-Meunier, B., Yager, R.R. (eds.) Information Processing and Management of Uncertainty in Knowledge-Based Systems. Communications in Computer and Information Science, vol. 444, pp. 364–373. Springer, Cham (2014)
19. Wilkin, T., Beliakov, G.: Weakly monotonic averaging functions. Int. J. Intell. Syst. **30**(2), 144–169 (2015)

Using Uninorms and Nullnorms to Modify Fuzzy Implication Functions

Isabel Aguiló, Jaume Suñer, and Joan Torrens[✉]

SCOPIA Research Group, Department of Mathematics and Computer Science,
University of the Balearic Islands, Crta. Valldemossa,
km. 7.5, 07122 Palma, Spain
{isabel.aguilo,jaume.sunyer,jts224}@uib.es

Abstract. In this comunication, some construction methods of fuzzy implication functions based on uninorms, nullnorms and fuzzy negations are presented. The main idea is to use these methods in order to obtain new implication functions from old ones in such a way that the obtained implication satisfies a desired property even if the old implication does not satisfy it. In this line, the paper focuses in the following three properties: the control of the decreasingness with respect to the first variable, the strong negation property and the property: $I(x, N(x)) = N(x)$. However, other properties could be also considered in the same way through the proposed methods.

Keywords: Fuzzy implication · Uninorm · Nullnorm · Construction method

1 Introduction

Among fuzzy logic operators, implication functions are one of the most recently studied because they generalize crisp conditional and consequently, they are used to model fuzzy conditionals and also in the inference process. Thus, implication functions become essential in fuzzy logic and approximate reasoning [11,13], but they are proved to be useful also in many other applications like fuzzy control, fuzzy subsethood measures, fuzzy indices, mathematical morphology, image processing, and so on, see [3–5,17] and the references therein.

Due to this great quantity of applications, many researchers have focused their efforts in the study of fuzzy implication functions from a pure theoretical point of view. In this topic, it is pointed out in [23] the necessity of having as many models and implication functions as possible in order to be able to modelize all possibilities, depending on the exact meaning of the fuzzy conditional to be modeled. In this sense, one of the main topics in the theoretic study of implication functions is the research of construction methods of new fuzzy implication functions from given ones. A lot of these methods have appeared in last years and many researchers have devoted their efforts to this topic. When they are constructed from a given implication function I, the interest of these methods lies in one or both of the following facts:

© Springer International Publishing AG 2018
V. Torra et al. (eds.), *Aggregation Functions in Theory and in Practice*,
Advances in Intelligent Systems and Computing 581, DOI 10.1007/978-3-319-59306-7_11

(i) It is important that the new implication function preserves as much properties as possible. This is the case for instance of the minimum and the maximum of two implications, any convex linear combination of them, the φ-conjugate or the N-reciprocal of an implication, see [4]. Other recent constructions in this direction are the threshold and vertical threshold generation methods [18,19], some algebraic operations between implications [24,25] and some others, see the recent survey [20].

(ii) Another possibility is that the construction method can modify a given implication I that do not satisfy some particular property, in order to obtain a new implication satisfying it. This is the case for instance of the contrapositivisation methods looking for implications satisfying the contraposition law, the classical ones that can be found in [4], and also some new contrapositivisations introduced in [1]. Other construction methods working in this direction were recently introduced in [26] looking for implication functions satisfying the ordering property.

The method presented in [21] (see also [22]) allows to construct fuzzy implication functions from a fuzzy negation N, but the method is in fact a particular case of a more general method involving a t-conorm S, a fuzzy negation N and a fuzzy implication function I. This last method was retrieved in [2] and generalized leading to the so-called FNI-implications constructed from a given implication I and using a disjunctive aggregation function F and a fuzzy negation N. Such implications were used in [2] to obtain new implications satisfying continuity and also the strong negation property.

The idea of this paper is to develop the FNI-method to obtain new implication functions satisfying some other properties like the control of the decreasingness with respect to the first variable, the strong negation property or the property: $I(x, N(x)) = N(x)$. Since the aggregation function used in the FNI-method must verify that $F(0,1) = 1$, we will do it by using some concrete aggregation functions with this property. In particular, we will use disjunctive uninorms and also nullnorms slightly modified in order to satisfy $F(0,1) = 1$. However, many other kinds of aggregation functions could be used and thus, we are dealing with a field of study with many possibilities and we simply highlight some of them in the current paper.

The paper is organized as follows. After this introduction, Sect. 2 is devoted to some preliminaries in order to make the paper as self-contained as possible. Section 3 presents the main results of the paper. In such section the construction method is investigated, and it is used to obtain new implication functions satisfying the three mentioned properties by dividing the reasoning in three subsections devoted to each one of these properties. Finally, the paper ends with Sect. 4 devoted to some conclusions and future work.

2   Preliminaries

We will suppose the reader to be familiar with the theory of t-norms, t-conorms and fuzzy negations (all necessary results and notations can be found in [14] and

[4]), and also with some basic facts about aggregation functions (that can be found in any of the excellent books [6,9,12]). We recall here only some facts on aggregation functions and fuzzy implications in order to establish the necessary notation that we will use along the paper and to make it as self-contained as possible.

Definition 1. *A binary operator* $F : [0,1] \times [0,1] \to [0,1]$ *is said to be an* aggregation function *if it is increasing in each variable and it satisfies* $F(0,0) = 0$ *and* $F(1,1) = 1$.

There are many kinds of aggregation functions and more details on them, their classes and their properties can be found in [6]. Let us recall here two well known families that will be used along the paper and that are specially important because they are generalizations of t-norms and t-conorms.

Definition 2. *A* uninorm *is an aggregation function which is associative, commutative and such that there exists an element* $e \in [0,1]$, *called* neutral element, *such that* $U(e,x) = x$ *for all* $x \in [0,1]$.

Evidently, a uninorm with neutral element $e = 1$ is a t-norm and a uninorm with neutral element $e = 0$ is a t-conorm. For any other value $e \in]0,1[$ the operation works as a t-norm in the $[0,e]^2$ square, as a t-conorm in $[e,1]^2$ and its values are between minimum and maximum in the set of points $A(e)$ given by

$$A(e) = [0,e[\times]e,1] \cup]e,1] \times [0,e[.$$

We will usually denote by $U \equiv \langle T_U, e, S_U \rangle$ a uninorm U with neutral element e and underlying t-norm and t-conorm, T_U and S_U. Any uninorm satisfies that $U(1,0) \in \{0,1\}$ and a uninorm U is called *conjunctive* if $U(1,0) = 0$ and *disjunctive* when $U(1,0) = 1$ (more details can be found for instance in [10] and also in the recent survey [15]).

Very related to uninorms are the so-called nullnorms [8] (also called t-operators in [16]).

Definition 3. *A function* $G : [0,1]^2 \to [0,1]$ *is called a* nullnorm *if it is an aggregation function which is associative, commutative and such that there exists* $k \in [0,1]$ *called* absorbing element *that verifies* $G(k,x) = k$ *for all* $x \in [0,1]$ *and*

$$G(0,x) = x \quad \text{for all } x \leq k \quad \text{and} \quad G(1,x) = x \quad \text{for all } x \geq k.$$

When $k = 0$ we obtain a t-norm and when $k = 1$ we obtain a t-conorm. In general, the absorbing element is always given by $k = G(1,0) = G(0,1)$. The structure of nullnorms can be found in [16] and it is given as follows.

Theorem 1. *Let* $G : [0,1] \times [0,1] \longrightarrow [0,1]$ *be a nullnorm with absorbing element* $G(1,0) = k \neq 0,1$. *Then*

$$G(x,y) = \begin{cases} kS\left(\frac{x}{k}, \frac{y}{k}\right) & \text{if } (x,y) \in [0,k]^2, \\ (1-k)T\left(\frac{x-k}{1-k}, \frac{y-k}{1-k}\right) + k & \text{if } (x,y) \in [k,1]^2 \\ k & \text{otherwise,} \end{cases}$$

where S *is a t-conorm and* T *is a t-norm.*

We will usually denote by $G \equiv \langle S_G, k, T_G \rangle$ a nullnorm G with absorbing element k and underlying t-norm and t-conorm, T_G and S_G.

Definition 4. *A decreasing function* $N : [0,1] \to [0,1]$ *is called a fuzzy negation if* $N(0) = 1$ *and* $N(1) = 0$. *A fuzzy negation* N *is called*

 (i) strict, *if it is strictly decreasing and continuous,*
 (ii) strong, *if it is an involution, i.e.,* $N(N(x)) = x$ *for all* $x \in [0,1]$.
 (iii) non-filling, *if it satisfies* $N(x) = 1$ *if and only if* $x = 0$.

Among fuzzy negations, we can highlight the classical fuzzy negation N_c given by $N_c(x) = 1 - x$ for all $x \in [0,1]$ which is a strong fuzzy negation.

Definition 5. *A binary operator* $I : [0,1] \times [0,1] \to [0,1]$ *is said to be a fuzzy implication function, or an implication, if it satisfies:*

(I1) $I(x,z) \geq I(y,z)$ *when* $x \leq y$, *for all* $z \in [0,1]$.
(I2) $I(x,y) \leq I(x,z)$ *when* $y \leq z$, *for all* $x \in [0,1]$.
(I3) $I(0,0) = I(1,1) = 1$ *and* $I(1,0) = 0$.

Note that, from the definition, it follows that $I(0,x) = 1$ and $I(x,1) = 1$ for all $x \in [0,1]$ whereas the symmetrical values $I(x,0)$ and $I(1,x)$ are not derived from the definition.

Definition 6. *Given a fuzzy implication function* I, *the function* $N_I(x) = I(x,0)$ *for all* $x \in [0,1]$ *is always a fuzzy negation, known as the* natural negation *of* I.

Among many other properties usually required for fuzzy implications we recall here some of the most important ones.

– The (Left) Neutrality Property:

$$I(1,y) = y \text{ for all } y \in [0,1]. \qquad (NP)$$

– The Consequent Boundary:

$$I(x,y) \geq y \text{ for all } y \in [0,1]. \qquad (CB)$$

– The Ordering Property:

$$I(x,y) = 1 \iff x \leq y \text{ for all } x, y \in [0,1]. \qquad (OP)$$

– The Identity Principle:

$$I(x,x) = 1 \text{ for all } x \in [0,1]. \qquad (IP)$$

– The Strong Negation Principle:

$$I(x,0) \text{ is a strong negation for all } x \in [0,1]. \qquad (SNP)$$

– The Continuity condition:

$$I \quad \text{is a continuous mapping} \qquad\qquad (CO)$$

– The Law of Contraposition with respect to a fuzzy negation N:

$$I(x,y) = I(N(y), N(x)) \text{ for all } x,y \in [0,1]. \qquad\qquad CP(N)$$

In [22] the following method to construct implication functions from t-conorms and fuzzy negations was presented.

$$I_{SNI}(x,y) = S(N(x), I(x,y)) \quad \text{ for all } x,y \in [0,1].$$

This method was recalled in [2] where it was also extended to general aggregation functions as follows.

Proposition 1. *Let F be an aggregation function such that $F(0,1) = 1$, N a fuzzy negation, and I an implication function. The function I_{FNI} given by:*

$$I_{FNI}(x,y) = F(N(x), I(x,y)) \quad \text{for all } x,y \in [0,1], \qquad\qquad (1)$$

is always an implication function.

Definition 7. *A fuzzy implication function I_{FNI} constructed through Eq. (1) will be called an FNI-implication. Whenever the aggregation function F is in fact a t-conorm S, I_{FNI} will be called a SNI-implication.*

According to Proposition 1, Eq. (1) gives a new method to construct implication functions that preserves some of the properties recalled before. Moreover, this method was also used in [2] to construct implication functions that satisfy continuity and (SNP) from implications not satisfying them.

Proposition 2. *Let F be a continuous disjunctive aggregation function and N a continuous fuzzy negation. Let I be an implication function which is continuous except at point $(0,0)$.*

(i) Then I_{FNI} satisfies (CO).
(ii) Moreover, if F has 0 as neutral element, N is strong, and I has N_{D_1} as natural negation, then I_{FNI} satisfies (SNP).

3   *FNI*-implications Derived from Uninorms and Nullnorms

In this section we want to follow in the line of Proposition 2 and we want to use FNI-implications to obtain new implication functions satisfying desired properties. We will do it by using uninorms and nullnorms as aggregation functions. Since from Proposition 1 the aggregation function should satisfy $F(0,1) = 1$, nullnorms are not adequate to generate FNI-implications. For this reason we introduce a slight modification of this kind of aggregations in order to satisfy the mentioned property.

Definition 8. *Let $G \equiv \langle S_G, k, T_G \rangle$ be a nullnorm. The following function*

$$G_d(x, y) = \begin{cases} 1 & \text{if } y = 1 \\ G(x, y) & \text{otherwise.} \end{cases}$$

will be called a d-nullnorm.

Similarly as for nullnorms, a d-nullnorm with $G_d(1,0) = k$ and underlying operators S_U and T_U will be denoted by $G_d \equiv \langle S_U, k, T_U \rangle_d$. It is clear from the definition that d-nullnorms, G_d, are aggregation functions such that $G_d(0,1) = 1$ and consequently they are suitable to be used in the construction method of FNI-implications. Thus, we will use the following notation from now on.

Definition 9. *An implication function derived from a disjunctive uninorm U, a fuzzy negation N and a given implication function I will be called a UNI-implication and will be denoted by I_{UNI}. Similarly, if the implication function is derived from N, I and a d-nullnorm G_d, it will be called a G_dNI-implication and will be denoted by I_{G_dNI}.*

From general results presented in [2] the following results can be easily proved.

Proposition 3. *Let N be a fuzzy negation, I an implication function, U a disjunctive uninorm, G_d a d-nullnorm and I_{UNI}, I_{G_dNI} the corresponding UNI and G_dNI-implications. The following items hold:*

(i) Both I_{UNI} and I_{G_dNI} preserve (IP).
(ii) I_{G_dNI} preserves (OP). If N is non-filling and U has trivial 1-region then I_{UNI} also preserves (OP).

On the contrary, note that UNI and G_dNI-implications never satisfy (NP) nor (CB) even when the initial implication I satisfies them. However the importance of these constructions does not fall in the properties that they preserve, but in the possibility to modify the given implication in order to obtain a new one satisfying a concrete desired property. In this direction we will devote our study to the control of the decreasingness with respect to the first variable, the strong negation property and the property: $I(x, N(x)) = N(x)$ and we will do it in a different section for each property.

Let us first present an example of how the UNI-implication method can slightly modify a given implication I.

Example 1. Let U be a disjunctive uninorm with neutral element $e \in]0, 1[$ and consider the fuzzy negation N given by:

$$N(x) = \begin{cases} 1 & \text{if } x = 0 \\ e & \text{if } 0 < x \leq \alpha \\ \frac{e}{1-\alpha}(1-x) & \text{if } x > \alpha, \end{cases}$$

where $\alpha < 1$ is a value near to 1. For any implication function I, the resulting I_{UNI} is an implication only modified in the region when $x > \alpha$. Thus, if for instance $e = 1/2$ and $\alpha = 3/4$, taking an implication function such that $I(x,y) \leq 1/2$ whenever $x > 3/4$ we obtain

$$I_{UNI}(x,y) = \begin{cases} I(x,y) & \text{if } x \leq 3/4 \\ \frac{1}{2}T_U(4 - 4x, 2I(x,y)) & \text{if } x > 3/4. \end{cases}$$

Note that the modification of I in the region $x > 3/4$ can be done even preserving continuity if the underlying t-norm T_U is continuous.

3.1   The Control of the Decreasingness with Respect to the First Variable

The control of the decreasingness with respect to the first variable (as well as the increasingness with respect to the second variable) of a fuzzy implication function I was already studied in [19] (respectively, in [18]), where the vertical threshold generation method of fuzzy implications was introduced and studied (respectively, the horizontal threshold generation method). This kind of implications are constructed from two given implications by rescaling their first variable depending on the vertical threshold $e \in]0, 1[$ (see [19]). In the same paper they were characterized as those implications satisfying $I(e,y) = e$ for all $y < 1$.

Thus, this kind of implications take values over e when $x < e$ and below e when $x > e$. We want to see in this case that UNI and G_dNI-implications can be used to control the decreasingness of I in the regions where $x < e$ and $x > e$. Specifically we have the following results.

Proposition 4. *Let $U \equiv \langle T_U, e, S_U \rangle$ be a disjunctive uninorm with neutral element $e \in]0,1[$ and N a fuzzy negation with fixed point e. If I is an implication function satisfying $I(e,y) = e$ for all $y < 1$ then the corresponding UNI-implication also satisfies $I_{UNI}(e,y) = e$ for all $y < 1$ and it is given by*

$$I_{UNI}(x,y) = \begin{cases} e + (1-e)S_U\left(\frac{N(x)-e}{1-e}, \frac{I(x,y)-e}{1-e}\right) & \text{if } x \leq e \\ eT_U\left(\frac{N(x)}{e}, \frac{I(x,y)}{e}\right) & \text{if } x > e. \end{cases} \tag{2}$$

Remark 1. Note that from the proposition above, I_{UNI} takes values over $I(x,y)$ in all points where $x < e$ and takes values below $I(x,y)$ in all points where $x > e$ allowing then a control of the decreasingness of the implication.

Moreover, since the values of $I_{UNI}(x,y)$ depend only of the underlying operators T_U and S_U, the above mentioned control can be done in a continuous way (at least in all points where I is continuous) taking T_U and S_U continuous.

Similarly, the control can be allowed by using d-nullnorms, but obtaining in this way the contrary effect (see Remark 2).

Proposition 5. *Let $G_d \equiv \langle S_U, k, T_U \rangle$ be a d-nullnorm with $G_d(1,0) = k \in \,]0,1[$ and N a fuzzy negation with fixed point k. If I is an implication function satisfying $I(k,y) = k$ for all $y < 1$ then the corresponding G_dNI-implication also satisfies $I_{GdNI}(k,y) = k$ for all $y < 1$ and it is given by*

$$I_{G_dNI}(x,y) = \begin{cases} 1 & \text{if } y = 1 \\ k + (1-k)T_U\left(\frac{N(x)-k}{1-k}, \frac{I(x,y)-k}{1-k}\right) & \text{if } y < 1 \text{ and } x \le k \\ kS_U\left(\frac{N(x)}{k}, \frac{I(x,y)}{k}\right) & \text{if } y < 1 \text{ and } x > k. \end{cases} \quad (3)$$

Remark 2. Note that from the proposition above, contrary to what happens with UNI-implications, the obtained G_dNI-implication, I_{G_dNI}, takes values below $I(x,y)$ in all points where $x < k$ and takes values over $I(x,y)$ in all points where $x > k$ allowing then the opposed effect to the one obtained by UNI-implications in the control of the decreasingness of the initial implication.

Example 2. Since the modified implications depend only on the underlying operators, any uninorm with the same associated T_U and S_U will get the same result. Thus given a fuzzy implication I satisfying $I(1/2,y) = 1/2$ for all $y < 1$ (many examples can be found in [19]), we can consider any uninorm with neutral element $1/2$ and with underlying operators T_U the product t-norm and S_U the probabilistic sum t-conorm. Consider also the classical fuzzy negation $N(x) = 1 - x$ (with fixed point $1/2$). Then the derived UNI-implication is given by:

$$I_{UNI}(x,y) = \begin{cases} 1 - 2x + 2xI(x,y) & \text{if } x \le 1/2 \\ 2I(x,y) - 2xI(x,y) & \text{if } x > 1/2, \end{cases}$$

which is a new implication satisfying $I_{UNI}(1/2,y) = 1/2$.

3.2 The Strong Negation Property

We want to deal in this section with the strong negation property (SNP). In this case UNI-implications can also be used to modify a given implication I obtaining a new implication satisfying this property. Specifically, we have the following result.

Proposition 6. *Consider $e \in\,]0,1[$ and let I be an implication function such that its natural negation N_I has fixed point e but it is not strong.*

(i) Suppose that N_I is strictly decreasing in $[0,e]$ and that $N_I^2(x) \le x$ for all $x \le e$. Take any disjunctive uninorm $U \equiv \langle T_U, e, S_U \rangle$ with neutral element e and underlying t-conorm $S_U = \max$, and the fuzzy negation N given by:

$$N(x) = \begin{cases} 1 & \text{if } x = 0 \\ e & \text{if } 0 < x < e, \\ N_I^{-1}(x) & \text{if } x \ge e. \end{cases} \quad (4)$$

Then the natural negation of the corresponding UNI-implication is given by

$$N_{I_{UNI}}(x) = I_{UNI}(x,0) = U(N(x), N_I(x)) \quad for\ all\ x \in [0,1], \qquad (5)$$

which is a strong negation and consequently I_{UNI} satisfies the (SNP) property.

(ii) Similarly, suppose that N_I is strictly decreasing in $[e,1]$ and that $N_I^2(x) \geq x$ for all $x \geq e$. Take any disjunctive uninorm $U \equiv \langle T_U, e, S_U \rangle$ with neutral element e and underlying t-norm $T_U = \min$, and the fuzzy negation N given by

$$N(x) = \begin{cases} N_I^{-1}(x) & if\ x \leq e \\ e & if\ e < x < 1, \\ 0 & if\ x = 1. \end{cases} \qquad (6)$$

Then the natural negation of the corresponding UNI-implication is given by Eq. (5), which is a strong negation and consequently I_{UNI} satisfies the (SNP) property.

Example 3. Let us consider the fuzzy negation given by

$$N_I(x) = \begin{cases} 1 - 2x & if\ x \leq 1/2 \\ 0 & if\ x > 1/2. \end{cases}$$

It is clear that N_I has fixed point $e = 1/3$ and that satisfies $N_I^2(x) \leq x$ for all $x \leq 1/3$. Note that there are many implication functions with N_I as natural negation like all (S, N)-implications derived from N_I and any t-conorm S and thus the previous proposition can be applied to any of these implication functions to obtain new ones satisfying (SNP).

Take for instance the (S, N)-implication given by $I(x, y) = \max(N_I(x), y)$ for all $x, y \in [0, 1]$. Take N the negation given by Eq. (4) with $e = 1/3$ and U any disjunctive idempotent uninorm with neutral element $e = 1/3$. Then the corresponding UNI-implication has natural negation given by

$$N_{I_{UNI}}(x) = \begin{cases} 1 - 2x & if\ x \leq 1/3 \\ 1/2(1 - x) & if\ x > 1/3. \end{cases}$$

which is clearly a strong negation. The three negations N_I, N and $N_{I_{UNI}}$ are depicted in Fig. 1.

3.3   The Property: $I(x, N(x)) = N(x)$

We devote this section to the property $I(x, N(x)) = N(x)$ introduced in [7] and lately studied due to its importance in the definition of fuzzy indices.

We will see that in this case an implication I can not be modified through UNI-implications to obtain a new implication satisfying this property, but G_dNI-implications can. On the contrary, we will see that UNI-implications in fact preserve this property. Specifically, we have the following results.

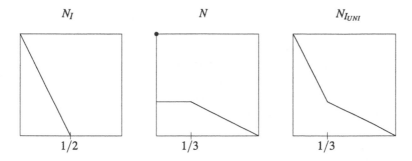

Fig. 1. The three fuzzy negations N_I, N and $N_{I_{UNI}}$ considered in Example 3.

Proposition 7. *Let $U \equiv \langle T_U, e, S_U \rangle$ be a disjunctive uninorm with neutral element $e \in]0,1[$ and underlying operators $T_U = \min$ and $S_U = \max$. Let N be a fuzzy negation with fixed point e and I an implication function satisfying $I(e,y) = y$ for all $y \in [0,1]$. If I satisfies the property $I(x, N(x)) = N(x)$ for all $x \in [0,1]$, the corresponding UNI-implication also satisfies this property.*

Proposition 8. *Let $G_d \equiv \langle S_U, k, T_U \rangle$ be a d-nullnorm with absorbent element $k \in]0,1[$ and underlying operators $T_U = \min$ and $S_U = \max$. Let N be a fuzzy negation with fixed point k and I an implication function satisfying $I(k,y) = y$ for all $y \in [0,1]$ and $I(x, N(x)) \neq 1$ for all x such that $N(x) < 1$. Then the corresponding G_dNI-implication satisfies the property $I_{G_dNI}(x, N(x)) = N(x)$ for all $x \in [0,1]$.*

Example 4. There are again many implication functions satsfying $I(e,y) = y$ for all $y \in [0,1]$ like all RU and (U,N)-implications derived from uninorms with neutral element e. Take for instance any non-filling fuzzy negation N and any disjunctive representable uninorm with neutral element e and additive generator h. Then the (U,N)-implication I derived from U and N is given by

$$I(x,y) = \begin{cases} h^{-1}(h(y) - h(x)) & \text{if } (x,y) \notin \{(1,0),(0,1)\} \\ 1 & \text{otherwise.} \end{cases}$$

It is clear that I satisfies $I(e,y) = y$ for all $y \in [0,1]$ and also satisfies $I(x, N(x)) \neq 1$ for all $x \in]0,1]$. By Proposition 7, if we take any idempotent d-nullnorm G_d with $G_d(0,1) = e$, then the corresponding G_dNI-implication satisfies $I_{G_dNI}(x, N(x)) = N(x)$ for all $x \in [0,1]$.

4   Conclusions

FNI-implications allow a way of constructing new implication functions from a given one using a disjunctive aggregation function F and a fuzzy negation N. This construction gives many possibilities to modify a given implication function in such a way that the resulting implication satisfies a desired property. In this paper, some of these possibilities have been highlighted by using uninorms and nullnorms (in fact a disjunctive modification of nullnorms) as disjunctive aggregation functions. Specifically, given a fuzzy implication function I, we have constructed new implication functions with good desired properties, like a control of the decreasingness in the first variable, the strong negation property (SNP), or the property $I(x, N(x)) = N(x)$ for all $x \in [0,1]$. In all these cases, an adequate uninorm or disjunctively modified nullnorm is used jointly with an adequate fuzzy negation N to obtain implications satisfying each one of these properties.

Acknowledgements. This paper has been supported by the Spanish Grant TIN2016-75404-P AEI/FEDER, UE.

References

1. Aguiló, I., Suñer, J., Torrens, J.: New types of contrapositivisation of fuzzy implications with respect to fuzzy negations. Inf. Sci. **322**, 223–226 (2015)
2. Aguiló, I., Suñer, J., Torrens, J.: A new look on fuzzy implication functions: FNI-implications. In: Carvalho, J.P., et al. (eds.) Information Processing and Management of Uncertainty in Knowledge-Based Systems. Communications in Computer and Information Science, vol. 610, pp. 375–386. Springer, Cham (2016)
3. Baczyński, M., Beliakov, G., Bustince, H., Pradera, A.: Advances in Fuzzy Implication Functions. Studies in Fuzziness and Soft Computing. Springer, Heidelberg (2013)
4. Baczyński, M., Jayaram, B.: Fuzzy Implications. Studies in Fuzziness and Soft Computing, vol. 231. Springer, Heidelberg (2008)
5. Baczyński, M., Jayaram, B., Massanet, S., Torrens, J.: Fuzzy implications: past, present, and future. In: Kacprzyk, J., Pedrycz, W. (eds.) Springer Handbook of Computational Intelligence, pp. 183–202. Springer, Heidelberg (2015)
6. Beliakov, G., Pradera, A., Calvo, T.: Aggregation Functions: A Guide for Practitioners. Studies in Fuzziness and Soft Computing, vol. 221. Springer, Heidelberg (2007)
7. Bustince, H., Burillo, P., Soria, F.: Automorphisms, negations and implication operators. Fuzzy Sets Syst. **134**, 209–229 (2003)
8. Calvo, T., De Baets, B., Fodor, J.: The functional equations of Frank and Alsina for uninorms and nullnorms. Fuzzy Sets Syst. **120**, 385–394 (2001)
9. Calvo, T., Mayor, G., Mesiar, R.: Aggregation Operators: New Trends and Applications. Studies in Fuzziness and Soft Computing. Physica-Verlag HD, Heidelberg (2002)
10. Fodor, J.C., Yager, R.R., Rybalov, A.: Structure of Uninorms. Int. J. Uncertain. Fuzziness Knowl.-Based Syst. **5**, 411–427 (1997)
11. Gottwald, S.: A Treatise on Many-Valued Logic. Research Studies Press, Baldock (2001)
12. Grabisch, M., Marichal, J.-L., Mesiar, R., Pap, E.: Aggregation Functions (Encyclopedia of Mathematics and Its Applications), 1st edn. Cambridge University Press, New York (2009)
13. Kerre, E.E., Huang, C., Ruan, D.: Fuzzy Set Theory and Approximate Reasoning. Wu Han University Press, Wu Chang (2004)

14. Klement, E.P., Mesiar, R., Pap, E.: Triangular Norms. Kluwer Academic Publishers, Dordrecht (2000)
15. Mas, M., Massanet, S., Ruiz-Aguilera, D., Torrens, J.: A survey on the existing classes of uninorms. J. Intell. Fuzzy Syst. **29**, 1021–1037 (2015)
16. Mas, M., Mayor, G., Torrens, J.: t-Operators. Int. J. Uncertain. Fuzziness Knowl.-Based Syst. **7**, 31–50 (1999)
17. Mas, M., Monserrat, M., Torrens, J., Trillas, E.: A survey on fuzzy implication functions. IEEE Trans. Fuzzy Syst. **15**(6), 1107–1121 (2007)
18. Massanet, S., Torrens, J.: Threshold generation method of construction of a new implication from two given ones. Fuzzy Sets Syst. **205**, 50–75 (2012)
19. Massanet, S., Torrens, J.: On the vertical threshold generation method of fuzzy implication and its properties. Fuzzy Sets Syst. **226**, 232–252 (2013)
20. Massanet, S., Torrens, J.: An overview of construction methods of fuzzy implications. In: [3], pp. 1–30 (2013)
21. Shi, Y., Van Gasse, B., Ruan, D., Kerre, E.: On a new class of implications in fuzzy logic. In: Hüllermeier, E., Kruse, R., Hoffmann, F. (eds.) Proceedings of IPMU 2010. CCIS, vol. 80, pp. 525–534. Springer, Heidelberg (2010)
22. Shi, Y., Van Gasse, B., Ruan, D., Kerre, E.: Fuzzy implications: classification and a new class. In: [3], pp. 31–53 (2013)
23. Trillas, E., Mas, M., Monserrat, M., Torrens, J.: On the representation of fuzzy rules. Int. J. Approx. Reason. **48**, 583–597 (2008)
24. Vemuri, N.R., Jayaram, B.: Representations through a monoid on the set of fuzzy implications. Fuzzy Sets Syst. **247**, 51–67 (2014)
25. Vemuri, N.R., Jayaram, B.: The ⋆-composition of fuzzy implications: closures with respect to properties, powers and families. Fuzzy Sets Syst. **275**, 58–87 (2015)
26. Zhang, W., Pei, D.: Two kinds of modifications of implications. In: Fan, T.-H., et al. (eds.) Quantitative Logic and Soft Computing 2016. Advances in Intelligent Systems and Computing, vol. 510, pp. 301–310. Springer, Cham (2017)

On the Aggregation of Zadeh's Z-Numbers Based on Discrete Fuzzy Numbers

Sebastia Massanet$^{(\boxtimes)}$, Juan Vicente Riera, and Joan Torrens

SCOPIA Research Group, Department of Mathematics and Computer Science, University of the Balearic Islands, Crta. Valldemossa, Km. 7.5, 07122 Palma, Spain
{s.massanet,jvicente.riera,jts224}@uib.es

Abstract. The accurate modelling of natural language is one of the main goals in the theory of computing with words. Based on this idea, Zadeh in 2011, introduced the concept of Z-number which has a great potential not only from the theoretical point of view but also for many possible applications such as in economics, decision analysis, risk assessment, etc. Recently, the authors proposed a new vision of Zadeh's Z-numbers based on discrete fuzzy numbers that simplifies the computations and maintains the flexibility of the original model from the linguistic point of view. Following with this novel interpretation, in this paper, algebraic structures in the set of Zadeh's Z-numbers are studied. In this framework, we propose a method to construct aggregation functions from couples of discrete aggregation functions. In particular, t-norms and t-conorms are built. Finally, an application to reach a final decision on a decision making problem is given.

Keywords: Zadeh's Z-numbers · Discrete fuzzy numbers · Aggregation functions

1 Introduction

Lattice theory is both a theoretical and applied field which has been widely developed since the publication of the seminal papers of Birkhoff [5], notably collected then into the book of Grätzer [13]. One of the most important characteristics of lattices is the possibility to interpret them on the one hand, as an algebraic structure with two underlying operations or on the other hand, as a set equipped with a partial order satisfying some conditions [12]. From a Fuzzy Logic point of view, both intepretations are quite interesting. If a lattice is interpreted as an algebraic structure, this structure can be understood as a possible semantics of a substructural logic [12,25]. Otherwise, if a lattice is interpreted as a partial ordered set, it can be used as a definition or valuation set of aggregation functions [1,11,18,20,37] or fuzzy connectives [8–10].

On the other hand, it is well-known that uncertainty is a common factor in a wide range of real life situations. Indeed, uncertainty appears in many decision making problems based on partial data or opinions expressed in natural language. It was from this fact that Zadeh introduced the idea of Computing With

© Springer International Publishing AG 2018
V. Torra et al. (eds.), *Aggregation Functions in Theory and in Practice*,
Advances in Intelligent Systems and Computing 581, DOI 10.1007/978-3-319-59306-7_12

Words (CWW) [35], as a computation based on *words*, or *perceptions*, or even sentences of the natural language, instead of the traditional computation based on numbers. This has motivated that computation with words have turned into a usual resource in the field of decision making. For this reason, several different linguistic models have been presented in the literature with the purpose of modelling experts' opinions. In [24] it is presented a systematic review process about multi-granular fuzzy linguistic model approaches (FLM) considering six different categories: Traditional multi-granular FLM based on fuzzy membership functions [19], Ordinal multi-granular FLM based on a basic Linguistic Term Set [14], Ordinal multi-granular FLM based on 2-tuple FLM [15], Ordinal multi-granular FLM based on hierarchical trees [17], Multi-granular FLM based on qualitative description spaces [32], and Ordinal multi-granular FLM based on discrete fuzzy numbers, or dfn for short [23,28,29]. All these linguistic frameworks are characterized by a robust algebraic background which allows the construction of aggregation functions and other logical operators. These operators are adapted to the framework where they have been defined and therefore, they can be successfully applied in decision making problems. Consequently, the proposal of linguistic models which fits as better as possible to the experts' opinions as well as the construction of the aggregation functions within these frameworks are always interesting challenges for researchers.

Following with the previous ideas, that is, the accurate modelling of natural language, Zadeh [36] in 2011, introduced the concept of Z-number as an ordered pair of fuzzy numbers (A, B). Thus, when a Z-number is associated with a real-valued uncertain variable X, the ordered triple (X, A, B) is referred to as a Z-valuation, where the first component A is interpreted not as a value of X, but as a restriction on the values which X can take; and the second one, B is referred to as certainty (sureness, confidence, reliability, probability, possibility...) about the value of A. For instance, the opinion *it is very likely that the investment risk is very low* can be modelled as the Z-valuation (investment risk, very low, very likely). Since then, many researchers have focused their studies on Z-numbers from different aspects (theoretical knowledge or practical applications) [2–4,26,34].

However, Zadeh [36] pointed out *"Problems involving computation with Z-numbers are easy to state but far from easy to solve"*. This complexity has led to the proposal of many approaches in the literature (see [2,26,34]). One of these approaches was presented in [22]. Similarly to [2], in this new vision of Zadeh's Z-numbers, Z-information is expressed as couples of discrete fuzzy numbers. However, the second component is not regarded from a probabilistic point of view but as a dfn-evaluation [23,28] that represents the sureness or confidence of the first component. This new approach not only increases the flexibility of the expert opinions, but it also eases the management and the operations between Z-valuations by using aggregation operators in the set of dfns [7,29–31]. In accordance with the above discussion and using the classical lattice theory [13], in this paper some lattice operators for this novel interpretation of Z-numbers will be defined allowing to construct a bounded distributive lattice in the set of

Z-numbers based on discrete fuzzy numbers. Moreover, using as definition and valuation set this new bounded lattice, several different aggregation functions useful for decision making problems where experts use this kind of linguistic valuations to express their opinions are presented.

2   Preliminaries

In this section we will present the main concepts related to discrete fuzzy numbers that will be used later.

By a fuzzy subset of $\mathbb{R}$, we mean a function $A : \mathbb{R} \to [0, 1]$. For each fuzzy subset A, let $A^\alpha = \{x \in \mathbb{R} : A(x) \geq \alpha\}$ for any $\alpha \in (0, 1]$ be its α-level set (or α-cut). By $supp(A)$, we mean the support of A, i.e., the set $\{x \in \mathbb{R} : A(x) > 0\}$. By A^0, we will denote the closure of $supp(A)$.

Definition 1 [33]. *A fuzzy subset A of $\mathbb{R}$ with membership mapping $A : \mathbb{R} \to [0, 1]$ is called a* discrete fuzzy number, *or dfn for short, if its support is finite, i.e., there exist $x_1, \ldots, x_n \in \mathbb{R}$ with $x_1 < x_2 < \cdots < x_n$ such that $supp(A) = \{x_1, \ldots, x_n\}$, and there are natural numbers s, t with $1 \leq s \leq t \leq n$ such that:*

1. *$A(x_i) = 1$ for all i with $s \leq i \leq t$. (core)*
2. *$A(x_i) \leq A(x_j)$ for all i, j with $1 \leq i \leq j \leq s$.*
3. *$A(x_i) \geq A(x_j)$ for all i, j with $t \leq i \leq j \leq n$.*

From now on, we will denote by L_n the finite chain $L_n = \{0, 1, \ldots, n\}$ and by $\mathcal{A}_1^{L_n}$ the set of discrete fuzzy numbers whose support is a subinterval of the finite chain L_n. Note that in this case, any α-cut is also a subinterval of L_n that will be denoted by A^α as well as the support of A which coincides with its closure, i.e., $supp(A) = A^0$. Moreover, given any $k \in L_n$, we will denote by 1_k the dfn in $\mathcal{A}_1^{L_n}$ whose support is the singleton $\{k\}$.

The following result holds for $\mathcal{A}_1^{L_n}$, but it is not true for the set of discrete fuzzy numbers in general (see [6]).

Theorem 1 [6]. *The triplet $(\mathcal{A}_1^{L_n}, MIN, MAX)$ is a bounded distributive lattice where $1_n \in \mathcal{A}_1^{L_n}$ and $1_0 \in \mathcal{A}_1^{L_n}$ are the maximum and the minimum, respectively, and where $MIN(A, B)$ and $MAX(A, B)$ are the discrete fuzzy numbers belonging to the set $\mathcal{A}_1^{L_n}$ such that they have the sets*

$$MIN(A, B)^\alpha = \{z \in L_n \mid \min(x_1^\alpha, y_1^\alpha) \leq z \leq \min(x_p^\alpha, y_k^\alpha)\} \text{ and}$$
$$MAX(A, B)^\alpha = \{z \in L_n \mid \max(x_1^\alpha, y_1^\alpha) \leq z \leq \max(x_p^\alpha, y_k^\alpha)\} \tag{1}$$

as α-cuts (and support) respectively, where $A, B \in \mathcal{A}_1^{L_n}$, with α cuts given by $[x_1^\alpha, x_p^\alpha]$ and $[y_1^\alpha, y_p^\alpha]$, respectively, for each $\alpha \in [0, 1]$.

Remark 1 [6]. Using these operations, we can define a partial order on $\mathcal{A}_1^{L_n}$ in the usual way:

$A \preceq B$ if and only if $MIN(A, B) = A$,

or equivalently,

$A \preceq B$ if and only if $MAX(A, B) = B$
for any $A, B \in \mathcal{A}_1^L$. Equivalently, we can also define the partial ordering in terms of α-cuts:

$A \preceq B$ if and only if $\min(A^\alpha, B^\alpha) = A^\alpha$
$A \preceq B$ if and only if $\max(A^\alpha, B^\alpha) = B^\alpha$

for all $\alpha \in [0, 1]$.

Aggregation functions defined on L_n have been extended to $\mathcal{A}_1^{L_n}$ (see for instance [7, 29]) according to the next result.

Theorem 2 [7, 29]. *Consider a binary aggregation function F on the finite chain L_n. The binary operation $\mathcal{F} : \mathcal{A}_1^{L_n} \times \mathcal{A}_1^{L_n} \longrightarrow \mathcal{A}_1^{L_n}$ which returns $\mathcal{F}(A, B)$, the discrete fuzzy number whose α-cuts are the sets*

$$\{z \in L_n \mid F(\min A^\alpha, \min B^\alpha) \leq z \leq F(\max A^\alpha, \max B^\alpha)\}$$

for each $\alpha \in [0, 1]$, is an aggregation function on $\mathcal{A}_1^{L_n}$.

Moreover, if F is a t-norm, a t-conorm, a uninorm or a nullnorm then so is its extension $\mathcal{F}$.

2.1   Linguistic Model Based on Discrete Fuzzy Numbers

The study of the lattice $\mathcal{A}_1^{L_n}$ as well as the construction of the aggregation functions we have already commented in the previous section were not only interesting from the algebraic point of view but also they have constituted the theoretical foundations for the multigranular computational linguistic model based on discrete fuzzy numbers [23]. Among the main advantages of this model, we want to highlight the following ones (see [16, 28] for more details):

(i) It allows to describe more accurately the experts' opinions when they express their valuations in decision making problems.
(ii) It eases the aggregation process of the linguistic information without need of any kind of previous transformation.

From the above discussion, we introduce the following definition.

Definition 2 [23]. *Let $L_n = \{0, \ldots, n\}$ be a finite chain. We call a dfn-evaluation to each discrete fuzzy number A belonging to $\mathcal{A}_1^{L_n}$.*

In Fig. 1, we have considered the lattice $\mathcal{A}_1^{L_6}$ with $L_6\{N, VL, L, M, H, VH, T\}$ where the letters refer to the linguistic terms *None, Very Low, Low, Medium, High, Very High* and *Total*. In this way, we can represent the expression *High* modelled by $A = \{0.5/2, 0.75/3, 1/4, 0.75/5\}$ and the expression "between *Low* and *High*" modelled by $B = \{0.5/1, 1/2, 1/3, 1/4, 0.25/5\}$. Note that this approach as we have previously mentioned is very flexible because another expert can give the same expressions *High* or "between Low and High" but now using other dfn-evaluations. For instance, $A' = \{0.8/2, 0.85/3, 1/4, 0.25/5\}$ and $B' = \{0.2/1, 1/2, 1/3, 1/4, 0.75/5\}$ represent the same linguistic expressions displayed in Fig. 1 but using other dfn-evaluations which can fit better the criterion of the second expert.

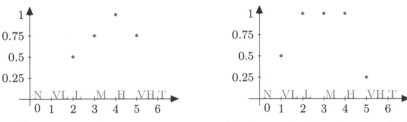

(a) A possible flexibilization of H (b) A possible flexibilization of "between Low and High"

Fig. 1. Different types of experts' evaluations using dfn-evaluations

2.2 A Review on Zadeh's Z-Numbers

In this section we recall the main concepts about this topic and we also analyse some of the most interesting ideas published in this framework.

Definition 3 [36]. *An ordered pair of fuzzy numbers (A, B) is a Z-number. A Z-number is associated with a real-valued uncertain variable, X, with the first component, A, playing the role of fuzzy restriction on X, while the fuzzy number, B is an imprecise estimation of reliability of A. The ordered triple (X, A, B) is referred as a Z-valuation and it is equivalent to an assignment statement, X is (A, B).*

Remark 2. When X is a random variable, X is (A, B) can be interpreted as $\text{Prob}(X$ is $A)$ is B where $\text{Prob}(X$ is $A)$ is the probability measure of the fuzzy event A in the sense of [36].

Operations with Z-Numbers: Let $Z_1 = (A_1, B_1)$ and $Z_2 = (A_2, B_2)$ be Z-numbers describing values of uncertain real-valued variables X_1 and X_2 respectively. The inference rule is represented as follows:

$$\frac{\begin{array}{c} Z_1 \ is \ (A_1, B_1) \\ Z_2 \ is \ (A_2, B_2) \end{array}}{Z_1 * Z_2 \ is \ (A_1 * A_2, B_1 \circ B_2)} \tag{2}$$

where $*$ represents an arithmetical operation and $A_1 * A_2$ is computed according to Zadeh's extension principle and $B_1 \circ B_2$ is computed applying the version of the extension principle which relates to probabilistic restrictions (for more details see [36]).

Note that the complexity of this operation is well known and usually yields a very complex non-linear variational problem (see [2,27]). To avoid or relax this complexity some different perspectives in the field of Z-numbers have been proposed to simplify the operations as well as the computational cost of their implementation [2,26,27,34].

Moreover, we highlight the following:

(i) Z-numbers can be conceptualized as a formidable tool in the design of discourse-oriented decision-making systems, risk assessments, etc.
(ii) It is necessary to find new linguistic models based on fuzzy sets to collect the main ideas established by Zadeh [36] in order to reduce the computational cost of the inference process.

It is clear that this second point is the central idea that generates all the above mentioned papers. Following this same line, we will recall in next section the recent interpretation of Z-numbers based on dfn in $\mathcal{A}_1^{L_n}$ stablished by the authors in [22] and we will propose a new algebraic framework that will allow us to construct formally aggregation functions whose values are this new kind of Z-numbers.

3   Zadeh's Z-Numbers Based on Dfn-Evaluations

In the previous section we have recalled that Zadeh's original concept can be a very appropriate tool to model the reasoning with words. However this idea presents some problems when we want to compute with Z-valuations. The different proposals previously presented in the literature consider the second component from a probabilistic point of view according to the original idea of Z-numbers. However, in the seminal paper [36], Zadeh also states that the second component, B, is a measure of reliability (certainty) of the first component, and closely related to certainty there are many concepts as: sureness, confidence, reliability, probability. That is, B can be interpreted from different points of view.

From this idea we introduced in [22] a new approach to Z-numbers based on couples of discrete fuzzy numbers in $\mathcal{A}_1^{L_n} \times \mathcal{A}_1^{L_m}$, where the second component represents the sureness or confidence of the first component avoiding in this way the probabilistic aspect considered in the previous approaches. Formally,

Definition 4 [22]. *Let us consider L_n and L_m two finite scales. An ordered pair of discrete fuzzy numbers (A, B) with $A \in \mathcal{A}_1^{L_n}$, $B \in \mathcal{A}_1^{L_m}$ is a (L_n, L_m)-discrete Z-number. An (L_n, L_m)-discrete Z-number is associated with an uncertain variable, X, with the first component, A, playing the role of fuzzy restriction on X, while the discrete fuzzy number B is an imprecise estimation of reliability of A. The ordered triple (X, A, B) is referred as a Z-valuation and it is equivalent to an assignment statement, X is (A, B).*

Example 1. Likely, the investment profit of this business will be very high can be interpreted as the Z-valuation $Z = $ (investment profit, Very High, Likely). For instance if we consider the linguistic term sets

$$S = \{Very\ Low(VL), Low(L), Neutral(N), High(H), Very\ High(VH)\},$$
$$S' = \{Impossible(I), Very\ Unlikely(VU), Unlikely(U), Maybe\ or \qquad (3)$$
$$Maybe\ Not(MN), Likely(Li), Very\ Likely(VLi), Sure(S)\}$$

to express the first and second components of Z respectively, the Z-valuation can be expressed by the couple $Z = (A, B)$ where

$$A = \{0.5/L, 0.6/N, 0.8/H, 1/VH\} = \{0.5/1, 0.6/2, 0.8/3, 1/4\} \in \mathcal{A}_1^{L_4},$$
$$B = \{0.7/MN, 1/Li, 0.6/VLi, 0.3/S\} = \{0.7/3, 1/4, 0.6/5, 0.3/6\} \in \mathcal{A}_1^{L_6}.$$

Let us consider the set $\mathcal{A}_1^{L_n} \times \mathcal{A}_1^{L_m} = \{Z = (A, B) \mid A \in \mathcal{A}_1^{L_n}, B \in \mathcal{A}_1^{L_m}\}$ of (L_n, L_m)-discrete Z-numbers. According to Theorem 1, the set $\mathcal{A}_1^{L_n}$ is a bounded distributive lattice for any finite chain L_n. Thus, using the well-known product lattice structure [13], we easily obtain the following result.

Proposition 1. *The triplet $(\mathcal{A}_1^{L_n} \times \mathcal{A}_1^{L_m}, \text{MIN}, \text{MAX})$ is a bounded distributive lattice where $1_{(n,m)} = (1_n, 1_m) \in \mathcal{A}_1^{L_n} \times \mathcal{A}_1^{L_m}$ and $0_{(n,m)} = (1_0, 1_0) \in \mathcal{A}_1^{L_n} \times \mathcal{A}_1^{L_m}$ are the maximum and the minimum, respectively, and where $\text{MIN}(Z_1, Z_2)$ and $\text{MAX}(Z_1, Z_2)$ denote the (L_n, L_m)-discrete Z-numbers belonging to the set $\mathcal{A}_1^{L_n} \times \mathcal{A}_1^{L_m}$ defined as follows*

$$\begin{aligned}\text{MIN}(Z_1, Z_2) &= (MIN(A_1, B_1), MIN(A_2, B_2)) \quad and \\ \text{MAX}(Z_1, Z_2) &= (MAX(A_1, B_1), MAX(A_2, B_2))\end{aligned} \tag{4}$$

where MIN and MAX stand for the operations defined in Theorem 1.

From the product lattice structure stated in Proposition 1, we can easily construct aggregation functions on $\mathcal{A}_1^{L_n} \times \mathcal{A}_1^{L_m}$ in the usual way (see [11]) as follows.

Proposition 2. *Consider $\mathcal{G}_1$ and $\mathcal{G}_2$ two aggregation functions on $\mathcal{A}_1^{L_n}$ and $\mathcal{A}_1^{L_m}$ respectively constructed according to Theorem 2. The binary operation on $\mathcal{A}_1^{L_n} \times \mathcal{A}_1^{L_m}$ defined as follows*

$$\begin{aligned}\mathbb{F} : (\mathcal{A}_1^{L_n} \times \mathcal{A}_1^{L_m}) \times (\mathcal{A}_1^{L_n} \times \mathcal{A}_1^{L_m}) &\longrightarrow \mathcal{A}_1^{L_n} \times \mathcal{A}_1^{L_m} \\ (Z_1 = (A_1, A_2), Z_2 = (B_1, B_2)) &\longmapsto \mathbb{F}(Z_1, Z_2)\end{aligned}$$

being $\mathbb{F}(Z_1, Z_2)$ the (L_n, L_m)-discrete Z-number

$$\mathbb{F}(Z_1, Z_2) = (\mathcal{G}_1(A_1, B_1), \mathcal{G}_2(A_2, B_2))$$

is an aggregation function on the poset $(\mathcal{A}_1^{L_n} \times \mathcal{A}_1^{L_m}, \leq, 0_{(n,m)}, 1_{(n,m)})$.

In particular, next result proves that many of the most usual properties of aggregation functions are preserved by the previous construction.

Proposition 3. *Let $\mathbb{F} : (\mathcal{A}_1^{L_n} \times \mathcal{A}_1^{L_m}) \times (\mathcal{A}_1^{L_n} \times \mathcal{A}_1^{L_m}) \longrightarrow \mathcal{A}_1^{L_n} \times \mathcal{A}_1^{L_m}$ be an aggregation function constructed according to Proposition 2 from $\mathcal{G}_1$ and $\mathcal{G}_2$ aggregation functions on $\mathcal{A}_1^{L_n}$ and $\mathcal{A}_1^{L_m}$ respectively. Then the following properties hold:*

1. *$\mathbb{F}$ is a commutative aggregation function if and only if $\mathcal{G}_1$ and $\mathcal{G}_2$ are commutative as well.*
2. *$\mathbb{F}$ is a associative aggregation function if and only if $\mathcal{G}_1$ and $\mathcal{G}_2$ are associative as well.*

3. $\mathbb{F}$ is an idempotent aggregation function if and only if $\mathcal{G}_1$ and $\mathcal{G}_2$ are idempotent as well.

4. $\mathbb{F}$ is a t-norm (t-conorm) if and only if $\mathcal{G}_1$ and $\mathcal{G}_2$ are t-norms (t-conorms) as well.

Example 2. Following with the Z-valuation $Z = (A, B)$ given in Example 1, if we consider another valuation Z'=(Investment profit, Low, $UnLikely$) modelled by the pair $Z' = (C, D)$ where

$$C = \{0.2/VL, 1/L, 0.8/N, 0.6/H\} \quad \in \mathcal{A}_1^{L_4},$$
$$D = \{0.3/VU, 1/U, 0.9/MN, 0.7/Li\} \in \mathcal{A}_1^{L_6},$$

respectively, these two Z-valuations can be aggregated using for instance the extensions of the kernel aggregation functions [21] with parameter $k = 2$ in L_4 and L_6, respectively, obtaining

$$\frac{\begin{array}{c} Investment\ Profit\ is\ \ (Very\ High,\ Likely) \\ Investment\ Profit\ is\ \ (Low,\ UnLikely) \end{array}}{Z = (High, Unlikely)} \tag{5}$$

where

$$High = \{0.2/VL, 0.6/L, 0.8/N, 1/H\}, \ Unlikely = \{0.3/VU, 1/U, 0.9/MN, 0.7/Li\}.$$

In the above computations we have considered the same parameter $k = 2$ for both aggregation functions. However, it is possible to consider different parameters for each linguistic scale. This parameter adjusts the degree of optimism of the aggregation process. Values of k close to 0 would lead to more optimistic aggregated opinions, while greater values would rise on less optimistic aggregated opinions. Thus, if we consider a more pessimistic parameter $k = 4$ for L_4 and a more optimistic parameter $k = 1$ for L_6 we obtain

$$\frac{\begin{array}{c} Investment\ Profit\ is\ \ (Very\ High,\ Likely) \\ Investment\ Profit\ is\ \ (Low,\ UnLikely) \end{array}}{Z = (Low, Likely)} \tag{6}$$

where

$$Low = \{0.2/VL, 1/L, 0.8/N, 0.6/H\}, \quad Likely = \{0.7/MN, 1/Li, 0.6/VLi, 0.7/S\}.$$

Remark 3. We wish to point out that it is possible to consider another vision of Zadeh's Z-numbers where the first component, similarly to the classical definition, is a fuzzy number but the second one, that expresses the sureness of this component, is represented as a dfn-evaluation. From this point of view, the complexity of the operations is also greatly reduced. In this way, the first components will be operated using the fuzzy arithmetic and the second components will be computed using an aggregation function constructed according to Proposition 2.

Concretely, let $Z_1 = (A_1, B_1)$ and $Z_2 = (A_2, B_2)$ be Z-numbers. The inference rule will be represented as follows:

$$Z_1 \text{ is } (A_1, B_1)$$
$$\frac{Z_2 \text{ is } (A_2, B_2)}{Z_1 * Z_2 \text{ is } (A_1 * A_2, \mathbb{F}(B_1, B_2))}$$

where $*$ represents a fuzzy arithmetical operation and $\mathbb{F}$ is an aggregation function on the set of discrete fuzzy numbers. For instance, the sentence *Next year, it is likely that the investment profit will be about 4 millions* can be interpreted as the Z-valuation $Z = $ (Investment profit, about 4 millions, *Likely*).

3.1   Application

In the previous section we have proposed a method to construct aggregation functions on $(\mathcal{A}_1^{L_n} \times \mathcal{A}_1^{L_m}, \text{MIN}, \text{MAX})$ from couples of aggregation functions $(\mathcal{G}_1, \mathcal{G}_2)$ defined on $\mathcal{A}_1^{L_n}$ and $\mathcal{A}_1^{L_m}$ respectively. Now, we propose to use this new kind of aggregation operators in order to obtain the final decision of a group of experts when their opinions are expressed through Z-valuations based on (L_n, L_m)-discrete Z-numbers.

Example 3. Let us suppose that the department of public works of a town hall analyses the feasibility of a work project. To this end, if the initial economic budget is accepted, the following two variables $X_1 = $ *time of delay in works already executed by this company* and $X_2 = $ *quality of the works previously executed* are taken in account.

The department will take a final decision using the global variable $X = $ {Project risk}, that will be obtained by merging the opinions given by three experts on the above two variables. To simplify the example we will suppose that all valuations have already been reduced to the linguistic scale (L_1, L_2) given by

$$L_1 = \{VL, L, M, H, VH\}, \qquad L_2 = \{VU, U, O, Li, VLi\},$$

where items in L_1 stand for *Very Low, Low, Moderate, High, Very High* and items in L_2 stand for *Very Unlikely, Unlikely, Occasional, Likely, Very Likely* respectively.

Thus, all the Z-valuations given by the experts in both components can be interpreted as discrete fuzzy numbers in $\mathcal{A}_1^{L_4}$. Let us suppose that they are given by

$$Z_{11} = (H = \{0.5/L, 0.8/M, 1/H, 0.1/VH\}, Li = \{0.2/U, 0.6/O, 1/Li, 0.8/VLi\})$$
$$Z_{21} = (M = \{0.8/VL, 0.9/L, 1/M, 0.2/H\}, U = \{0.4/VU, 1/U, 0.2/O\})$$
$$Z_{31} = (H = \{0.6/L, 0.8/M, 1/H, 0.4/VH\}, VLi = \{0.2/O, 0.4/Li, 1/VLi\})$$

for the variable X_1 and

$$Z_{12} = (H = \{0.3/L, 0.7/M, 1/H, 0.6/VH\}, VLi = \{0.3/O, 0.6/Li, 1/VLi\})$$
$$Z_{22} = (VH = \{0.3/L, 0.4/M, 0.5/H, 1/VH\}, O = \{0.4/VU, 0.7/U, 1/O, 0.6/Li\})$$
$$Z_{32} = (L = \{0.3/VL, 1/L, 0.4/M, 0.3/H\}, U = \{0.4/VU, 1/U, 0.7/O, 0.6/Li\})$$

for the variable X_2. These valuations are collected in Table 1.

Table 1. Z-valuations expressed by the three experts on $L = (L_1, L_2)$.

	Variables	
	X_1 = Time of delay	X_2 = Quality
E_1	$Z_{11} = (High, Likely)$	$Z_{12} = (High, Very\ Likely)$
E_2	$Z_{21} = (Moderate, UnLikely)$	$Z_{22} = (Very\ High, Occasional)$
E_3	$Z_{31} = (High, Very\ Likely)$	$Z_{32} = (Low, UnLikely)$

Once we have obtained the valuations of each variable, we compute for each one a representative final valuation obtained by the aggregation of the opinions given by the experts on that variable. In this way, to symplify the operations, let us choose the same aggregation function in all cases. In particular, we will consider $\mathcal{G}$ the extension to $\mathcal{A}_1^{L_4}$ of the kernel aggregation function on L_4 with parameter $k = 2$ (see [21]). Thus, we get the global evaluations for the variables X_1 and X_2

$$\mathbb{G}_1(Z_{11}, Z_{21}, Z_{31}) = (\{0.8/VL, 0.9/L, 1/M, 0.1/VH\}, \{0.2/VU, 0.4/U, 1/O\}),$$
$$\mathbb{G}_2(Z_{12}, Z_{22}, A_{32}) = (\{0.2/VL, 0.4/L, 1/M\}, \{0.3/VU, 0.6/U, 1/O, 0.6/Li\}),$$

respectively. Finally, the global valuation is obtained aggregating the two previous Z-valuations by using the same aggregation function:

$$\mathbb{G}(\mathbb{G}_1(Z_{11}, Z_{21}, Z_{31}), \mathbb{G}_2(Z_{12}, Z_{22}, A_{32})) =$$
$$(\{0.8/VL, 0.9/L, 1/M\}, \{0.3/VU, 0.6/U, 1/O\}) = (Moderate, Occasional).$$

Then it is occasional that the risk is moderate.

4   Conclusions and Future Work

In [22], the authors presented a new approach on Z-numbers based on discrete fuzzy numbers with support in a finite chain L_n. Following with this investigation, in this article we have studied algebraic structures in the set of Z-numbers. Concretely, a bounded distributive lattice is constructed. We have proposed a method to build aggregation functions on this lattice from couples of discrete aggregation functions. Moreover, we have shown that these new aggregation functions preserve similar properties than the initials ones. In particular, we have proved that if we consider a couple of discrete t-norms or t-conorms this new aggregation function so is. Finally, we have proposed an application to reach a final decision on a decision making problem.

As a future work, we want to use these aggregations functions in the solution of group decision making problems.

Acknowledgments. This paper has been partially supported by the Spanish Grant TIN2016-75404-P AEI/FEDER, UE.

References

1. Akif, M., Karaal, F., Mesiar, R.: Medians and nullnorms on bounded lattices. Fuzzy Sets Syst. **289**, 74–81 (2016)
2. Aliev, R., Alizadeh, A., Huseynov, O.: The arithmetic of discrete Z-numbers. Inf. Sci. **290**, 134–155 (2015)
3. Aliev, R.A., Huseynov, O.H., Aliyev, R.R., Alizadeh, A.A.: The Arithmetic of Z-Numbers: Theory and Applications. World Scientific Publishing, River Edge (2015)
4. Aliev, R.A., Mraiziq, D., Huseynov, O.H.: Expected utility based decision making under Z-information and its application. Comput. Intell. Neurosci. Article ID 364512, 11 pages (2015)
5. Birkhoff, G.: Lattices and their applications. Bull. Am. Math. Soc. **44**, 793–800 (1940)
6. Casasnovas, J., Riera, J.V.: Lattice properties of discrete fuzzy numbers under extended min and max. In: Proceedings of IFSA-EUSFLAT, Lisbon, pp. 647–652 (2009)
7. Casasnovas, J., Riera, J.V.: Extension of discrete t-norms and t-conorms to discrete fuzzy numbers. Fuzzy Sets Syst. **167**(1), 65–81 (2011)
8. Cornelis, C., Deschrijver, G., Kerre, E.: Implication in intuitionistic fuzzy and interval-valued fuzzy set theory: construction, classification, application. Int. J. Approx. Reason. **35**, 55–95 (2004)
9. De Baets, B.: Model implicators and their characterization. In: Steele, N. (ed.) Proceedings of ISFL 95, First ICSC International Symposium on Fuzzy Logic ICSC, pp. A42–A49. Academic Press (1995)
10. De Baets, B.: Coimplicators, the forgotten connectives. Tatra MT. **12**, 229–240 (1997)
11. De Baets, B., Mesiar, R.: Triangular norms on product lattices. Fuzzy Sets Syst. **104**(1), 61–75 (1999)
12. Galatos, N., Jipsen, P., Kowalski, T.: Residuated Lattices: An Algebraic Glimpse at Substructural Logics. Studies in Logic and the Foundations of Mathematics. Elsevier, Amsterdam (2007)
13. Grätzer, G.: General Lattice Theory. Academic Press, Cambridge (1978)
14. Herrera, F., Herrera-Viedma, E., Martínez, L.: A fusion approach for managing multi-granularity linguistic term sets in decision making. Fuzzy Sets Syst. **114**, 43–58 (2000)
15. Herrera, F., Martínez, L.: A 2-tuple fuzzy linguistic representation model for computing with words. IEEE Trans. Fuzzy Syst. **8**(6), 746–752 (2000)
16. Herrera-Viedma, E., Riera, J.V., Massanet, S., Torrens, J.: Some remarks on the fuzzy linguistic model based on discrete fuzzy numbers. In: Angelov, P., et al. (eds.) Intelligent Systems 2014. AISC, pp. 319–330. Springer International Publishing, Cham (2015)
17. Huynh, V., Nakamori, Y.: A satisfactory-oriented approach to multiexpert decision-making with linguistic assessments. IEEE Trans. Syst. Man Cybernet. **35**, 184–196 (2005)
18. Jenei, S., De Baets, B.: On the direct decomposability of t-norms on product lattices. Fuzzy Sets Syst. **139**, 699–707 (2003)
19. Jiang, Y., Fan, Z., Ma, J.: A method for group decision making with multigranularity linguistic assessment information. Inf. Sci. **178**, 1098–1109 (2008)
20. Lizasoain, I., Moreno, C.: OWA operators defined on complete lattices. Fuzzy Sets Syst. **224**, 36–52 (2013)

21. Mas, M., Monserrat, M., Torrens, J.: Kernel aggregation functions on finite scales. Constructions from their marginals. Fuzzy Sets Syst. **241**, 27–40 (2014)
22. Massanet, S., Riera, J.V., Torrens, J.: A new vision of Zadeh's Z-numbers. In: Carvalho, J., Lesot, M.J., Kaymak, U., Vieira, S., Bouchon-Meunier, B., Yager, R. (eds.) Information Processing and Management of Uncertainty in Knowledge-Based Systems. IPMU 2016. Communications in Computer and Information Science, vol. 611. Springer, Cham (2016)
23. Massanet, S., Riera, J.V., Torrens, J., Herrera-Viedma, E.: A new linguistic computational model based on discrete fuzzy numbers for computing with words. Inf. Sci. **258**, 277–290 (2014)
24. Morente-Molinera, J., Pérez, I., Ureña, M., Herrera-Viedma, E.: On multi-granular fuzzy linguistic modeling in group decision making problems: a systematic review and future trends. Knowl.-Based Syst. **74**, 49–60 (2015)
25. Ono, H.: Substructural logics and residuated lattices - an introduction. In: Hendricks, V.F., Malinowski, J. (eds.) Trends in Logic: 50 Years of Studia Logica, pp. 193–228. Kluver Academic Publishers, Netherlands (2003)
26. Pal, S.K., Banerjee, R., Dutta, S., Sarma, S.: An insight into the Z-number approach to CWW. Fundamentae Informaticae **124**, 197–229 (2013)
27. Patel, P., Khorasani, E.S., Rahimi, S.: Modeling and implementation of Z-numbers. Soft. Comput. **20**(4), 1341–1364 (2016)
28. Riera, J.V., Massanet, S., Herrera-Viedma, E., Torrens, J.: Some interesting properties of the fuzzy linguistic model based on discrete fuzzy numbers to manage hesitant fuzzy linguistic information. Appl. Soft Comput. **36**, 383–391 (2015)
29. Riera, J.V., Torrens, J.: Aggregation of subjective evaluations based on discrete fuzzy numbers. Fuzzy Sets Syst. **191**, 21–40 (2012)
30. Riera, J.V., Torrens, J.: Aggregation functions on the set of discrete fuzzy numbers defined from a pair of discrete aggregations. Fuzzy Sets Syst. **241**, 76–93 (2014)
31. Riera, J.V., Torrens, J.: Using discrete fuzzy numbers in the aggregation of incomplete qualitative information. Fuzzy Sets Syst. **264**, 121–137 (2015)
32. Roselló, L., Sánchez, M., Agell, N., Prats, F., Mazaira, F.: Using consensus, distances between generalized multi-attribute linguistic assessments for group decision-making. Inf. Fusion **32**, 65–75 (2011)
33. Voxman, W.: Canonical representations of discrete fuzzy numbers. Fuzzy Sets Syst. **118**(3), 457–466 (2001)
34. Yager, R.: On Z-valuations using Zadeh's Z-numbers. Int. J. Intell. Syst. **27**, 259–278 (2012)
35. Zadeh, L.: Fuzzy logic = computing with words. IEEE Trans. Fuzzy Syst. **4**, 103–111 (1996)
36. Zadeh, L.: A note on Z-numbers. Inf. Sci. **9**(1), 43–80 (2011)
37. Zhang, D.: Triangular norms on partially ordered sets. Fuzzy Sets Syst. **181**, 2923–2932 (2005)

Aggregation over Property-Based Preference Domains

Marta Cardin$^{(\boxtimes)}$

Department of Economics, Università Ca' Foscari Venezia, Venice, Italy
mcardin@unive.it

Abstract. We provide an axiomatic characterization of preorders that are defined with respect to a set of properties. Moreover, it is proven that property-based posets are in natural correspondence with topological spaces. This paper propose also a characterization and a generalization of a Sugeno-type integral in our framework.

1 Introduction

Decision problems are characterized by a plurality of points of view. We have to consider the different dimensions from which the alternatives can be viewed in a multi-attribute decision model or the preferences of voters in a social choice problem. In order to solve a decision problem we have to compare and rank a set of alternatives and each alternative is often defined by its attributes or properties.

We consider the model of abstract Arrowian aggregation introduced in [17] that represents a decision problem in terms of a set of Boolean properties specifying for every alternative a list of properties that are satisfied.

A *property space* is a pair $(X; \mathscr{H})$ where X is a non-empty set and $\mathscr{H}$ is a collection of non-empty subsets of X that is closed under complementation and that separates points(i.e. if $x, y \in X$ and $x \neq y$ there exists $H \in \mathscr{H}$ such that $x \in H$ and $y \notin H$). The elements of $\mathscr{H}$ are referred to as properties and if $x \in H$ we say that x has property represented by the subset H.

The "property space" model has received attention in the literature on judgement aggregation for studying the problem of aggregating sets of logically interconnected propositions. Moreover, it provides a general framework for representing preferences and then aggregation of preferences (see [17]).

The structure of a state property systems was introduced in the context of the foundations of quantum mechanics. In [1] a physical entity is represented by a mathematical model that considers its set of states, its set of properties, and a relation of "actuality of a certain property for a certain state". This model contains a complete lattice of properties of the physical entity. In [2] it is shown that the lattice can be viewed as the lattice of closed sets of a closure space.

We have two goals in this paper. Our principal goal is to introduce a general framework to study property spaces. We do not consider a finite property space and we consider also "non classical" properties. In Sect. 3 the categorical

© Springer International Publishing AG 2018
V. Torra et al. (eds.), *Aggregation Functions in Theory and in Practice*,
Advances in Intelligent Systems and Computing 581, DOI 10.1007/978-3-319-59306-7_13

equivalence between the description of a partially ordered set by means of objects and properties and the representation of the corresponding topological space is studied while in Sect. 4 an algebraic characterization is considered.

We then focus on aggregation operators over property-based domains and our second goal is to characterize some important class of aggregation functionals. Section 5 contains such results. Finally in Sect. 6 we briefly address some conclusions and future work.

2   The Model

First we recall some basic notions in lattice and ordered set theory. More detailed introduction to this subject can be found in e.g., Caspard, Leclerc and Monjardet [5], Davey and Priestley or Grätzer [14].

A *partially ordered set* (*poset* for short) $(P, \leq)$ is a set P with a reflexive, antisymmetric and transitive binary relation $\leq$. We will write $(x, y) \in R$ as $x \leq y$ (or equivalently, $y \geq x$) and we will use $x > y$ to mean that $x \leq y$ and $x \neq y$.

The word "partial" indicates that there's no guarantee that all elements can be compared to each other i.e. we don't know that for all $x, y \in P$, at least one of $x \leq y$ and $x \geq y$ holds. A poset in which this is guaranteed is a *totally ordered set*.

A relation that is reflexive and transitive is said to be a *preorder*. This is a rather general concept, as every partial order and every equivalence order is a preorder. Given a poset P and a set $S \subseteq P$, when S has a supremum p with respect to $\leq$ we say that p is the *join* of S and write $\bigvee S = \bigvee_{x \in S} x = p$. Similarly when S has an infimum q with respect to $\leq$ and we say q is the *meet* of S and write $\bigwedge S = \bigwedge_{x \in S} x = q$.

A lattice is a poset in which every pair of elements (and thus every finite subset) has both a meet and a join. Every lattice L constitutes a partially ordered set endowed with the partial order $\leq$ such that for every $x, y \in L$, write $x \leqslant y$ if $x \wedge y = x$ or, equivalently, if $x \vee y = y$. If for every $x, y \in L$, we have $x \leq b$ or $y \leq x$, then L is said to be a *chain*. A lattice L is said to be *bounded* if it has a least and a greatest element, denoted by 0 and 1, respectively.

A lattice L is said to be *distributive*, if for every $x, y, z \in L$,

$$x \vee (y \wedge z) = (x \vee y) \wedge (x \vee z) \quad \text{or, equivalently,} \quad x \wedge (y \vee z) = (x \wedge y) \vee (x \wedge z).$$

Clearly, every chain is distributive. A lattice L is said to be *complete* if $\bigwedge I = \bigwedge_{x \in I} x$ and $\bigvee I = \bigvee_{x \in I} x$ exist for every $I \subseteq L$. Clearly, every complete lattice is also bounded.

In this paper we adopt the following property space formulation which focuses on a partially ordered set.

A poset $(X, \leq)$ is a *property-based poset* if there exists a complete lattice L and a function $f \colon X \to \mathscr{P}(L)$ such that we have:

(i) $0 \notin f(x)$ for every element $x \in X$
(ii) if for every $i \in I$, $a_i \in f(x)$ then $\bigwedge_{i \in I} a_i \in f(x)$

(iii) if $x, y \in X$ then $x \leq y$ if and only if $f(x) \subseteq f(y)$
(iv) if $x_i \in X$ for every $i \in I$ and there exists $\wedge_{i \in I} x_i$ then $f(\wedge_{i \in I} x_i) = \cap_{i \in I} f(x_i)$
 (v) if $x \in X$ and $a, b \in L$ and $a \leq b$, if $a \in f(x)$ then $b \in f(x)$
(vi) if $a, b \in L$ and $x \in X$ if $a \vee b \in f(x)$ then $a \in f(x)$ or $b \in f(x)$.

The set X is the set of alternatives while the lattice L consists of properties of elements in X. The statement "the element x satisfies the property a" is formally expressed by the formula $a \in f(x)$. We demand that the set of properties is a complete lattice. So the set of properties is a partially ordered set with the meaning of the partial order relation is the following. If $a, b \in L$ then $a \leq b$ means that whenever a is satisfied for an element x then also property b is satisfied for x. Moreover, it is easy to prove that the function f is injective. Requirement (ii) expresses that if for an alternative x all the properties a_i are satisfied, this implies that for x also the infimum property $\bigwedge_{i \in I} a_i$ is satisfied. The property $\bigwedge_{i \in I} a_i$ is the property that is satisfied if and only if all of the properties a_i are satisfied. Hence the infimum represents the logical "and".

We say that a property $a \in L$ is a *classical property* if there exists a property b such that $a \vee b = 1$ and $a \wedge b = 0$. Note that a classical property a is such that $a \in f(x)$ if and only if $b \notin f(x)$ and so classical properties are "yes" or "no" properties.

The concept of property space studied in [17] consider only a finite set of elements and only classical properties. In [1–3] it is considered the definition of property state system where if $a, b \in L$, $a \leq b$ if and only if for every $x \in X$ when $a \in f(x)$ then $b \in f(x)$.

3   A Topological Characterization of Property-Based Posets

In this section we will prove that a property-based poset can be seen as a topological space and that conversely every topological space is a property-based poset. With this equivalence a concept which can be defined using closed set on a topological space can be translated in a concept for a property-based poset.

If $(X, \leq)$ is a property-based poset we introduce the following map

$$c \colon L \to \mathscr{P}(X), \quad a \mapsto c(a) = \{x \in X \mid a \in f(x)\}.$$

Proposition 1. *If $(X, \leq)$ is a property-based poset the set $c(L)$ is the family of closed sets of a T_0-topology defined on the set X.*

Proof. Since $1 \in f(x)$ for every $x \in X$, we have $c(1) = X$. As $0 \notin f(x)$ for every $x \in X$, we have $c(0) = \emptyset$ and then X and $\emptyset$ are elements of $c(L)$. If we consider the sets $c(a_i) \in c(L)$ such that $i \in I$ then we get $\cap_{i \in I} c(a_i) = c(\wedge_{i \in I} a_i) \in c(L)$. Then the intersection of arbitrary elements in $c(L)$ is an element in $c(L)$.

If $a, b \in L$ and $x \in c(a \vee b)$ then $a \vee b \in f(x)$. By requirement [(vi)] in the definition of a property-based poset we have that $a \in f(x)$ or $b \in f(x)$. So we get that $x \in c(a)$ or $x \in c(b)$ which shows that $x \in c(a) \cup c(b)$. We can conclude that $c(a) \cup c(b) \subseteq c(a \vee b)$ and this shows that $c(a) \cup c(b) = c(a \vee b)$ since the other inclusion is obvious hence $c(a) \cup c(b) \in c(L)$.

The topology defined above satisfies the T_0 separation axiom since if $x \neq y$ then there exists $a \in L$ such that $a \in f(x)$ and $a \notin f(y)$ (or vice versa) and then the closed set $c(a)$ separates the points x and y. $\qquad \square$

This result proves that to a property-based poset naturally corresponds a topological space $(X, \mathcal{T})$ on the set of alternatives, where the properties are represented by the closed subsets. The following proposition proves that we can also associate a property-based poset with any topological space.

Proposition 2. *Let $\mathcal{C}$ be the set of closed set of a topological space X that satisfies T_0 separation axiom. Then the set $\mathcal{C}$ is a complete lattice with respect to set inclusion and the function $f : X \to \mathscr{P}(\mathcal{C})$ such that*

$$f(x) = \{C \in \mathcal{C} : x \in C\}$$

defines a property-based poset structure on the set X.

Proof. It is easy to prove that $\mathcal{C}$ is a lattice with respect to set inclusion with minimal element $\emptyset$ and maximal element X. If $(C_i)_{i \in I}$ is a family of elements of $\mathcal{C}$ then we have that $\bigwedge_{i \in I} C_i = \bigcap_{i \in I} C_i$ and $\bigvee_{i \in I} C_i = \overline{\bigcup_{i \in I} C_i}$ (where $\overline{\bigcup_{i \in I} C_i}$ is the closure of the set $\bigcup_{i \in I} C_i$) and so we can say that $\mathcal{C}$ is a complete lattice.

The relation defined in X by $x \leq y$ if and only if $f(x) \subseteq f(y)$ is clearly a preorder on X. Furthermore, the preorder is a partial order since the topology is satisfies T_0 separation axiom and closed sets separate points in X. Clearly $\emptyset \notin f(x)$ for every element $x \in X$. If for every $i \in I$, $C_i \in f(x)$ then $x \in C_i$ for every $i \in I$ and $x \in \bigcap_{i \in I} C_i$. As a consequence $\bigwedge_{i \in I} a_i = \bigcap_{i \in I} C_i \in f(x)$. We can also prove that if there exists $\bigwedge_{i \in I} x_i$ then $f(\bigwedge_{i \in I} x_i) = \{C \in \mathcal{C} : x_i \in C \text{ for every } i \in I\} = \bigcap_{i \in I} f(x_i)$ so Requirement(iv) is satisfied. Let C, D be elements of $\mathcal{C}$ such that $C \leq D$. Then $C \subseteq D$ and so if $C \in f(x)$ for $x \in X$ we can say that $D \in f(x)$. Finally we prove that Requirement(vi) is verified. Suppose that C, D are elements of $\mathcal{C}$ such that $f(x) \in C \bigvee D$ then $x \in C \bigvee D = C \bigcup D$ and then we have that or $x \in C$ or $x \in D$. We have proved that or $C \in f(x)$ or $D \in f(x)$. $\qquad \square$

We can show that the classical properties correspond to the subset of the associated topological space that are closed and open.

Proposition 3. *Let $(X, \leq)$ be a property-based poset and $(X, \mathcal{T})$ the corresponding topological space. For every $a \in L$, a is classical if and only if the set $c(a)$ is a set that is open and closed with respect to the topology on X.*

Proof. If a property $a \in L$ is a classical property there exists a property b such that $a \vee b = 1$ and $a \wedge b = 0$ then it is easy to prove that $c(b) = C(c(a))$ (where $C(c(a))$ is the complement of the set $c(a)$) and then and $c(b)$ are closed and open sets. To prove the converse we consider the set $c(a)$, $a \in L$, and we suppose that $c(a)$ is an open and closed set. Then there exists an element $c(b) \in c(L)$ such that $c(b) = C(c(a))$. We have that $a \wedge b = 0$ since there exists no $x \in X$ such that $x \in c(a)$ and $x \in c(b)$. We can also prove that $X = c(a) \cup c(b)$ and then we get that $a \vee b = 1$. $\square$

4   An Algebraic Characterization of Property-Based Posets

We introduce some definitions that will be needed in this section. A *filter* of a poset P is a subset F of P such that

(i) if $x \in F$ and $x \leq y$ then $y \in F$,
(ii) if $x, y \in F$ there is $z \in F$ such that $z \leq x$ and $z \leq y$.

Sets satisfying Condition (i) of a filter are called upsets. The dual notation is that of an *ideal*. If $a \in P$ we define the *principal filter* generated by x as $\uparrow x = \{y \in L : y \geq x\}$. It is easy to prove that $\uparrow x$ is a filter for every $x \in L$. It can be proved that in a finite lattice each filter and each ideal are principal.

In a lattice L a filter F is an upset such that if $x, y \in F$ then $x \wedge y \in F$.

A *proper filter* is a filter that is neither empty nor the whole lattice while a *prime filter* is a proper filter F such that whenever $\bigvee_{i \in I} x_i$ is defined in P for a finite set I we have $x_i \in F$ for some $i \in I$. Prime ideals are defined dually.

If $(X, \leq)$ is a property-based poset with respect to a complete lattice L we can define a relation in L by

$$a \trianglelefteq b \quad \text{if and only if} \quad c(a) \subseteq c(b)$$

and we want to characterize this relation with some properties studied in [10].

A binary relation $R \subseteq L \times L$ in a lattice L is said to be *monotone* if when $x \leq y$ then $(x, y) \in R$. Let us say that a binary relation $R \subseteq L \times L$ in a lattice L is *compatible* whenever it preserves the join and the meet i.e.

(i) if $(x, y) \in R$ then $(x \wedge z, y \wedge z) \in R$ for each $z \in L$;
(ii) if $(x, y) \in R$ then $(x \vee z, y \vee z) \in R$ for each $z \in L$.

Proposition 4. *Let $(X, \leq)$ be a property-based poset on a complete lattice L. Then the relation in L defined by*

$$a \trianglelefteq b \quad \text{if and only if} \quad c(a) \subseteq c(b)$$

is a monotone and compatible preorder.

If $\trianglelefteq$ is a monotone and compatible preorder on a complete and distributive lattice L there exists complete and distributive lattice M and a function $f : L \to \mathscr{P}(M)$ such that for every $x \in L$

$$a \trianglelefteq b \quad \text{if and only if} \quad a \in f(x) \quad \text{implies that} \quad b \in f(x).$$

Proof. It is straightforward to prove the first statement.

Conversely if $\trianglelefteq$ is a monotone and compatible preorder on a complete and distributive lattice L we consider the complete and distributive lattice $\mathscr{P}$ of prime filters of L. By Theorem 2 in [4] there exists a subset $\mathscr{Q} \subseteq \mathscr{P}$ such that

$$a \trianglelefteq b \quad \text{if and only if} \quad a \in F \quad \text{then} \quad b \in F \quad \text{for every} \quad F \in \mathscr{Q}.$$

Hence if $M = \mathscr{P}$ and we define the function $f \colon L \to \mathscr{P}(M)$ such that $f(x) = \{F \in \mathscr{Q} : x \in F\}$ we can prove the result. $\qquad\square$

The following result characterizes posets that are property-based. First we introduce some definitions.

A *weak filter* of a poset P is a subset F of P such that

(i) if $x \in F$ and $x \leq y$ then $y \in F$,
(ii) if there exists $\bigwedge S$ with $\emptyset \subset S \subseteq F$ then $\bigwedge S \in F$

A *prime weak filter* is a weak filter F, $\emptyset \subset P \subset P$ such that whenever $\bigvee_{i \in I} x_i$ is defined in P for a finite set I we have $x_i \in F$ for some $i \in I$. Note that (prime) filters in a poset are (prime) weak filters.

A set $\mathscr{S} \subseteq \mathscr{P}(P)$ where P is a poset is *separating* if when $x, y \in P$ with $x \neq y$ there exists $S \in \mathscr{S}$ such that $x \in S$ and $y \notin S$ or vice versa.

Proposition 5. *A poset $(X, \leq)$ is a property-based poset if and only if the set of prime weak filter $\mathscr{F}$ is closed under finite union and arbitrary union and separating.*

Proof. If $(X, \leq)$ is a property-based poset the set $\{c(a) : a \in L\}$ is the set of prime weak filter $\mathscr{F}$ of X. Conversely if the set of prime weak filter $\mathscr{F}$ is closed under finite union and arbitrary union and separating $\mathscr{F}$ define a T_0-topology on the set X hence by Proposition 2 $(X, \leq)$ is a property-based poset. $\qquad\square$

5 Aggregation Functional over Property-Based Posets

Aggregation operators are mathematical functions that are used to combine several inputs into a single representative outcome; see [13] for a comprehensive overview on aggregation theory. Aggregation operators play an important role in several fields such as decision sciences, computer and information sciences, economics and social sciences. There are a large number of different aggregation operators that differ on the assumptions on the inputs and about the information that we want to consider in the model. One of the most important aggregation functional making sense in a qualitative framework is Sugeno integral that is a very useful non-linear functional in several applications in mathematics, economics and decision making (see [6–8]). The aim of this section is to present a Sugeno-type integral representation for aggregation operators defined on property-based posets.

Our framework is very general, we do not assume that the set X is finite or that the map $f \colon L \to \mathscr{P}(X)$ is surjective. Moreover, we consider the case in which there are more than one equivalent solutions and also the case in which the only solution is the element if it exists with no properties.

Let $N = \{1, \ldots, n\}$ be a set of individuals with $n \geq 2$ and X a property-based poset. We define an *aggregation functional* as a map $F \colon X^n \to \mathscr{P}(X)$. We consider now some of the properties that an aggregation functional $F \colon X^n \to \mathscr{P}(X)$ may or may not satisfy:

Monotonicity. If $F(x_1, \ldots x_i \ldots, x_n) \in c(a)$ for $a \in L$ and $y_i \in c(a)$ then $F(x_1, \ldots y_i \ldots, x_n) \in c(a)$.

Independence. If $F(x_1, \ldots x_n) \in c(a)$ for $a \in L$ and for all $i \in N$, $x_i \in c(a)$ if and only if $y_i \in c(a)$ we have that $F(y_1, \ldots y_n) \in c(a)$.

Strong Independence. If $F(x_1, \ldots x_n) \in c(a)$ for $a, b \in L$ and for all $i \in N$, $x_i \in c(a)$ if and only if $y_i \in c(b)$ we have that $F(y_1, \ldots y_n) \in c(b)$.

Monotonicity states that if the final outcome has a property a and the voters' supports for this property increase then the resulting final outcome has property a as well. Independence is deciding for each property where the final outcome has this property.

It is easy to prove that a functional F is monotone and independent if and only if If $F(x_1, \ldots x_n) \in c(a)$ for and for all $i \in N$, if $x_i \in c(a)$ then $y_i \in c(a)$ we have that $F(y_1, \ldots y_n) \in c(a)$ and that a functional F is monotone and strongly independent if and only if if $F(x_1, \ldots x_n) \in c(a)$ and for all $i \in N$, if $x_i \in c(a)$ then $y_i \in c(b)$ we have that $F(y_1, \ldots y_n) \in c(b)$.

If a is an element of L we define the set $N(x, a) = \{i \in N : x_i \in c(a)\}$.

Proposition 6. *If $(X, \leq)$ is a property-based poset and $F \colon X^n \to \mathscr{P}(X)$ a monotone and independent aggregation functional then there exists for every $a \in L$ a family $\mathscr{F}_a$ of subsets of N such that*

$$F(x) = \bigcap \{c(a) : N(x, a) \in \mathscr{F}_a\}.$$

Proof. We say that a set A is a-decisive if there exists $x \in X^n$ such that $N(x, a) = A$ and $F(x) \in c(a)$. Being F monotone and independent a set is a-decisive if and only if for every $x \in X^n$ such that $N(x, a) = A$, $F(x) \in c(a)$. For every $a \in L$ let $\mathscr{F}_a$ the family of a-decisive subsets of N. Hence for every $x \in X^n$, $F(x) \in c(a)$ if and only if $N(x, a) \in \mathscr{F}_a$. Note that we can have $\mathscr{F}_a = \emptyset$ for some $a \in L$. So we have proved that $F(x) = \bigcap \{c(a) : N(x, a) \in \mathscr{F}_a\}$. $\square$

Proposition 7. *If $(X, \leq)$ is a property-based poset and $F \colon X^n \to \mathscr{P}(X)$ a monotone and strongly independent aggregation functional then there exists a family $\mathscr{F}$ of subsets of N such that*

$$F(x) = \bigcup_{A \in \mathscr{F}} \{c(a) : x_i \in c(a) \text{ for every } i \in A\}.$$

Proof. It can be easily proved that if the functional F is monotone and strongly idempotent then the set of a-decisive subset of N does not depend on the property a. Let $\mathscr{F} = \mathscr{F}_a$ for every $a \in L$.

If $x \in X^n$ we have that $F(x) \in c(a)$ if and only if $N(x, a) = A$ for some $A \in \mathscr{F}$ hence if and only if $x_i \in c(a)$ for every $i \in A$ for some $A \in \mathscr{F}$. Then the statement follows at once. $\square$

6   Concluding Remarks

In this paper we have introduced a general framework for studying preferences representation. Our framework is abstract and lattice-theoretic and the crucial operations being the joining and meet of two properties. It appears that there are many connections between the work presented here with the results of [1–3,15–17]. Applications of these types of results can be found in [9,11,17]. There are however many opportunities for much more detailed research in this area in particular from the point of view of aggregation theory.

References

1. Aerts, D.: Foundations of quantum physics: a general realistic and operational approach. Int. J. Theor. Phys. **38**, 289358 (1999)
2. Aerts, D., Colebunders, E., van der Voorde, A., van Steirteghem, B.: State property systems and closure spaces: a study of categorical equivalence. Int. J. Theor. Phys. **38**, 359–385 (1999)
3. Aerts, D., van der Voorde, A., Deses, D.: Connectedness applied to closure spaces and state property systems. J. Electr. Eng. **52**, 1821 (2001)
4. Cardin, M.: Benchmarking over distributive lattices. In: Carvalho, J., Lesot, M.J., Kaymak, U., Vieira, S., Bouchon-Meunier, B., Yager, R. (eds.) IPMU 2016. Communications in Computer and Information Science, pp. 117–125. Springer, Cham (2016)
5. Caspard, N., Leclerc, B., Monjardet, B.: Finite Ordered Sets. Encyclopedia of Mathematics and Its Applications. Cambridge University Press, Cambridge (2012)
6. Couceiro, M., Marichal, J.-L.: Polynomial functions over bounded distributive lattices. J. Multiple-Valued Log S **18**, 247–256 (2012)
7. Couceiro, M., Marichal, J.L.: Characterizations of discrete Sugeno integrals as lattice polynomial functions. In: Proceedings of the 30th Linz Seminar on Fuzzy Set Theory (LINZ 2009), pp. 17–20 (2009)
8. Couceiro, M., Marichal, J.L.: Characterizations of discrete Sugeno integrals as polynomial functions over distributive lattices. Fuzzy Set Syst. **161**, 694–707 (2010)
9. Chambers, C.P., Miller, A.D.: Scholarly influence. J. Econ. Theory **151**(1), 571–583 (2014)
10. Chambers, C.P., Miller, A.D.: Benchmarking. Working Paper (2015)
11. Daniëls, T., Pacuit, E.: A general approach to aggregation problems. J. Logic Comput. **19**(3), 517–536 (2009)
12. Davey, B.A., Priestley, H.A.: Introduction to Lattices and Order. Cambridge University Press, New York (2002)
13. Grabisch, M., Marichal, J.L., Mesiar, R., Pap, E.: Aggregation Functions. Encyclopedia of Mathematics and Its Applications. Cambridge University Press, Cambridge (2009)
14. Grätzer, G.: General Lattice Theory. Birkhäuser Verlag, Berlin (2003)
15. Leclerc, B., Monjardet, B.: Aggregation and residuation. Order **30**, 261–268 (2013)
16. Monjardet, B.: Arrowian characterization of latticial federation consensus functions. Math. Soc. Sci. **20**, 51–71 (1990)
17. Nehring, K., Puppe, C.: Abstract Arrowian aggregation. J. Econ. Theory **145**, 467–494 (2010)

Generalization of Czogała-Drewniak Theorem for n-ary Semigroups

Gergely Kiss[1]([✉]) and Gabor Somlai[2]

[1] Université du Luxembourg, 2, l'Université L-4365, Avenue de l'Universite,
Esch-sur-Alzette, Luxembourg
gergely.kiss@uni.lu
[2] Eötvös Loránd University, 1/C, HU-1117, Pázmány Péter sétány,
Budapest, Hungary
zsomlei@cs.elte.hu

Abstract. We investigate n-ary semigroups as a natural generalization of binary semigroups. We refer it as a pair (X, F_n), where X is a set and an n-associative function $F_n : X^n \to X$ is defined on X. We show that if F_n is idempotent, n-associative function which is monotone in each of its variables, defined on an interval $I \subset \mathbb{R}$ and has a neutral element, then F_n is combination of the minimum and maximum operation. Moreover we can characterize the n-ary semigroups (I, F_n) where F_n has the previous properties.

1 Introduction

A function $F_n : X^n \to X$ is called *n-associative* if for every $x_1, \ldots, x_{2n-1} \in X$ and for every $1 \le i \le n-1$ we have

$$F_n(F_n(x_1, \ldots, x_n), x_{n+1}, \ldots, x_{2n-1})$$
$$= F_n(x_1, \ldots, x_i, F_n(x_{i+1}, \ldots, x_{i+n}), x_{i+n+1}, \ldots, x_{2n-1}).$$

Throughout this paper we assume that the underlying sets X are partially ordered sets (poset). However, some of the results only work for totally ordered sets. In our main results we investigate n-ary semigroups on arbitrary nonempty subintervals of the real numbers.

A set X endowed with an n-associative function $F_n : X^n \to X$ is called an *n-ary semigroup* and is denoted by (X, F_n). We say that (X, F_n) is a *totally (partially) order based n-ary semigroup* for emphasizing that X is totally (partially) ordered. Clearly, we obtain a generalisation of associative functions, which are the 2-associative functions using our terminology. The main purpose of this paper is to describe a class of n-ary semigroups. An n-ary semigroup is called *idempotent* if $F_n(a, \ldots, a) = a$ for all $a \in X$. On a partially ordered set X we can define monotonicity of a function F_n. An n-associative function is called *monotone in the i'th variable* if for every $a_1, \ldots, a_{i-1}, a_{i+1}, \ldots, a_n$ the 1-variable functions $f_i(x) := F_n(a_1, \ldots, a_{i-1}, x, a_{i+1}, \ldots, a_n)$ are all order-preserving or all are order-reversing. An n-associative function is called *monotone* if it is monotone

© Springer International Publishing AG 2018
V. Torra et al. (eds.), *Aggregation Functions in Theory and in Practice*,
Advances in Intelligent Systems and Computing 581, DOI 10.1007/978-3-319-59306-7_14

in each of its variables. Further we say that an n-associative function has *neutral element* denoted by $e \in X$ if for every $x \in X$ and $1 \leq i \leq n$ we have $F(e, \ldots, e, x, e, \ldots, e) = x$, where x is substituted into the i'th coordinate.

Finally, we say that an n-ary semigroup (X, F_n) is *conservative* (or it is said to be *quasitrivial*) if for every $x_1, \ldots, x_n \in X$ we have $F_n(x_1, \ldots, x_n) \in \{x_1, \ldots, x_n\}$. Such an n-variable function F_n is called a *choice function*. One might also say that F_n preserves all subsets of X. Ackerman [1] investigated conservative semigroups and also gave a characterization of them.

If we take $n = 2$ we get the binary version of the definitions introduced above. The pair (X, F_2) is called a *semigroup*, where X is a set and the binary function $F_2 : X^2 \to X$ is (2-)associative.

2 Preliminary Results

In this section we collect the previously known results that we use in order to prove our main results.

2.1 Binary Case

Let $I \subset \mathbb{R}$ be a not necessarily bounded, nonempty interval and $\overline{I}$ be the closure of I. We also use the standard terminology of the extended reals $\overline{\mathbb{R}} = \mathbb{R} \cup \{\pm\infty\}$. Let $g : \overline{I} \to \overline{I}$ be a decreasing function. For every $x \in I$ let $g(x - 0)$ and $g(x + 0)$ denote the limit of g at x from the left and from the right, respectively. On the boundary we take the one sided limit of g. We denote by Γ_g the *completed graph* of g, which is a subset of $\overline{I}^2$ obtained by extending the graph of the function g in the following way. If $x \in I$ is a discontinuity point of g, then we add a vertical line segment between the points $(x, g(x - 0))$ and $(x, g(x + 0))$ to extend the graph of g. Formally,

$$\Gamma_g = \{(x, y) \in \overline{I}^2 : g(x + 0) \leq y \leq g(x - 0)\}.$$

On the infimum and the supremum of $\overline{I}$, the extended graph Γ_g defined with the sets

$$\{(\inf \overline{I}, y) \in \overline{I}^2 : g(\inf \overline{I} + 0) \leq y \leq \sup \overline{I}\},$$
$$\{(\sup \overline{I}, y) \in \overline{I}^2 : \inf \overline{I} \leq y \leq g(\sup \overline{I} - 0)\},$$

respectively. It is easy to show that Γ_g is a closed set. We call Γ_g *(id-)symmetric* if Γ_g is symmetric to the line $x = y$. These definitions was introduced in [10,11].

The following theorem gives a description of idempotent, monotone, (2-ary) semigroups with neutral elements. These semigroups were first investigated by Czogała and Drewniak [2], where the authors only dealt with closed, bounded subintervals of $\mathbb{R}$ but the statement holds for any nonempty interval as it was mentioned in [6]. On the other hand, instead of monotonicity it was assumed that the binary function is monotone increasing. However, Lemma 4 shows that monotonicity implies monotone increasingness in this case.

Theorem 1. *Let I be an arbitrary nonempty real interval. If a function $F_2 :$ $I^2 \to I$ is associative, idempotent, monotone and has a neutral element $e \in I$, then there exists a monotone decreasing function $g : \overline{I} \to \overline{I}$, with $g(e) = e$, such that for every $x, y \in I$*

$$F_2(x, y) = \begin{cases} \min(x, y), & if \ y < g(x) \\ \max(x, y), & if \ y > g(x) \\ \min(x, y) \ or \ \max(x, y), & if \ y = g(x). \end{cases}$$

Now we present a full characterization of idempotent, monotone increasing, (2-ary) semigroups with neutral elements. First this was proved by Martin, Mayor and Torrens [10]. The statement of their theorem contained a small error in its condition, but essentially it was correct. In the original paper [10] the results worked on the closed unit interval $[0, 1]$ and there was given the following condition for g, instead of the symmetry of Γ_g. The function $g : [0, 1] \to [0, 1]$ satisfies

$$\inf\{y : g(y) = g(x)\} \le (g \circ g)(x) \le \sup\{y : g(y) = g(x)\} \text{ for all } x \in [0, 1]. \quad (1)$$

It was proved in [11] that Theorem 2 holds if F_2 is commutative and shown that condition (1) is not equivalent to the (id)-symmetry of Γ_g. Recently, Theorem 2 was reproved in an alternative way in [5] for any nonempty subinterval of $\mathbb{R}$.

From now on, we denote $(g \circ g)(x)$ by $g^2(x)$.

Theorem 2. *Let $I \subseteq \mathbb{R}$ be an arbitrary, nonempty interval. A function $F_2 \colon I^2 \to I$ is associative, idempotent, monotone and has a neutral element $e \in I$ if and only if there exists a decreasing function $g : \overline{I} \to \overline{I}$ with $g(e) = e$ $(e \in I)$ such that the completed graph Γ_g is (id)-symmetric and for every $x, y \in I$*

$$F_2(x, y) = \begin{cases} \min(x, y), & if \ y < g(x) \ or \ y = g(x) \ and \ x < g^2(x) \\ \max(x, y), & if \ y > g(x) \ or \ y = g(x) \ and \ x > g^2(x) \quad (2) \\ \min(x, y) \ or \ \max(x, y), & if \ y = g(x) \ and \ x = g^2(x). \end{cases}$$

Moreover, $F_2(x, y) = F_2(y, x)$ except perhaps the set of points $(x, y) \in I^2$ satisfying $y = g(x)$ and $x = g^2(x) = g(y)$.

2.2   n-ary Case

An important construction of n-ary semigroups is the following. Let (X, F_2) be a binary semigroup. Let $F_n := \underbrace{F_2 \circ F_2 \circ \ldots \circ F_2}_{n-1}$, where

$$F_n(x_1, \ldots, x_n) = \underbrace{F_2 \circ F_2 \circ \ldots \circ F_2}_{n-1}(x_1, \ldots, x_n)$$
$$= F_2(x_1, F_2(x_2, \ldots, F_2(x_{n-1}, x_n))).$$

The last equality is one of the possible evaluation of the composition. By associativity any evaluation gives the same value.

We obtain an n-associative function $F_n \colon X^n \to X$ and an n-ary semigroup (X, F_n). In this case we say that (X, F_n) *is derived* from the binary semigroup (X, F_2). Generally we simply say that F_n is derived from F_2.

Dudek and Mukhin [3] have found the exact condition when an n-ary semigroup (X, F_n) is derived from a binary one.

Proposition 1 [3]. *If (X, F_n) is an n-ary semigroup with a neutral element e, then F_n can be derived from a binary semigroup denoted by F_2, where*

$$F_2(a, b) = F_n(a, e, \ldots, e, b). \tag{3}$$

As an application of Proposition 1 the authors of [3] obtained that X is an n-ary group which is derived from a group if and only if it contains a neutral element. An n-ary semigroup (X, F_n) is called an *n-ary group* if for $i \in \{1, \ldots, n\}$ and every $n - 1$ elements $x_1, \ldots, x_{i-1}, x_{i+1}, \ldots, x_n$ in X and every $a \in X$ there exists a unique $b \in X$ with $F_n(x_1, \ldots, x_{i-1}, b, x_{i+1}, \ldots, x_n) = a$. It is easy to see from the definition that ordinary groups are exactly the 2-ary groups. Clearly, a function F_n derived from a semigroup F_2 is n-associative but not every n-ary semigroup can be obtained in this way. We can easily construct n-ary groups which are not derived from binary groups if n is odd. Indeed, let $G_n(x_1, \ldots, x_n) = \sum_{i=1}^{n}(-1)^i x_i$. It is easy to verify that G_n is n-associative and we obtain an n-ary group. Moreover G_n is clearly monotone and there is no neutral element for G_n. For further examples see [9].

3   From n-ary to Binary Semigroups

The main purpose of this section is to derive properties from the n-ary semigroup to the corresponding binary semigroup and vice versa. The results of this section are also preparations for proving Theorem 3.

The following lemma is an easy consequence of the definitions.

Lemma 1. *Let $(X, \leq)$ be a partially ordered set, (X, F_2) be a semigroup and F_n be derived from F_2. If F_2 has any of the following properties*

1. *monotonicity,*
2. *idempotent,*
3. *has a neutral element,*

then the n-associative F_n also has.

From now on we focus on the possible reverse of the cases of Lemma 1.

First we investigate the *neutral element* property. By Proposition 1, if F_n has a neutral element, then F_n is derived from F_2 which is defined by Eq. (3).

Remark 1. By the definition (3) of F_2, if e is a neutral element of F_n, then e is also a neutral element of F_2. Indeed, $F_2(e, a) = F_n(e, \ldots, e, a) = a = F_n(a, e, \ldots, e) = F_2(a, e)$ for every $a \in X$.

For *monotonicity* the following statement have been proved for more general settings. On the other hand, by Remark 2, it turns out that this weaker condition implies that F_n is monotone in each of its variables.

Lemma 2. *Let $F_n : X^n \to X$ be an n-associative function on the partially ordered set X. Assume F_n is idempotent and monotone in the first and the last coordinates and derived from an associative function F_2. Then F_2 is monotone.*

Remark 2. As a consequence of Lemmas 1 and 2 we have that if F_n is n-associative, idempotent and monotone in the first and the last variables on a poset X and derived from F_2, then F_n is monotone in each of its variables.

We can verify *idempotency* only for totally ordered sets. In Example 1 we show that this requirement is essential.

Lemma 3. *Let $F_n : X^n \to X$ be an n-associative function on a **totally ordered** set. Assume F_n is idempotent and monotone in each variable and derived from an associative function F_2. Then F_2 is idempotent as well.*

Example 1. For $k \geq 3$ we construct a k-ary semigroup (X, F_k), which is derived from a non-idempotent semigroup (X, F_2), where F_2 is monotone in both of its variables and have a neutral element.

Let $X = \{m, M\} \cup Z_{k-1}$, where Z_{k-1} is the cyclic group of order $k - 1$. We define a partial ordering on X in the following way. M and m are the largest and smallest elements of X, respectively. The elements of Z_{k-1} are mutually incomparable but they are all larger than m and smaller than M. The set X endowed with this partial ordering is a modular lattice. Further we build up an associative function F_2:

$$F_2(x, y) = \begin{cases} M, & \text{if } x = M \text{ or } y = M \\ m, & \text{if } x = m \text{ or } y = m \text{ and } x, y < M \\ xy, & \text{if } x, y \in Z_{k-1}. \end{cases}$$

It is easy to verify that F_2 is associative and monotone increasing in both of its variables. The identity element e of Z_{k-1} is the neutral element of (X, F_2). One can define F_{k-1} and F_k as before. By Lemma 1 the functions F_{k-1} and F_k are $(k - 1)$- and k-associative functions, respectively. Both of them are monotone having neutral element. Finally, it is easy to check that F_{k-1} is not idempotent since $F_{k-1}(a, \ldots, a) = e$ for every $a \in Z_{k-1}$ while $F_k(x, \ldots, x) = x$ for every $x \in X$. Since F_{k-1} is non-idempotent, F_2 cannot be idempotent by Lemma 1 (ii).

We note that the cyclic group Z_{k-1} might be substituted by any nontrivial group whose exponent divides $k - 1$.

Remark 3. We note that for distributive lattices the statement of Lemma 3 seems true, but a potential proof would be basically different from the one of Lemma 3.

The following easy lemma provides that monotonicity implies *monotone increasingness* for partially order based, idempotent semigroups.

Lemma 4. *Let (X, F_2) be a partially order based semigroup, where $F_2 : X^2 \to X$ is idempotent and monotone in each variable, then F_2 is monotone increasing in each variable.*

Remark 4. Now we obtain some examples showing that we cannot omit any of the conditions of Lemma 4.

1. Let $F_2(x, x) = x$ for $x \in \mathbb{R}$ and $F_2(x, y) = 0$ if $x, y \in \mathbb{R}, x \neq y$. Then F_2 is associative and idempotent, but not monotone in each variable.
2. Let $F_2(x, y) = 2x - y$ for $x, y \in \mathbb{R}$. Then F_2 is idempotent and monotone in each variable, but not associative and clearly not monotone increasing.
3. Let $F_2(x, y) = -x$, if $x, y > 0$, and $F_2(x, y) = 0$ otherwise. Then F_2 is associative, since $F_2(x, F_2(y, z)) = F_2(F_2(x, y), z)) = 0$ and F_2 is monotone decreasing in each variable but F_2 is not idempotent.

Corollary 1. *If (X, F_n) is a totally order based n-ary semigroup, where F_n is idempotent and monotone in the first and in the last variables and derived from F_2, then F_n is monotone increasing in each variable. Moreover, F_k is monotone increasing for every $k \geq 2$.*

Using the results of this section we get the following proposition.

Proposition 2. *Let (X, F_n) be a totally order based n-ary semigroup, which is monotone, idempotent and has a neutral element. Then F_n is derived from a binary semigroup (X, F_2), where F_2 is also monotone idempotent and it also has a neutral element. Moreover, F_n is monotone increasing in each variables.*

As a consequence of Proposition 2 we can prove the following.

Lemma 5. *Let (X, F_n) be a totally order based n-ary semigroup derived from (X, F_2), where F_2 is idempotent, associative, monotone increasing and have a neutral element on X. Then*

$$F_n(a, y_1, \dots, y_{n-2}, b) = F_2(a, b)$$
$$F_n(b, y_1, \dots, y_{n-2}, a) = F_2(a, b)$$

for every $a \leq y_1, \dots, y_{n-2} \leq b$.

4   Main Results

If (X, F_n) is an n-ary semigroup having a neutral element e, then one can assign a semigroup by $F_2(a, b) = F_n(a, e, \dots, e, b)$ for every $a, b \in X$ as it was defined in Eq. (3). This operation will be denoted by $\mathscr{F}$. One of our main theoretic result is the following:

Theorem 3. *For any totally ordered set X the operation $\mathscr{F}$ creates bijection between the set of idempotent, monotone, associative functions on X having neutral elements and the set of n-associative, idempotent, monotone functions on X having neutral elements.*

We get the following as an easy consequence of our investigation.

Theorem 4. *Let I be a nonempty interval. For $n \geq 2$ let $F_n : I^n \to I$ be n-associative, monotone increasing, idempotent n-ary semigroup and has a neutral element $e \in I$. Then F_n is conservative.*

Applying Theorems 2 and 3 we can obtain a practical method to calculate the value of $F_n(a_1, \ldots, a_n)$ for any $a_1, \ldots, a_n \in I$, where $I \subset \mathbb{R}$ is a nonempty interval.

For every decreasing function $g : \bar{I} \to \bar{I}$ a pair $(a, b) \in I^2$ is called *critical* if $g(a) = b$ and $g(b) = a$. By Theorem 2 and Lemma 4, for every idempotent, monotone semigroup (X, F_2) with neutral element there exists a unique decreasing function g satisfying (2). Theorem 2 shows also that F_2 commutes in every non-critical pair $(x, y) \in I^2$ (i.e. $F_2(x, y) = F_2(y, x)$). Since for a critical pair (a, b) the value of $F_2(a, b)$ and $F_2(b, a)$ can be independently chosen from g we have two cases. A pair (a, b) is called *extra-critical* if critical and $F_2(a, b) \neq F_2(b, a)$. We note that being critical or extra-critical are both symmetric relations.

Finally, in order to simplify notation and give a compact way to express a value of F_n we introduce the following. The set of entries $\{a_1, \ldots, a_n\}$ of F_n is denoted by A. The smallest and largest element of A is denoted by c and d, respectively. Further there exist $1 \leq i \leq j \leq n$ such that $a_i, a_j \in \{c, d\}$ and $a_k \notin \{c, d\}$ for every $1 \leq k < i$ and $j < k \leq n$. We write $e_1 = a_i$ and $e_2 = a_j$.

Theorem 5. *Let $F_n : I^n \to I$ be an n-associative, idempotent function with neutral element. Assume that F_n is monotone in its first and last coordinates. If (c, d) is not an extra-critical pair, then $F_n(a_1, \ldots, a_n) = F_2(c, d)$. If (c, d) is an extra-critical pair, then $F_n(a_1, \ldots, a_n) = F_2(e_1, e_2)$.*

Now we point out three important consequences of Theorem 5. First we generalise Czogala-Drewniak's theorem (Theorem 1) as follows.

Theorem 6. *Let I be an arbitrary nonempty real interval. If a function $F_n : I^n \to I$ is n-associative, idempotent, monotone and has a neutral element $e \in I$, then there exits a monotone decreasing function $g : \bar{I} \to \bar{I}$ with $g(e) = e$ $(e \in I)$ such that Γ_g is symmetric and*

$$F_n(a_1, \ldots, a_n) = \begin{cases} c, & \text{if } c < g(d) \\ d, & \text{if } c > g(d) \\ c \text{ or } d, & \text{if } c = g(d), \end{cases}$$

where c and d denote the minimum and the maximum of set $A = \{a_1, \ldots, a_n\} \subset \mathbb{R}$, respectively.

We note that a generalization of Theorem 2 is essentially stated in Theorem 5. In [11] the authors investigated idempotent uninorms, which are idempontent, associative, commutative, monotone functions with a neutral element and defined on $[0,1]$. We introduce *n-ary uninorms*, which are n-associative, commutative, monotone functions with neutral element. Here we show a generalization of [11, Theorem 3] for n-ary operations.

Theorem 7. *An n-ary operator U_n is an idempotent n- ary uninorm on $[0,1]$ with neutral element $e \in [0,1]$ if and only if there exists a decreasing function $g : [0,1] \rightarrow [0,1]$ with fixed point e and with symmetric graph Γ_g such that*

$$U_n(a_1, \ldots, a_n) = \begin{cases} c & \text{if } c < g(d)) \text{ or } d < g(c) \\ d & \text{if } c > g(d) \text{ or } d > g(c) \\ c \text{ or } d & \text{if } c = g(d) \text{ and } d = g(c), \end{cases} \tag{4}$$

where c and d are as in Theorem 6. Moreover, if (c,d) is a critical pair ($c = g(d), d = g(c)$), then the value of $U_n(a_1, \ldots, a_n)$ can be chosen to be c or d arbitrarily and independently from other critical pairs.

One may extend the concept of associativity for string functions ([4,8]). Let us define

$$X^* = \bigcup_{n \in \mathbb{N}} X^n$$

to be the space of finite length words over the alphabet X. A multivariate function $F : X^* \rightarrow X$ is *associative* if it satisfies

$$F(\mathbf{x}, \mathbf{x}') = F(F(\mathbf{x}), F(\mathbf{x}'))$$

for all $\mathbf{x}, \mathbf{x}' \in X^*$. It is easy to check that $F|_{X^n}$ is n-associative for every $n \in \mathbb{N}$. Idempotency, monotonicity and the neutral element properties of F can be defined as they hold for every $n \in \mathbb{N}$.

Theorem 8. *Let I be a nonempty real interval. Then $F : I^* \rightarrow I$ is associative, idempotent, monotone and has a neutral element if and only if there is a decreasing function $g : \overline{I} \rightarrow \overline{I}$ with symmetric completed graph Γ_g such that $F|_{X^2}$ satisfies (2). Furthermore F must be monotone increasing in each variable.*

Acknowledgements. The results based on the article [7], which is an extended version of the current paper. This research is partly supported by the internal research project R-AGR-0500-MRO3 of University of Luxembourg.

References

1. Ackerman, N.L.: A characterization of quasitrivial n-semigroups, preprint
2. Czogala, E., Drewniak, J.: Associative monotonic operations in fuzzy set theory. Fuzzy Sets Syst. **12**, 249–269 (1984)
3. Dudek, W.A., Mukhin, V.V.: On n-ary semigroups with adjoint neutral element. Quasigroups Relat. Syst. **14**, 163–168 (2006)
4. Grabisch, M., Marichal, J.-L., Mesiar, R., Pap, E.: Aggregation Functions. Encyclopedia of Mathematics and its Applications, vol. 127. Cambridge University Press, Cambridge (2009)

5. Devillet, J., Kiss, G., Marichal, J.-L., Teheux, B.: On reflexive, monotone functions that are associative and bisymmetric, in preparation

6. Fodor, J.: An extension of Fung-Fu's theorem. Int. J. Uncertainty Fuzziness Knowl.-Based Syst. **4**(3), 235–243 (1996)

7. Kiss, G., Somlai, G.: Characterization of n-associative, monotone, idempotent functions on an interval which have neutral elements, Semigroup forum, in press. arXiv:1609.00279

8. Lehtonen, E., Marichal, J.-L., Teheux, B.: Associative string functions. Asian-Eur. J. Math. **7**(4), 1450059 (2014)

9. Marichal, J.-L., Mathonet, P.: A description of n-ary semigroups polynomial-derived from integral domains. Semigroup Forum **83**(2), 241–249 (2011)

10. Martin, J., Mayor, G., Torrens, J.: On locally internal monotonic operations. Fuzzy Sets Syst. **137**, 27–42 (2003)

11. Ruiz-Aguilera, D., Torrens, J., De Baets, B., Fodor, J.: Some remarks on the characterization of idempotent uninorms. In: Hüllermeier, E., Kruse, R., Hoffmann, F. (eds.) Computational Intelligence for Knowledge-Based Systems Design, pp. 425–434. Springer, Heidelberg (2010)

On Idempotent Discrete Uninorms

Miguel Couceiro[1], Jimmy Devillet[2], and Jean-Luc Marichal[2(✉)]

[1] LORIA, CNRS - Inria Nancy Grand Est - Université de Lorraine,
BP 239, 54506 Vandoeuvre-lès-Nancy, France
miguel.couceiro@inria.fr
[2] Mathematics Research Unit, University of Luxembourg, Maison du Nombre,
6, Avenue de la Fonte, 4364 Esch-sur-Alzette, Luxembourg
{jimmy.devillet,jean-luc.marichal}@uni.lu

Abstract. In this paper we provide two axiomatizations of the class of idempotent discrete uninorms as conservative binary operations, where an operation is conservative if it always outputs one of its input values. More precisely we first show that the idempotent discrete uninorms are exactly those operations that are conservative, symmetric, and nondecreasing in each variable. Then we show that, in this characterization, symmetry can be replaced with both bisymmetry and existence of a neutral element.

1   Introduction

Aggregation functions defined on linguistic scales (i.e., finite chains) have been intensively investigated for about two decades; see, e.g., [1–4,6–11,13,14]. Among these functions, discrete fuzzy connectives (such as discrete uninorms) are associative binary operations that play an important role in fuzzy logic.

This short paper focuses on characterizations of the class of idempotent discrete uninorms. Recall that a discrete uninorm is a binary operation on a finite chain that is associative, symmetric, nondecreasing in each variable, and has a neutral element.

A first characterization of the class of idempotent discrete uninorms was given by De Baets et al. [1, Theorem 3]. This characterization reveals that any idempotent discrete uninorm is a combination of the minimum and maximum operations. In particular, such an operation is *conservative* in the sense that it always outputs one of the input values.

The outline of this paper is as follows. After presenting some preliminary results on conservative operations in Sect. 2, we show in Sect. 3 that the idempotent discrete uninorms are exactly those operations that are conservative, symmetric, and nondecreasing in each variable. This new characterization is very simple and requires neither associativity nor the existence of a neutral element. In Sect. 4 we provide an alternative characterization of this class in terms of the bisymmetry property. More specifically, we show that the idempotent discrete uninorms are exactly those operations that are conservative, bisymmetric, nondecreasing in each variable, and have neutral elements.

© Springer International Publishing AG 2018
V. Torra et al. (eds.), *Aggregation Functions in Theory and in Practice*,
Advances in Intelligent Systems and Computing 581, DOI 10.1007/978-3-319-59306-7_15

2 Preliminaries

In this section we present some basic definitions and preliminary results.

Let X be an arbitrary nonempty set and let $\Delta_X = \{(x, x) \mid x \in X\}$.

Definition 1. An operation $F\colon X^2 \to X$ is said to be

- *idempotent* if $F(x, x) = x$ for all $x \in X$.
- *conservative* if $F(x, y) \in \{x, y\}$ for all $x, y \in X$.
- *associative* if $F(F(x, y), z) = F(x, F(y, z))$ for all $x, y, z \in X$.

An element $e \in X$ is said to be a *neutral element* of F (or simply a *neutral element*) if $F(x, e) = F(e, x) = x$ for all $x \in X$. In this case we easily show by contradiction that such a neutral element is unique. The points (x, y) and (u, v) of X^2 are said to be *connected for F* (or simply *connected*) if $F(x, y) = F(u, v)$. We observe that "being connected" is an equivalence relation. The point (x, y) of X^2 is said to be *isolated for F* (or simply *isolated*) if it is not connected to another point in X^2.

Remark 1. Conservativeness was introduced in Pouzet et al. [12]. This condition is also called "local internality" in Martín et al. [5].

Lemma 1. *Let $F\colon X^2 \to X$ be an idempotent operation. If the point $(x, y) \in X^2$ is isolated, then it lies on Δ_X, that is, $x = y$.*

Remark 2. We observe that idempotency is necessary in Lemma 1. Indeed, consider the operation $F\colon X^2 \to X$, where $X = \{a, b\}$, defined as $F(x, y) = a$, if $(x, y) = (a, b)$, and $F(x, y) = b$, otherwise. Then (a, b) is isolated and $a \neq b$. The contour plot of F is represented in Fig. 1. Here and throughout, connected points are joined by edges. To keep the figures simple we sometimes omit the edges obtained by transitivity.

Fig. 1. A non-idempotent operation

The following lemma provides an easy test for the existence of a neutral element of a conservative operation.

Lemma 2. *Let $F\colon X^2 \to X$ be a conservative operation and let $e \in X$. Then e is a neutral element if and only if (e, e) is isolated.*

Corollary 1. *Any isolated point (x, y) of a conservative operation $F\colon X^2 \to X$ is unique and lies on Δ_X. Moreover, $x = y$ is a neutral element.*

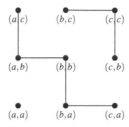

Fig. 2. An operation with no neutral element

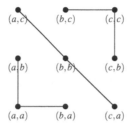

Fig. 3. An operation with no isolated point

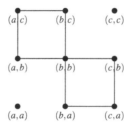

Fig. 4. An operation with two isolated points

Remark 3. Lemma 2 no longer holds if conservativeness is relaxed into idempotency. Indeed, by simply taking $X = \{a, b, c\}$ we can easily construct an idempotent operation with an isolated point on Δ_X and no neutral element (see Fig. 2). Also, it is easy to construct an idempotent operation with a neutral element and no isolated point (see Fig. 3). It is also noteworthy that there are idempotent operations with more than one isolated point (see Fig. 4).

3   Main Results

We now focus on characterizations of the class of idempotent discrete uninorms. These operations are defined on finite chains. Without loss of generality we will only consider the n-element chains $L_n = \{1, \ldots, n\}$, $n \geq 1$, endowed with the usual ordering relation $\leq$.

Recall that an operation $F \colon L_n^2 \to L_n$ is said to be *nondecreasing in each variable* if $F(x, y) \leq F(x', y')$ whenever $x \leq x'$ and $y \leq y'$.

Definition 2 (see, e.g., [1]). A *discrete uninorm* on L_n is an operation $U: L_n^2 \to L_n$ that is associative, symmetric, nondecreasing in each variable, and has a neutral element.

A characterization of the class of idempotent discrete uninorms is given in the following theorem. Although this characterization is somewhat intricate, it shows, together with Lemma 3 below, that any idempotent discrete uninorm is conservative.

Theorem 1 (see [1, Theorem 3]). *An operation $F: L_n^2 \to L_n$ with a neutral element $1 < e < n$ is an idempotent discrete uninorm if and only if there exists a nonincreasing map $g: [1, e] \to [e, n]$ (nonincreasing means that $g(x) \geq g(y)$ whenever $x \leq y$), with $g(e) = e$, such that*

$$F(x,y) = \begin{cases} \min\{x,y\}, & \text{if } y \leq \overline{g}(x) \text{ and } x \leq \overline{g}(1), \\ \max\{x,y\}, & \text{otherwise}, \end{cases}$$

where $\overline{g}: L_n \to L_n$ is defined by

$$\overline{g}(x) = \begin{cases} g(x), & \text{if } x \leq e, \\ \max\{z \in [1,e] \mid g(z) \geq x\}, & \text{if } e \leq x \leq g(1), \\ 1, & \text{if } x > g(1). \end{cases}$$

We now show that the idempotent discrete uninorms are exactly those operations that are conservative, symmetric, and nondecreasing in each variable (see Theorem 2).

First consider the following lemma, which actually holds on arbitrary, not necessarily finite, chains.

Lemma 3. *If $F: L_n^2 \to L_n$ is idempotent, nondecreasing in each variable, and has a neutral element $e \in L_n$, then $F|_{[1,e]^2} = \min$ and $F|_{[e,n]^2} = \max$.*

Proposition 1. *If $F: L_n^2 \to L_n$ is conservative, symmetric, and nondecreasing in each variable, then it is associative and it has a neutral element.*

For $n = 2$ and $n = 3$, the possible operations $F: L_n^2 \to L_n$ that are conservative, symmetric, and nondecreasing in each variable have contour plots depicted in Figs. 5 and 6, respectively.

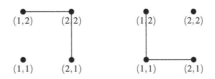

Fig. 5. Possible operations when $n = 2$

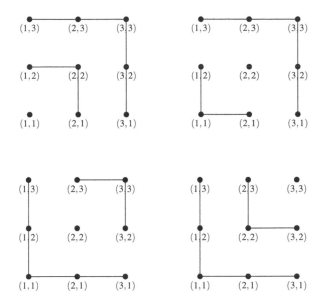

Fig. 6. Possible operations when $n = 3$

Remark 4

(a) The existence of a neutral element in Proposition 1 is no longer guaranteed if the chain is not finite. For instance, the real operation $F\colon [0,1]^2 \to [0,1]$ defined by $F(x,y) = \min\{x,y\}$, if $x,y \in [0,\frac{1}{2})^2$, and $F(x,y) = \max\{x,y\}$, otherwise, is conservative, symmetric, and nondecreasing in each variable, but it does not have a neutral element.

(b) We observe that conservativeness cannot be relaxed into idempotency in Proposition 1. For instance the operation $F\colon L_3^2 \to L_3$ whose contour plot is depicted in Fig. 2 is idempotent, symmetric, and nondecreasing in each variable, but one can show that it is not associative and it has no neutral element.

(c) We also observe that each of the conditions of Proposition 1 is necessary. Indeed, we give in Fig. 7 an operation that is conservative and symmetric but that is not nondecreasing in each variable. We also give in Fig. 8 an operation that is conservative and nondecreasing in each variable but not symmetric. Finally, we give in Fig. 9 an operation that is symmetric and nondecreasing in each variable but not conservative. None of these three operations is associative and none has a neutral element.

Theorem 2. *An operation $F\colon L_n^2 \to L_n$ is conservative, symmetric, and nondecreasing in each variable if and only if it is an idempotent discrete uninorm. Moreover, there are exactly 2^{n-1} such operations.*

Remark 5. Theorem 2 enables us to provide a graphical characterization of the idempotent discrete uninorms in terms of their contour plots. Indeed, denoting

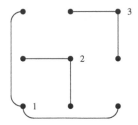

Fig. 7. An operation that fails to be nondecreasing in each variable

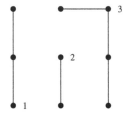

Fig. 8. An operation that fails to be symmetric

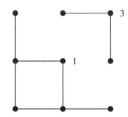

Fig. 9. An operation that fails to be conservative

by L an arbitrary n-element chain, we observe that the restriction $F|_{L'}$ of any idempotent discrete uninorm $F: L^2 \to L$ to any subchain L' obtained by removing one of the endpoints of L is also an idempotent discrete uninorm. Moreover, the operation F (or equivalently its contour plot) can be retrieved from $F|_{L'}$ by connecting all the points of $L^2 \setminus L'^{\,2}$. It follows that all the idempotent discrete uninorms can be constructed recursively in terms of their contour plots.

4   Bisymmetric Operations

In this section we provide a characterization of the class of idempotent discrete uninorms in terms of the bisymmetry (or mediality) property.

Definition 3. An operation $F: X^2 \to X$ is said to be *bisymmetric* if

$$F(F(x, y), F(u, v)) = F(F(x, u), F(y, v))$$

for all $x, y, u, v \in X$.

Proposition 2. *Let $F\colon X^2 \to X$ be a conservative operation that has a neutral element. Then F is bisymmetric if and only if it is associative and symmetric.*

Combining Proposition 2 with Theorem 2, we can easily derive the following alternative characterization of idempotent discrete uninorms.

Theorem 3. *An operation $F\colon L_n^2 \to L_n$ is conservative, bisymmetric, nondecreasing in each variable, and has a neutral element if and only if it is an idempotent discrete uninorm.*

Acknowledgements. The authors thank Gergely Kiss for fruitful discussion and valuable remarks. This research is partly supported by the internal research project R-AGR-0500-MRO3 of the University of Luxembourg.

References

1. De Baets, B., Fodor, J., Ruiz-Aguilera, D., Torrens, J.: Idempotent uninorms on finite ordinal scales. Int. J. Uncertainty Fuzziness Knowl.-Based Syst. **17**(1), 1–14 (2009)
2. De Baets, B., Mesiar, R.: Discrete triangular norms. In: Rodabaugh, S., Klement, E.P. (eds.) Topological and Algebraic Structures in Fuzzy Sets, A Handbook of Recent Developments in the Mathematics of Fuzzy Sets. Trends in Logic, pp. 389–400. Kluwer Academic Publishers, Dordrecht (2003)
3. Fodor, J.: Smooth associative operations on finite ordinal scales. IEEE Trans. Fuzzy Syst. **8**, 791–795 (2000)
4. Li, G., Liu, H.-W., Fodor, J.: On weakly smooth uninorms on finite chain. Int. J. Intell. Syst. **30**, 421–440 (2015)
5. Martín, J., Mayor, G., Torrens, J.: On locally internal monotonic operations. Fuzzy Sets Syst. **137**, 27–42 (2003)
6. Mas, M., Mayor, G., Torrens, J.: t-Operators and uninorms on a finite totally ordered set. Int. J. Intell. Syst. **14**, 909–922 (1999)
7. Mas, M., Monserrat, M., Torrens, J.: On bisymmetric operators on a finite chain. IEEE Trans. Fuzzy Syst. **11**, 647–651 (2003)
8. Mas, M., Monserrat, M., Torrens, J.: On left and right uninorms on a finite chain. Fuzzy Sets Syst. **146**, 3–17 (2004)
9. Mas, M., Monserrat, M., Torrens, J.: Smooth t-subnorms on finite scales. Fuzzy Sets Syst. **167**, 82–91 (2011)
10. Mayor, G., Suñer, J., Torrens, J.: Copula-like operations on finite settings. IEEE Trans. Fuzzy Syst. **13**, 468–477 (2005)
11. Mayor, G., Torrens, J.: Triangular norms in discrete settings. In: Klement, E.P., Mesiar, R. (eds.) Logical, Algebraic, Analytic, and Probabilistic Aspects of Triangular Norms, pp. 189–230. Elsevier, Amsterdam (2005)
12. Pouzet, M., Rosenberg, I.G., Stone, M.G.: A projection property. Algebra Universalis **36**(2), 159–184 (1996)
13. Ruiz-Aguilera, D., Torrens, J.: A characterization of discrete uninorms having smooth underlying operators. Fuzzy Sets Syst. **268**, 44–58 (2015)
14. Su, Y., Liu, H.-W.: Discrete aggregation operators with annihilator. Fuzzy Sets Syst. **308**, 72–84 (2017)

On the F-partial Order and Equivalence Classes of Nullnorms

Emel Aşıcı[✉]

Department of Software Engineering, Faculty of Technology,
Karadeniz Technical University, 61830 Trabzon, Turkey
emelkalin@hotmail.com

Abstract. In this paper, we define the set $I_F^{(x)}$, denoting the set of all incomparable elements with arbitrary but fixed $x \in (0,1)$ according to F-partial order and this set is deeply investigated. Then, an equivalence relation on the class of nullnorms induced by a F-partial order is defined and discussed. Finally, we give an answer to a recently posed open problem.

1 Introduction

Nullnorms and t-operators were introduced in [6,17], respectively, which are also generalizations of the notions of t-norms and t-conorms. And then in [18], it is pointed out that nullnorms and t-operators are equivalent since they have the same block structures in $[0,1]^2$. Namely, if a binary operator F is a nullnorm then it is also a t-operator and vice versa.

In [19], a natural order for semigroups was defined. Similarly, in [13], a partial order defined by means of t-norms on a bounded lattice was introduced

$x \preceq_T y :\Leftrightarrow T(\ell, y) = x$ for some $\ell \in L$,

where L is a bounded lattice, $x, y \in L$ and T is a t-norm on L. This partial order $\preceq_T$ is called a T-partial order on L.

In [1], with the help of any t-norm T on $[0,1]$, a family of t-norms on $[0,1]$, $(T_\lambda)_{\lambda \in (0,1)}$ was constructed. If T was a divisible t-norm, then it was obtained that $([0,1], \preceq_{T_\lambda})$ was a lattice. The nullnorms and t-norms were also studied by many other authors [2,7,10,11,14,16,18,20,21].

The present paper is organized as follows. We shortly recall some basic notions in Sect. 2. In Sect. 3, we define a set $I_F^{(x)}$, denoting the set of all incomparable elements with arbitrary but fixed $x \in (0,1)$ according to $\preceq_F$. In Sect. 4, we define an equivalence on the class of nullnorms on a bounded lattice $(L, \leq, 0, 1)$ and we determine the equivalence class of some special nullnorms. In [3], the following open problem was proposed: Given an $(a, b) \subsetneq (0,1)$, can we find an F such that $K_F = (a, b)$? In Sect. 5, we give an answer to an this open problem in [3].

V. Torra et al. (eds.), *Aggregation Functions in Theory and in Practice*,
Advances in Intelligent Systems and Computing 581, DOI 10.1007/978-3-319-59306-7_16

2   Notations, Definitions and a Review of Previous Results

Definition 1 [9]. Let $(L, \leq, 0, 1)$ be a bounded lattice. A *triangular norm* T (briefly t-norm) is a binary operation on L which is commutative, associative, monotone and has neutral element 1.

Definition 2 [15]. Let $(L, \leq, 0, 1)$ be a bounded lattice. A *triangular conorm* S (briefly t-conorm) is a binary operation on L which is commutative, associative, monotone and has neutral element 0.

Example 1 [15]. Well-known triangular norms and triangular conorms are:
$T_M(x, y) = min(x, y)$
$T_P(x, y) = x.y$
$T_L(x, y) = max(x + y - 1, 0)$
$$T_D(x, y) = \begin{cases} 0, & (x, y) \in [0, 1)^2 \\ min(x, y), & \text{otherwise} \end{cases}$$
$S_M(x, y) = max(x, y)$
$S_P(x, y) = x + y - x.y$
$S_L(x, y) = min(x + y, 1)$
$$S_D(x, y) = \begin{cases} 1, & (x, y) \in (0, 1]^2 \\ max(x, y), & \text{otherwise} \end{cases}$$

Also, t-norms on a bounded lattice $(L, \leq, 0, 1)$ are defined in similar way, and then extremal t-norms $T_\wedge$ and T_W on L is defined as follows, respectively:
$T_\wedge(x, y) = x \wedge y$
$$T_W(x, y) = \begin{cases} x & , if\ y = 1 \\ y & , if\ x = 1 \\ 0 & , \text{otherwise} \end{cases}$$
Similarly it can be defined the t-conorms $S_\vee$ and S_W.

Especially we obtained that $T_W = T_D$ and $T_\wedge = T_M$ for $L = [0, 1]$.

Definition 3 [7]. A t-norm T on L is *divisible* if the following condition holds:

$$\forall x, y \in L \quad \text{with} \quad x \leq y \quad \text{there is a} \quad z \in L \quad \text{such that} \quad x = T(y, z).$$

A basic example of a non-divisible t-norm on an arbitrary lattice L (i.e., card $L > 3$) is the weakest t-norm T_D. Trivially, the infimum $T_\wedge$ is divisible: $x \leq y$ is equivalent to $x \wedge y = x$.

Proposition 1 [8]. *Let T be a t-norm on $[0, 1]$. T is divisible if and only if T is continuous.*

Definition 4 [5]. Given a bounded lattice $(L, \leq, 0, 1)$ and $a, b \in L$, if a and b are incomparable, in this case we use the notation $a \parallel b$.

Definition 5 [5]. Given a bounded lattice $(L, \leq, 0, 1)$ and $a, b \in L$, $a \leq b$, a subinterval $[a, b]$ of L is defined as

$[a, b] = \{x \in L \mid a \leq x \leq b\}$

Similarly, $[a, b) = \{x \in L \mid a \leq x < b\}$, $(a, b] = \{x \in L \mid a < x \leq b\}$ and $(a, b) = \{x \in L \mid a < x < b\}$.

Definition 6 [6]. Let $(L, \leq, 0, 1)$ be a bounded lattice. A commutative, associative, non-decreasing in each variable function $F : L^2 \to L$ is called a nullnorm if there is an element $a \in L$ such that $F(x, 0) = x$ for all $x \leq a$, $F(x, 1) = x$ for all $x \geq a$.

It can be easily obtained that $F(x, a) = a$ for all $x \in L$. So $a \in L$ is the zero (absorbing) element for F.

Consider the set $\mathscr{F}$ of all nullnorms on L with the following order: For $F_1, F_2 \in \mathscr{F}$,

$F_1 \leq F_2 \Leftrightarrow F_1(x, y) \leq F_2(x, y)$ for all $(x, y) \in L^2$.

$D_a = [0, a) \times (a, 1] \cup (a, 1] \times [0, a)$ for $a \in L \backslash \{0, 1\}$.

Proposition 2 [12]. *Let $(L, \leq, 0, 1)$ be a bounded lattice, $a \in L \setminus \{0, 1\}$ and F be a nullnorm with zero element a on L. Then,*

(i) $S^* = F \mid_{[0,a]^2} : [0, a]^2 \to [0, a]$ *is a t-conorm on $[0, a]$.*
(ii) $T^* = F \mid_{[a,1]^2} : [a, 1]^2 \to [a, 1]$ *is a t-norm on $[a, 1]$.*

Definition 7 [13]. Let L be a bounded lattice, T be a t-norm on L. The order defined as following is called a T- *partial order* (triangular order) for t-norm T:

$$x \preceq_T y :\Leftrightarrow T(\ell, y) = x \text{ for some } \ell \in L.$$

Definition 8 [3]. Let L be a bounded lattice, S be a t-conorm on L. The order defined as following is called an S-*partial order* for t-conorm S:

$$x \preceq_S y :\Leftrightarrow S(\ell, x) = y \text{ for some } \ell \in L.$$

Definition 9 [3]. Let $(L, \leq, 0, 1)$ be a bounded lattice and F be a nullnorm with zero element a on L. Define the following relation, for $x, y \in L$, as

$$x \preceq_F y :\Leftrightarrow \begin{cases} \text{if } x, y \in [0, a] \text{ and there exist } k \in [0, a] \text{ such that } F(x, k) = y \text{ or} \\ \text{if } x, y \in [a, 1] \text{ and there exist } \ell \in [a, 1] \text{ such that } F(y, \ell) = x \text{ or,} \\ \text{if } (x, y) \in L^* \text{ and } x \leq y. \end{cases}$$

$$(1)$$

Where $I_a = \{x \in L \mid x \parallel a\}$ and $L^* = [0, a] \times [a, 1] \cup [0, a] \times I_a \cup [a, 1] \times I_a \cup [a, 1] \times [0, a] \cup I_a \times [0, a] \cup I_a \times [a, 1] \cup I_a \times I_a$.

Proposition 3 [3]. *The relation $\preceq_F$ defined in (1) is a partial order on L.*

Note: The partial order $\preceq_F$ in (1) is called F-partial order on L.

Proposition 4 [3]. *Let* $(L, \leq, 0, 1)$ *be a bounded lattice and* F *be a nullnorm on* L. *If* $x \preceq_F y$ *for any* $x, y \in L$, *then* $x \leq y$.

Proposition 5 [3]. *Let* $(L, \leq, 0, 1)$ *be a bounded lattice and* F *be a nullnorm with zero element* a. *Then*, $(L, \preceq_F)$ *is a bounded partially ordered set.*

Remark 1 [3]. Let $(L, \leq, 0, 1)$ be a bounded lattice and F be a nullnorm with zero element a. The order $\preceq_F$ coincides with the order $\preceq_T$ ($\preceq_S$), when $a = 0$ ($a = 1$).

Definition 10 [3]. Let F be a nullnorm on $[0, 1]$ and let K_F be defined by

$$K_F = \{x \in [0, 1] | \quad \text{for some} \quad y \in [0, 1], \quad [x < y \quad \text{and} \quad x \not\preceq_F y] \quad \text{or}$$
$$[y < x \quad \text{and} \quad y \not\preceq_F x]\}.$$

3   About the Set $I_F^{(x)}$ Consisting All Incomparable Elements with Any $x \in (0, 1)$ According to $\preceq_F$

Definition 11. Let F be a nullnorm on $[0, 1]$ with zero element a and let $I_F^{(x)}$ for $x \in (0, 1)$ be defined by

$$I_F^{(x)} = \{y_x \in (0, 1) \mid [x < y_x \text{ and } x \not\preceq_F y_x] \text{ or } [y_x < x \text{ and } y_x \not\preceq_F x]\}$$

After that we will use the notation $I_F^{(x)}$ to denote the set of all incomparable elements with $x \in (0, 1)$ according to $\preceq_F$.

Note: It is clear that $x \neq a$, by the definition of $\preceq_F$-partial order.

Example 2. Consider the nullnorm $F^{(\vee)} : [0, 1]^2 \to [0, 1]$ with zero element $a \in (0, 1)$ defined by

$$F^{(\vee)}(x, y) = \begin{cases} max(x, y), & (x, y) \in [0, a]^2 \\ a, & (x, y) \in [a, 1]^2 \cup D_a \\ y, & x = 1 \\ x, & y = 1 \end{cases}$$

Then,

(a) $I_{F^{(\vee)}}^{(x)} = \{y_x \in (a, 1) \mid x \neq y_x\}$ for $x \in (a, 1)$ and
(b) $I_{F^{(\vee)}}^{(x)} = \varnothing$ for $x \leq a$.

Example 3. Consider the nullnorm $F^{(\wedge)} : [0, 1]^2 \to [0, 1]$ with zero element $a \in (0, 1)$ defined by

$$F^{(\wedge)}(x, y) = \begin{cases} min(x, y), & (x, y) \in [a, 1]^2 \\ a, & (x, y) \in (0, a]^2 \cup D_a \\ y, & x = 0 \\ x, & y = 0 \end{cases}$$

Then,

(a) $I_{F(\wedge)}^{(x)} = \{y_x \in (0, a) \mid x \neq y_x\}$ for $x \in (0, a)$ and
(b) $I_{F(\wedge)}^{(x)} = \varnothing$ for $x \geq a$.

Example 4. Consider the function $F := F_{(T^n M, S, \frac{1}{5})} : [0,1]^2 \to [0,1]$ defined as follows:

$$F_{(T^n M, S, \frac{1}{5})}(x, y) = \begin{cases} S(x, y), & (x, y) \in [0, \frac{1}{5}]^2 \\ \frac{1}{5}, & ((x, y) \in [\frac{1}{5}, 1]^2 \text{ and } x + y \leq 1) \text{ or } (x, y) \in D_{\frac{1}{5}} \\ min(x, y), & \text{otherwise} \end{cases}$$

where S is a continuous t-conorm on $[0, \frac{1}{5}]$. Then, the function F is a nullnorm with zero element $\frac{1}{5}$. Then,

(a) $I_F^{(x)} = \{y_x \in (\frac{1}{5}, 1 - x] \mid x \neq y_x\}$ for $x \in (\frac{1}{5}, \frac{4}{5})$
(b) $I_F^{(x)} = \varnothing$ for $x \leq \frac{1}{5}$ or $x \geq \frac{4}{5}$.

Proposition 6. *Let F be a nullnorm on $[0,1]$. Then $K_F = \bigcup_{x \in [0,1]} I_F^{(x)}$.*

Proposition 7. *Let F_1 and F_2 be two uninorms on $[0,1]$. If for all $x \in [0,1]$, $I_{F_1}^{(x)} = I_{F_2}^{(x)}$, then the set K_{F_1} is equal to the set K_{F_2}.*

Remark 2. The converse of Proposition 7 may not be true.

Proposition 8. *Let F be a nullnorm on $[0,1]$. If K_F is a non-empty set, then K_F is infinite.*

4   The Equivalence Classes Obtained from $\preceq_F$

The above introduced $\preceq_F$-partial order allows us to introduce the next equivalence relation on the class of all nullnorms on $(L, \leq, 0, 1)$.

Definition 12. Let $(L, \leq, 0, 1)$ be a given bounded lattice. Define a relation α_F on the class of all nullnorms on $(L, \leq, 0, 1)$ by $F_1 \alpha_F F_2$ if and only if the F_1-partial order coincides with the F_2-partial order.

Lemma 1. *The relation α_F given in Definition 12 is an equivalence relation.*

Definition 13. For a given nullnorm F on a bounded lattice $(L, \leq, 0, 1)$, we denote by $\overline{F}$ the α_F equivalence class linked to F, i.e.,

$$\overline{F} = \{F' \mid F' \alpha_F F\}.$$

Proposition 9. *Consider the weakest nullnorm $F^{(\vee)} : [0,1]^2 \to [0,1]$ with zero element $a \in (0,1)$ of Example 2. Then, the equivalence class of the nullnorm $F^{(\vee)}|_{[0,a]^2}$ is the set of all divisible t-conorms on $[0, a]$ and the equivalence class of the nullnorm $F^{(\vee)}|_{[a,1]^2}$ is the nullnorm $F^{(\vee)}|_{[a,1]^2}$.*

Note: If $(x, y) \in D_a$, then we have that $F(x, y) = a$ for all nullnorms with zero element a. So, in this case all nullnorms with zero element a are equivalent.

Corollary 1. *The equivalence class of the weakest t-norm T_D on $[0, 1]$ only consist of the t-norm T_D.*

Proposition 10. *Consider the greatest nullnorm $F^{(\wedge)} : [0, 1]^2 \to [0, 1]$ with zero element $a \in (0, 1)$ of Example 3. Then, the equivalence class of the nullnorm $F^{(\wedge)} \mid_{[a,1]^2}$ is the set of all divisible t-norms on $[a, 1]$ and the equivalence class of the nullnorm $F^{(\wedge)} \mid_{[0,a]^2}$ is the nullnorm $F^{(\wedge)} \mid_{[0,a]^2}$.*

Corollary 2. *The equivalence class of the weakest t-conorm S_D on $[a, 1]$ only consist of the t-norm S_D.*

Proposition 11 [6]. *Let $a \in (0, 1)$. A binary operation F is a nullnorm with zero element a if and only if there exists triangular norm T and triangular conorm S such that*

$$
F(x, y) = \begin{cases} \phi^{-1}(S(\phi(x), \phi(y))), & (x, y) \in [0, a]^2 \\ \varphi^{-1}(T(\varphi(x), \varphi(y))), & (x, y) \in [a, 1]^2 \\ a, & \text{otherwise} \end{cases}
$$

where $\phi : [0, a] \to [0, a]$ and $\varphi : [a, 1] \to [a, 1]$ are linear bijection such that $\phi(x) = \frac{x}{a}$ and $\varphi(x) = \frac{x-a}{1-a}$.

Proposition 12. *Let F be a nullnorm on $[0, 1]$ with zero element a and $\phi : [0, a] \to [0, a]$ be a linear bijection and S be a t-conorm on $[0, a]$. The following statements are equivalent.*

(i) S and $F \mid_{[0,a]^2}$ are in the same equivalence class.
(ii) ϕ is order-preserving with respect to $\preceq_S$.

Proposition 13. *Let F be a nullnorm on $[0, 1]$ with zero element a and $\varphi : [a, 1] \to [a, 1]$ be a linear bijection and and T be a t-norm on $[a, 1]$. The following statements are equivalent.*

(i) T and $F \mid_{[a,1]^2}$ are in the same equivalence class.
(ii) φ is order-preserving with respect to $\preceq_T$.

5   An Answer to an Open Problem

Let $(L, \leq, 0, 1)$ be a bounded lattice and F be a nullnorm on L. In [3], it has been shown that K_F need not be $(0, 1)$ but any sub-interval of $(0, 1)$ and the following problem has been posed by Asici [3].

Given an $(a, b) \subsetneq (0, 1)$, can we find an F such that $K_F = (a, b)$? In the following theorem, we give an answer to this open problem.

Theorem 1. *Let $\{0,1\} \subseteq A \subseteq [0,1]$ be an arbitrary set. If there exists a family $((u_i, v_i))_{i \in I}$ and $((y_j, z_j))_{j \in I}$ be pairwise disjoint open sub-intervals of $[0,1]$ such that*

$$\bigcup_{i \in I} ((u_i, v_i)) \cup \bigcup_{j \in I} (y_j, z_j) \subseteq [0,1] \setminus A$$

where I is a finite or countably infinite index set. Then there is a nullnorm F such that A coincides with the set of all comparable elements of $[0,1]$ with respect to $\preceq_F$.

Proof. To proof this theorem, we defined a nullnorm F on $[0,1]$ with zero element a. Then we obtained that $A = [0,1] \setminus K_F$, that is, A is the set of all comparable elements of $[0,1]$ with respect to $\preceq_F$. So, we showed that $K_F = [0,1] \setminus A$. Thus, we proved that given an $(a,b) \subsetneq (0,1)$, we can find an F such that $K_F = (a,b)$.

6   Conclusions

We have discussed and investigated some properties of F-partial order. Also, we have defined that the set of all incomparable elements with arbitrary but fixed x element of $(0,1)$ according to the F-partial order and we have investigated some properties of this set. Then, we have defined an equivalence on the class of nullnorms on a bounded lattice L and we have determined the equivalence class of some special nullnorms. Finally, we have given an answer to an open problem posed in [1].

Acknowledgement. In this paper, the full proofs are contained in [4].

References

1. Aşıcı, E., Karaçal, F.: On the T-partial order and properties. Inf. Sci. **267**, 323–333 (2014)
2. Aşıcı, E., Karaçal, F.: Incomparability with respect to the triangular order. Kybernetika **52**, 15–27 (2016)
3. Aşıcı, E.: An order induced by nullnorms and its properties. Fuzzy Sets Syst. doi:10.1016/j.fss.2016.12.004
4. Aşıcı, E.: The equivalence classes of nullnorms. Working paper
5. Birkhoff, G.: Lattice Theory, 3rd edn. American Mathematical Society, Providence (1967)
6. Calvo, T., De Baets, B., Fodor, J.: The functional equations of Frank and Alsina for uninorms and nullnorms. Fuzzy Sets Syst. **120**, 385–394 (2001)
7. Casasnovas, J., Mayor, G.: Discrete t-norms and operations on extended multisets. Fuzzy Sets Syst. **159**, 1165–1177 (2008)
8. De Baets, B., Mesiar, R.: Triangular norms on the real unit square. In: Proceedings of the 1999 EUSFLAT-ESTYLF Joint Conference, Palma de Mallorca, Spain, pp. 351–354 (1999)

9. Çaylı, G.D., Karaçal, F., Mesiar, R.: On a new class of uninorms on bounded lattices. Inf. Sci. **367–368**, 221–231 (2016)
10. Drewniak, J., Drygaś, P., Rak, E.: Distributivity between uninorms and nullnorms. Fuzzy Sets Syst. **159**, 1646–1657 (2008)
11. Drygaś, P.: A characterization of idempotent nullnorms. Fuzzy Sets Syst. **145**, 455–461 (2004)
12. Karaçal, F., Ince, M.A., Mesiar, R.: Nullnorms on bounded lattice. Inf. Sci. **325**, 227–236 (2015)
13. Karaçal, F., Kesicioğlu, M.N.: A T-partial order obtained from t-norms. Kybernetika **47**, 300–314 (2011)
14. Kesicioğlu, M.N., Karaçal, F., Mesiar, R.: Order-equivalent triangular norms. Fuzzy Sets Syst. **268**, 59–71 (2015)
15. Klement, E.P., Mesiar, R., Pap, E.: Triangular Norms. Kluwer Academic Publishers, Dordrecht (2000)
16. Liang, X., Pedrycz, W.: Logic-based fuzzy networks: a study in system modeling with triangular norms and uninorms. Fuzzy Sets Syst. **160**, 3475–3502 (2009)
17. Mas, M., Mayor, G., Torrens, J.: t-Operators. Int. J. Uncertain. Fuzz Knowl.-Based Syst. **7**, 31–50 (1999)
18. Mas, M., Mayor, G., Torrens, J.: The distributivity condition for uninorms and t-operators. Fuzzy Sets Syst. **128**, 209–225 (2002)
19. Mitsch, H.: A natural partial order for semigroups. Proc. Am. Math. Soc. **97**, 384–388 (1986)
20. Saminger, S.: On ordinal sums of triangular norms on bounded lattices. Fuzzy Sets Syst. **157**, 1403–1416 (2006)
21. Xie, A., Liu, H.: On the distributivity of uninorms over nullnorms. Fuzzy Sets Syst. **211**, 62–72 (2013)

A Generalization of the Gravitational Search Algorithm

Humberto Bustince[1][(✉)], Maria Minárová[2], Javier Fernandez[1],
Mikel Sesma-Sara[1], Cedric Marco-Detchart[1], and Javier Ruiz-Aranguren[1]

[1] Public University of Navarra and Institute of Smart Cities,
Campus Arrosadia s/n, 31006 Pamplona, Spain
{bustince,fcojavier.fernandez,mikel.sesma,cedric.marco}@unavarra.es,
jruizaranguren@gmail.com
[2] Slovak University of Technology, Radlinského 11, 813 68 Bratislava 1, Slovakia
maria.minarova@stuba.sk

Abstract. In this work we propose a generalization of the gravitational search algorithm where the product in the expression of the gravitational attraction force is replaced by more general functions. We study some conditions which ensure convergence of our proposal and we show that we recover a wide class of aggregation functions to replace the product.

1 Introduction

The development of algorithms inspired in the behaviour of nature has a long history in artificial intelligence. In a short and non-exhaustive summary, we can recall evolutionary algorithms [1] or perceptron and neural networks [5].

Regarding physical laws, it is worth to mention that by the year 1977, Wright proposed an adaptation of Newton's Law of Gravitation for clustering problems [8]. More recently, in [7], an optimization algorithm which makes use of the gravitational law was proposed, in order to approach to the maximum (or the minimum) of a given function. The idea for this algorithm comes from two of the most relevant physical laws in the Newtonian framework. A simplified (and modified) form of the gravitational law, on the one hand, and the second law of dynamics, on the other hand. Roughly speaking, the Gravitational Search Algorithm (GSA) considers each possible solution of the optimization problem as a particle in a dynamical system which evolves under the sole action of a simplified gravitational attraction in a space of as many dimensions as the arity of the function to be minimized.

Besides, the gravitational law has also been used in image processing in the problem of edge detection. For instance, in [6] it is shown how the classical expression of the Gravitational Law can be modified in order to improve results of the edge detection algorithms.

In this work, we propose to replace the product in the classical expression of the gravitational attracting force in order to define a more general form of the Gravitational Search Algorithm. We mainly focus in the theoretical aspects of

© Springer International Publishing AG 2018
V. Torra et al. (eds.), *Aggregation Functions in Theory and in Practice*,
Advances in Intelligent Systems and Computing 581, DOI 10.1007/978-3-319-59306-7_17

these modified algorithms, providing some easy conditions to ensure convergence and analyzing what functions which are commonly used in the fuzzy framework can be considered to replace the product.

The structure of this paper is the following: In Sect. 2 we present some preliminary results and we discuss the original Gravitational Search Algorithm. In Sect. 3 we propose our modified version of the Gravitational Search Algorithm. In Sect. 4 we make some considerations about the convergence of the algorithm. We finish with some conclusions and references.

2   Preliminaries

2.1   Mathematical Concepts and Notations

We will denote by $\mathbf{X} = (x^1, \ldots, x^k)$ a vector in the Euclidean space $\mathbb{R}^k$. By $\|\mathbf{X}\|$, we denote the Euclidean L^2 norm of the vector $\mathbf{X}$, i.e.

$$\|\mathbf{X}\| = \sqrt{(x^1)^2 + \cdots + (x^k)^2}$$

When we say that a sequence of vectors $\{\mathbf{X}_n\}$ converges to some other vector $\mathbf{X}_0$, we mean that it does so in the Euclidean norm. Note that $\mathbf{X}_n \to \mathbf{X}_0$ as $n \to \infty$ if and only if $x_n^d \to x_0^d$ as $n \to \infty$ for every $d \in \{1, \ldots, n\}$.

2.2   The Gravitational Search Algorithm

We now briefly describe the original Gravitational Search Algorithm (GSA) [7]. The goal of the algorithm is to find an optimum (either a maximum or a minimum) of a given fitness function $fit : \mathbb{R}^n \to \mathbb{R}$. The search of the optimum is done in an iterative way. Assume that at some iteration t we have N particles; that is, N points $\mathbf{X}_1(t), \ldots, \mathbf{X}_N(t) \in \mathbb{R}^n$. Each of these particles represents a possible solution for our minimization problem. For each of these particles, a velocity vector $\mathbf{V}_i \in \mathbb{R}^n$ ($i \in \{1, \ldots, N\}$) is also provided.

In order to calculate the corresponding values for the iteration $t + 1$ from the values of iteration t, the GSA does as follows.

1. Assignation of masses to each of the points at iteration t.
    (i)   We evaluate the value of the function fit at each of the points $\mathbf{X}_i$. In particular, we denote:

$$best(t) = \min_{j \in \{1, \ldots, N\}} fit(\mathbf{X}_j(t))$$

$$worst(t) = \max_{j \in \{1, \ldots, N\}} fit(\mathbf{X}_j(t))$$

whenever we are minimizing the fitness function, and just opposite in case we are maximizing the fitness function.

(ii) We assign a relative mass $m_i(t)$ to the particle $\mathbf{X}_i(t)$ as follows.

$$m_i(t) = \frac{fit(\mathbf{X_i}(t)) - worst(t)}{best(t) - worst}$$

Note that in this way the particle(s) with the best fitness is (are) given a mass equal to 1, whereas the particle(s) with the worst fitness is (are) given a mass equal to 0.

(iii) We finally assign a mass M_i to the particle $\mathbf{X}_i(t)$ just normalizing the relative mass $m_i(t)$, i.e.,

$$M_i(t) = \frac{m_i(t)}{\sum_{j=1}^{N} m_j(t)}$$

Note that the particle(s) with the worst fitness get a mass equal to 0.

2. Calculation of gravitational forces at iteration t.

Following the gravitational law model, for each particle we are going to measure the effect of the "gravitational force" of all the other particles of the system acting over it. So, for every $i \in \{1, \dots, N\}$:

(i) We define the force $\mathbf{F}_{ij}(t)$ of particle j acting over particle i ($i \neq j$) as

$$\mathbf{F}_{ij}(t) = G(t) \frac{M_i(t) M_j(t)}{R_{ij}(t) + \varepsilon} (\mathbf{X}_j(t) - \mathbf{X}_i(t)) \tag{1}$$

where R_{ij} is the Euclidean (L^2) distance between $\mathbf{X}_j(t)$ and $\mathbf{X}_i(t)$ and $\varepsilon > 0$ is a small constant which is introduced to avoid possible indeterminacies. Observe that the distance is not squared, contrary to the case of the Newtonian Law of Gravitation. $G(t)$ is a positive constant which may vary from iteration to iteration. It is usually required that $\lim_{t \to \infty} G(t) = 0$.

(ii) We calculate the total force acting over particle $\mathbf{X}_i(t)$. To do so, we build a vector $(w_1, \dots, w_N) \in [0,1]^N$, where each w_j is a uniformly distributed random number in the interval $[0, 1]$. With this vector at hand, we define the total force acting over particle $\mathbf{X}_i(t)$ as

$$\mathbf{F}_i(t) = \sum_{j=1, j \neq i}^{n} w_j \mathbf{F}_{ij}(t).$$

Note that the random vector plays a significant role, and, in particular, it can make the effect of particles with small mass (and hence a bad fitness) very relevant.

(iii) Now we can calculate the corresponding acceleration $\mathbf{a}_i(t)$ of the particle $\mathbf{X}_i(t)$ as:

$$\mathbf{a}_i(t) = \frac{\mathbf{F}_i(t)}{M_i(t)}.$$

3. Calculation of the new positions
   (i) The new velocity of the particle $\mathbf{X}_i(t)$ is calculated as

$$\mathbf{V}_i(t+1) = p_i \mathbf{V}_i(t) + \mathbf{a}_i(t)$$

   where $p_i \in [0,1]$ is a random number calculated according to a uniform distribution of probability.
   (ii) Finally, the new position of the particle is calculated as

$$\mathbf{X}_i(t+1) = \mathbf{X}_i(t) + \mathbf{V}_i(t+1)$$

   (iii) We check the stopping criterion of the algorithm. If fulfilled, finish, otherwise go back to the step (1).

Particles and velocities for the first iteration are chosen randomly. The process is iterated until some stopping condition is satisfied.

3   A Modified Proposal of the GSA Algorithm

In this section, we propose a modification of the Gravitational Search Algorithm. In particular, our idea is to replace the product in Eq. (1) by another appropriate function and to analyse what is the effect of such a change in the behaviour of the algorithm. This approach is inspired by the studies carried out in [6], where, in the framework of edge detection, the replacement of the product by other, more general functions led to a significant improvement in the results.

So we are going to define the total force $F_i(t)$ acting over particle i within the tth iteration:

$$\mathbf{F}_{ij}(t) = G(t) \frac{H(M_i(t), M_j(t))}{R_{ij}(t) + \varepsilon} (\mathbf{X_i}(t) - \mathbf{X_j}(t)) \tag{2}$$

for some appropriate function $H : [0,1]^2 \to [0,1]$.
Then the total force acting over i^{th} particle is

$$\mathbf{F}_i(t) = \sum_{\substack{j=1 \\ i \neq j}}^{N} w_j \mathbf{F}_{ij}(t) \tag{3}$$

with $w_j \in [0,1]$ being random numbers specifying the influence of the jth particle. These numbers are drawn from a uniform distribution on $[0,1]$. Accordingly, herein the randomness reflects the fact that sometimes the particle with very good fitness has very small influence.

Analogously to the original GSA proceeding, see [7], the first approximation of MGSA pseudocode can be proposed:

1. Generate initial population in $t = 1$, of N various particles, $N \geq 2$, each one with its initial particular D dimensional position vector $\mathbf{X}_i(t)$. Take an initial value $G_0 > 0$ for $G(t)$.

2. Evaluate fitness for each particle $fit(\mathbf{X}_i(t)) : \mathbb{R}^n \to \mathbb{R}$
3. Update $G(t), best(t), worst(t)$. In principle, it is usually required that $G(t)$ is a decreasing functions such that $\lim_{t \to \infty} G(t) = 0$. For instance, we can assume that:

$$G(t) = G_0 \frac{1}{t} \tag{4}$$

Then, for each t, the following functions are defined:

$$best(t) = \min_{j \in \{1...,N\}} fit(\mathbf{X}_j(t)) \tag{5}$$

$$worst(t) = \max_{j \in \{1...,N\}} fit(\mathbf{X}_j(t)) \tag{6}$$

It is clear that for each $j : best(t) \le fit(\mathbf{X}_j(t)) \le worst(t)$.
If $|best(t) - worst(t)| < crit$ or $t \ge t_{max}$ then finish.
4. Being $best(t) < worst(t)$, which means **there exist at least two different values of fitness function**, we can establish mass $M_i(t)$ for each i:

$$m_i(t) = \frac{fit(\mathbf{X}_i(t)) - worst(t)}{best(t) - worst(t)} \tag{7}$$

$$M_i(t) = \frac{m_i(t)}{\sum_{j \in \{1...,N\}} m_j(t)} \tag{8}$$

It is apparent that in any $t : 0 \le m_i(t) \le 1, 0 \le M_i(t) \le 1$.
Moreover $\sum_{i \in \{1...,N\}} M_i(t) = 1$, which follows directly from its definition,

$$\sum_{i \in \{1...,N\}} M_i(t) = \sum_{i \in \{1...,N\}} \frac{m_i(t)}{\sum_{j \in \{1...,N\}} m_j(t)} = \frac{\sum_{i \in \{1...,N\}} m_i(t)}{\sum_{j \in \{1...,N\}} m_j(t)} = 1 \tag{9}$$

5. Calculate the gravitational force in t^{th} ITERATION:
For every $i \in \{1, \ldots, N\}$:
(i) The force $\mathbf{F}_{ij}(t)$ of particle j acting over particle i ($i \ne j$) is

$$\mathbf{F}_{ij}(t) = G(t) \frac{H(M_i(t)M_j(t))}{R_{ij}(t) + \varepsilon} (\mathbf{X}_j(t) - \mathbf{X}_i(t)) \tag{10}$$

and, the total force acting over particle $\mathbf{X}_i(t)$ is

$$\mathbf{F}_i(t) = \sum_{j=1, j \ne i}^{n} w_j \mathbf{F}_{ij}(t).$$

with $(w_1, \ldots, w_N) \in [0; 1]^N$ being a vector, where each w_j is a uniformly distributed random number in the interval $[0, 1]$.

(ii) Hence, the corresponding acceleration $\mathbf{a}_i(t)$ of the particle $\mathbf{X}_i(t)$ is:

$$\mathbf{a}_i(t) = \frac{\mathbf{F}_i(t)}{M_i(t)}.$$

6. Calculate the new positions.
   (i) The new velocity of the particle $\mathbf{X}_i(t)$ is calculated as

   $$\mathbf{V}_i(t+1) = p_i \mathbf{V}_i(t) + \mathbf{a}_i(t)$$

   where $p_i \in [0, 1]$ is a random number calculated according to a uniform distribution of probability.
   (ii) Finally, the new position of the particle is calculated as

   $$\mathbf{X}_i(t+1) = \mathbf{X}_i(t) + \mathbf{V}_i(t+1)$$

   (iii) Go back to the step (2)

4 Study of the Convergence of the Generalized Gravitational Search Algorithm

Mathematically, both the gravitational search algorithm and the modified version can be seen as two instances of dynamical systems. The particle i at iteration $t+1$ is obtained from particle i at iteration t by means of the expression:

$$\mathbf{X}_i(t+1) = \mathbf{X}_i(t) + \mathbf{V}_i(t+1)$$

Taking into account the expression for the velocity, this identity is the same as

$$\mathbf{X}_i(t+1) = \mathbf{X}_i(t) + p_i(t)\mathbf{V}_i(t) + \mathbf{a}_i(t) \tag{11}$$

So, from standard analysis, we can provide the following lemma.

Lemma 1. *The sequence $\{\mathbf{X}_i(t)\}_t$ converges (in the usual Euclidean metric) if and only if*

$$\lim_{t \to \infty} p_i(t)\mathbf{V}_i(t) + \mathbf{a}_i(t) = \mathbf{0}$$

Proof. Rewriting Eq. 11 as

$$\mathbf{X}_i(t+1) - \mathbf{X}_i(t) = p_i(t)\mathbf{V}_i(t) + \mathbf{a}_i(t)$$

the result follows from the fact that $\{\mathbf{X}_i(t)\}_t$ converges if and only if it is a Cauchy sequence.

With this Lemma at hand, we can provide the following general convergence theorem.

Theorem 1. *The sequence $\{\mathbf{X}_i(t)\}_t$ converges if and only if $\mathbf{a}_i(t) \to \mathbf{0}$ as $t \to \infty$.*

Proof. Assume first that $\mathbf{a}_i(t) \to \mathbf{0}$ as $t \to \infty$. In this setting, since

$$\mathbf{V}_i(t+1) - p_i(t)\mathbf{V}_i(t) = \mathbf{a}_i(t)$$

it follows that

$$\|\mathbf{V}_i(t)\| \to 0$$

and from Lemma 1, we get convergence.

Conversely, assume that $\mathbf{a}_i(t)$ does not converge to $\mathbf{0}$ as $t \to \infty$. Note that if the sequence $\{\mathbf{X}_i(t)\}_t$ converges, it must also hold that $\mathbf{V}_i(t) \to \mathbf{0}$. But from

$$\mathbf{a}_i(t) = \mathbf{V}_i(t+1) - p_i(t)\mathbf{V}_i(t)$$

if $\mathbf{a}_i(t)$ does not converge, $\mathbf{V}_i(t)$ can not converge, either.

Taking into account these results, we also get a convergence criterion in terms of the velocity which was already discussed in [3].

Corollary 1. *Assume that* $\lim_{t\to\infty} p_i(t) \neq 1$. *The sequence* $\{\mathbf{X}_i(t)\}_t$ *converges if and only if* $\mathbf{V}_i(t) \to \mathbf{0}$ *as* $t \to \infty$.

Proof. It is straight from the previous theorem.

Remark 1. From a physical point of view, each acceleration component magnitude is going down to zero, hence the entire system is tending to the steady state.

So in order to analyse convergence, we must study the behaviour of acceleration along the time.

Using this lemma, we can state the following result.

Theorem 2. *The sequence* $\{\mathbf{X}_i(t)\}_t$ *converges if and only if*

$$\lim_{t\to\infty} G(t) \sum_{i\neq j} w_j \frac{1}{M_i(t)} \frac{H(M_i(t), M_j(t))}{R_{ij}(t) + \varepsilon}(x_j^d(t) - x_i^d(t)) = 0 \qquad (12)$$

for every $i \in \{1, \ldots, n\}$. *In the case of* $M_i(t) = 0$ *we take* $H(Mi, Mj)/Mi = 0$ *due to fact that the particle with worst fitness has no influence within the* t^{th} *iteration.*

Proof. It is just necessary to rewrite the acceleration in terms of the masses.

We recall now the following well-known result that will be useful for our subsequent estimations.

Lemma 2. *For every* $\mathbf{X} = (x^1, \ldots, x^N) \in \mathbb{R}^n$ *and for every* $d \in \{1, \ldots, n\}$, *the following inequality holds.*

$$\frac{|x^d|}{\|\mathbf{X}\|} \leq 1 \qquad (13)$$

Proof. It follows from a straightforward calculation.

Taken into account this Lemma, we get the following Corollary.

Corollary 2. *The sequence* $\{\mathbf{X}_i(t)\}_t$ *converges if*

$$\lim_{t\to\infty} G(t) \sum_{i\neq j} \frac{1}{M_i(t)} H\big(M_i(t), M_j(t)\big) = 0 \tag{14}$$

for every $i \in \{1,\ldots,n\}$, *having* $H(M_i(t), M_j(t))/M_i(t) = 0$ *in case of* $M_i(t) = 0$.

Proof. First of all, note that the dth component of the acceleration vector for the ith particle is given by

$$a_i^d(t) = \frac{F_i^d(t)}{M_i(t)} = \frac{1}{M_i(t)} \sum_{\substack{j\in\{1\ldots,N\}\\i\neq j}} w_j G(t) \frac{H\big(M_i(t), M_j(t)\big)}{R_{ij}(t) + \varepsilon} (x_j^d(t) - x_i^d(t)) \tag{15}$$

Then, from Lemma 2, we see that

$$|a_i^d(t)| \leq G(t) \sum_{\substack{j\in\{1\ldots,N\}\\i\neq j}} \frac{1}{M_i(t)} H\big(M_i(t), M_j(t)\big). \tag{16}$$

So, if the right hand side term of (16) tends to zero as t goes to infinity, so does the acceleration.

Remark 2. 1. Note that as soon as the function $\frac{H(x,y)}{x}$ is bounded, if we assume that $G(t) \to 0$ as $t \to \infty$ and since we are dealing with a finite number of particles, we get convergence. This is in particular the case of the original gravitational search algorithm, as, from $H(x,y) = xy$ it follows that

$$\frac{H(x,y)}{x} = \begin{cases} y & \text{if } xy > 0 \\ 0 & \text{in other case.} \end{cases}$$

2. However, even if $\frac{H(x,y)}{x}$ is not bounded, we may also have convergence of the algorithm.

We are now going to get an easier to handle condition for convergence. From now on, and to make the text more easily readable, we will write the shortened forms m_i, M_i, etc. instead of $m_i(t), M_i(t)$, etc., whenever possible.

First of all, note that the condition $\sum_{\substack{j\in K\\i\neq j}} \frac{1}{M_i} H\big(M_i, M_j\big) \leq 1$ from (16) can be rewritten as

$$\sum_{i\neq j} H(M_i, M_j) \leq M_i \tag{17}$$

However, we can also expect that

$$\sum_{j\neq i} H(M_j, M_i) \leq M_j \tag{18}$$

In order to ensure the fulfillment of (17) and (18) simultaneously, we require that the strictest of both inequalities holds. So in next consideration we can just put $min\{M_i, M_j\}$ on the right hand side of (18). So in the following consideration, without loss of generality, we can assume $M_i \leq M_j$ (17).

Lemma 3. *For every function $H \in \mathcal{M}$ and for every $M_i \in [0,1[$, the following statements are equivalent:*

1. $\sum_{i \neq j} H(M_i, M_j) \leq M_i$.
2. $\sum_{i \neq j} H(M_i, M_j) \leq \sum_{i \neq j} M_j \frac{M_i}{1 - M_i}$.

Proof. It is enough, taking into account (9), to observe the following:

$$M_i = M_i \sum_{\substack{j \in K}} M_j = \sum_{\substack{j \in K \\ i \neq j}} M_i M_j + M_i^2 = \sum_{\substack{j \in K \\ i \neq j}} M_i M_j + \sum_{\substack{j \in K \\ i \neq j}} M_i^2 M_j + \cdots + \sum_{\substack{j \in K \\ i \neq j}} M_i^m M_j + \cdots$$

$$= \sum_{\substack{j \in K \\ i \neq j}} M_j \sum_{m=1}^{\infty} M_i^m = \sum_{\substack{j \in K \\ i \neq j}} M_j \frac{M_i}{1 - M_i},$$

$$(19)$$

so the result follows.

Taking into account our previous discussion, we can provide the following convergence criterion for the MGSA, whose proof is now straightforward.

Theorem 3. *Let $H \in \mathcal{M}$. Then, if*

$$H(x, y) \leq \frac{xy}{1 - \min(x, y)} \tag{20}$$

for every $x, y \in [0, 1]$, the MGSA converges.

With this criterion, many aggregation functions can be used to replace the product. For instance.

Proposition 1. *Let T be a t-norm [4]. If we take $H = T$ in the MGSA, then the algorithm converges.*

Proof. It is enough to note that the minimum satisfies Eq. 20 and, for every other t-norm $T, T(x, y) \leq \min(x, y)$ for every $x, y \in [0, 1]$.

5   Conclusions

In this work we have presented the first discussion of a generalization of the Gravitational Search Algorithm. Although this study is still preliminary, it has already shown that the product can be replaced by a large class of aggregation functions, including t-norms. In future works, we intend to expand this analysis to get full convergence criteria which allow us to identify every possible aggregation function to be used. We will also develop an experimental analysis of the new functions to be considered.

Acknowledgements. This work was supported by Spanish Research Project TIN-77356-P (AEI/FEDER, UE) and by projects APVV-14-0013 and VEGA-1/0420/15.

References

1. Ashlock, D.: Evolutionary Computation for Modeling and Optimization. Springer, New York (2005)
2. Bustince, H., Fernandez, J., Mesiar, R., Montero, J., Orduna, R.: Overlap functions. Nonlinear Anal. Theory Methods Appl. **72**, 1488–1499 (2010)
3. Ghorbani, F., Neamabadi-Pour, H.: On the convergence analysis of gravitational search algorithm. J. Adv. Comput. Res. **3**(2), 45–51 (2012)
4. Grabisch, M., Marichal, J.-L., Mesiar, R., Pap, E.: Aggregation Functions. Cambridge University Press, Cambridge (2009)
5. Haykin, S.: Neural Networks: A Comprehensive Foundation. Prentice Hall, Upper Saddle River (1999)
6. Lopez-Molina, C., Bustince, H., Fernandez, J., Couto, P., De Baets, B.: A gravitational approach to edge detection based on triangular norms. Pattern Recogn. **43**, 3730–3741 (2010)
7. Rashedi, E., Neamabadi-Pour, H., Sariazdi, S.: GSA: a gravitational search algorithm. Inf. Sci. **179**(13), 2232–2248 (2009)
8. Wright, W.E.: Gravitational clustering. Pattern Recogn. **9**, 151–166 (1977)

On Stability of Families for Improper Aggregation Operators

Pablo Olaso[1([⊠])], Karina Rojas[2], Daniel Gómez[3], and Javier Montero[2]

[1] Facultad de Ciencias Economicas y Empresariales,
Universidad Complutense de Madrid, Campus Somosaguas,
Pozuelo de Alarcon, Spain
polasore@ucm.es
[2] Facultad de Ciencias Matematicas, Universidad Complutense de Madrid,
Plaza de Ciencias 3, Ciudad Universitaria, Madrid, Spain
{krpatuelli,monty}@mat.ucm.es
[3] Facultad de Estadistica, Universidad Complutense de Madrid,
Avda. Puerta de Hierro s/n, Ciudad Universitaria, Madrid, Spain
dagomez@estad.ucm.es

Abstract. This work extends the notion of consistency in terms of stability for *Families of Aggregation Operators* (*FAO*), as defined in previous works. The notion of stability proposed in this work, not only extends the previous one, but it can be applied to a wider set of *FAOs*, particularly, to those that we name here as *Family of Improper Aggregation Operators* (*FIAO*), or improper *FAOs*. When the aggregated value cannot be considered as a new item from the input, the present definition of consistency cannot be applied. This is usual in several areas, namely in the development of social, economic and political indexes, as far as the aggregation process typically yield a new and different concept from the input elements.

1   Introduction

An aggregation operator is usually defined as a real function $A : [0,1]^n \to [0,1]$, such that for n items in $[0,1]$, yields an aggregation value in the same interval [1,2,8,9,13]. This definition can be extended (see for example [18]) by considering a *Family of Aggregation Operators* $\{A_n\}_{n \in N}$, which aggregates a collection of items of any length n. It is also referred to as extended aggregation function by other authors [2,16]. Many properties have been studied in relation to single aggregation operators A_n, in contrast, few efforts have been dedicated to study the relations between these operators as members of a family of aggregation functions. As it has been pointed out in some previous works [1,8–10], most commonly assumed properties (e.g.continuity) represent desirable features related to each aggregation function A_n, but they do not imply the consistency of the FAO as a whole. In this sense, no relation is being imposed among the members of a given family of operators. In this context, in the mentioned work [18] it was studied the relation that must exist between $\{A_n\}$ and $\{A_m\}$ in order

© Springer International Publishing AG 2018
V. Torra et al. (eds.), *Aggregation Functions in Theory and in Practice*,
Advances in Intelligent Systems and Computing 581, DOI 10.1007/978-3-319-59306-7_18

to ensure consistency when the input cardinality changes (from n data to m). Thus, some properties of consistency among operators from the same FAO are defined, that guarantee a robust process of aggregation under such cardinality changes (e.g. missing data situations). This concept of consistency in $FAOs$ was named *Stability*.

In [18], the stability of some of the most commonly used $FAOs$ has been studied (minimum, maximum, median, arithmetic mean, geometric mean, harmonic mean, owa, weighted mean and product). It was also considered the structure of the input, widening the scope of the notion of stability for cases of unstructured data, lineally structured and hierarchically structured data [15, 19]. In addition, for each of those kind of input data new definitions of stability were introduced, each one less restrictive, starting with the strict stability, followed by the *asymptotically strict stability*, the *almost sure strict stability* and, finally, the definition of *instability* [14, 17, 18].

Stability in $FAOs$ relays in the concept of continuity, as small input changes should imply small output changes. In particular, for a collection of n items $(x_1, \ldots, x_n)$, when a new item x_{n+1} is added, such that x_{n+1} is *close* to the aggregation of the previous n items $A_n(x_1, \ldots, x_n)$, then $A_{n+1}(x_1, \ldots, x_n, x_{n+1})$ should be either *close* to $A_n(x_1, \ldots, x_n)$. Naturally arises the need to study the symmetry of the $FAOs$, since for those nonsymmetric operators, the position of x_{n+1} is relevant for the result of the process. Notice that from this point of view, the notion of *Self-identity* from Yager [21] can be regarded as a particular case of the stability property for symmetric $FAOs$. Stability property has been used in different problems as weights determination for weighted average mean [3]; data missing problems [4, 11, 12, 20] or index [19] among many others.

Nevertheless, this stability definition cannot be applied to any FAO since classical definition assumes that the output of the aggregation process $A_n(x_1, \ldots, x_n)$ can be considered as a new input. The stability equation establishes that $A_{n+1}(x_1, \ldots, x_n, A_n(x_1, \ldots, x_n)) = A_n(x_1, \ldots, x_n)$. But, does this equation always make sense? In some contexts this could be weird. Although from a mathematical point of view all values are in $[0, 1]$, their meaning could be quite different.

Based on this idea, we define the concept of proper aggregation function as an aggregation function ϕ_n in which the meaning of value $\phi_n(x_1, \ldots, x_n)$ could be consider as a new input in the aggregation process (as for example the mean, the minimum or maximum). After that, and aiming to extend the notions of stability for any class of aggregation functions, in this work we propose a new definition of stability that follows the same ideas of the original definition.

Moreover, the definition of stability proposed in this work is richer and, in some cases, more restrictive than the previous one, as we shall see below.

2   Preliminary

In this section, the most important stability properties for $FAOs$ are reminded, for more about them, we refer the reader to [18].

2.1   Strict Stability for *FAOs*

Definition 1. Let $\{A_n : [0,1]^n \to [0,1], n \in N\}$ be a family of aggregation operators. Then, it is said that:

1. $\{A_n\}_n$ is an R-strictly stable family if

$$A_n(x_1, x_2, \ldots, x_{n-1}, A_{n-1}(x_1, x_2, \ldots, x_{n-1})) = A_{n-1}(x_1, x_2, \ldots, x_{n-1}) \quad (1)$$

   holds $\forall n \geq 3$ and $\forall \{x_n\}_{n \in N}$ in $[0,1]$

2. $\{A_n\}_n$ is an L-strictly stable family if

$$A_n(A_{n-1}(x_1, x_2, \ldots, x_{n-1}), x_1, x_2, \ldots, x_{n-1}) = A_{n-1}(x_1, x_2, \ldots, x_{n-1}) \quad (2)$$

   holds $\forall n \geq 3$ and $\forall \{x_n\}_{n \in N}$ in $[0,1]$

3. $\{A_n\}_n$ is an RL-strictly stable family if both properties hold simultaneously.

Let us observe that the notion of self-identity as defined in [21] is the first part of this definition of stability. If the *FAO* is symmetric, both *L-* (Yager's self-identity) and *R-stability* definitions, and consequently, *LR-stability*, are equivalent. But this is not true in general if such symmetry is not satisfied.

For example, let us analyze self-identity in the *backward inductive extension* $\{A_n^b\}_{n \in N}$ and *forward inductive extension* $\{A_n^f\}_{n \in N}$ [3] of any binary aggregation operator, defined for $n > 2$ as $A_n^b = L_2(x_1, L_2(\ldots, L_2(x_{n-1}, x_n) \ldots)$ *for* $n > 2$, and $A_n^f = L_2(\ldots, (L_2(L_2(x_1, x_2), x_3)), \ldots, x_n)$, for $n > 2$, where L_2 is a binary aggregation operator, i.e. $L_2 : [0,1]^2 \to [0,1]$.

It can be proven that the family of aggregation functions $\{A_n^f\}_{n \in N}$ satisfies self-identity if L_2 is idempotent, i.e., $A_n(x, \ldots, x) = x$, for all $n \in N$ and $x \in [0,1]$ (see also [3]). Nevertheless, the family $\{A_n^b\}_{n \in N}$ does not satisfy self-identity since the order in which this family aggregates the information is inverse (i.e. from right to left). In our opinion, the family $A_n^b = L_2(x_1, L_2(\ldots, L_2(x_{n-1}, x_n)))$ *for* $n > 2$ should be consistent in the sense of stability when the information is aggregated from right to left.

Moreover, it is important to note that in some cases, the data should be introduced in the k-th position instead of just left or right. For instance, stability can be studied for an aggregation process in which there are missing values at any position (see [5–7, 15]).

3   A Generalization of Stability for Any FAOs

As mentioned above, the notion of consistency of an aggregation process has been introduced in previous works based on the property of stability of a *FAO*, a property that has been studied for different data input structures, and also less restrictive definitions have been introduced, be it strict, asymptotic and in probability. However, such stability is based on the equivalency for a single value, it does not take into account a range of neighboring values. Besides, it cannot be used in any situation, for there are problems in which there is no sense in feeding the input data with the aggregated output since both have completely different meanings.

Therefore, a distinction has to be established between both kinds of families that has to be concerned with their use rather than their mathematical properties. In this sense, let X be a data set and a FAO $\{A_n\}$, both involved in an aggregation process, if the aggregated value can be regarded as a new element of the original set X, we will say that $\{A_n\}$ is a *Family of Proper Aggregation Operators* ($FPAO$) or a proper FAO. Likewise, we will say that $\{A_n\}$ is a *Family of Improper Aggregation Operators* ($FIAO$) or an improper FAO if it is not a Proper FAO. Notice that any FAO may be proper or improper depending on the context they are being used. What we want to stress here is that in many practical problems the need to consider Improper $FAOs$ arises.

Taking this into account, the consistency properties defined in [18] only apply to proper $FAOs$, since for an improper FAO $\{A_n : [0,1]^n \longrightarrow [0,1], n \in N\}$ the expression $A_{n+1}(x_1, \ldots, x_n, A_n(x_1, \ldots, x_n))$ has no sense. The concept must then be reformulated, and to do that let us retake the consistency principle that guided the previous definition in order to ensure a robust aggregation process: "*small input changes must yield small output changes*". Let us begin for what "small changes" mean.

Previous idea of stability establish that if the new input that has to be aggregate coincides with the aggregation of the previous ones (i.e. $x_{n+1} = A_n(x_1, \ldots, x_n)$), then the aggregation of $n+1$ items $A_{n+1}(x_1, \ldots, x_n, A_n)$ should coincide with A_n. Since the expression $A_{n+1}(x_1, \ldots, x_n, A_n)$ in general has not sense, the new stability can be reformulated as follows: if x_{n+1} is close to $x_1, \ldots, x_n$, then A_{n+1} should be close to A_n. The concept of proximity or closeness between a point and a set of points is widely used in clustering techniques under the name of similarity or, equivalently, dissimilarity. So, given a value $x_k \in [0,1]$ and a data set $X = \{x_1, \ldots, x_n\}$ also in $[0,1]$, let us denote by $\mathscr{D}_n$ the set of dissimilarity measures $\mathscr{D}_n : [0,1]x[0,1]^n \longrightarrow [0,1]$ for any n. Therefore, given x_a and x_b in $[0,1]$, let us say that x_a is closer to X than x_b with respect to $\mathscr{D}$, if $\mathscr{D}_{|X|}(x_a, X) < \mathscr{D}_{|X|}(x_b, X)$.

We will see more about dissimilarities below, but now we are able to introduce a new definition of stability more general than strict stability as defined so far, and based on the principle of continuity. This new definition should be of applicability to any kind of FAO, including the improper $FAOs$, and should "contain" the previous concept of stability at least in the sense that the latter implies the former for some particular dissimilarity measures. This definition is based in the principle of consistency as has been presented above: given a data set and the aggregation of its elements, the process of aggregating a new value outside the given set but close enough to it should yield a value close itself to the previous aggregation. The definition is as follows:

Definition 2. Let $\mathscr{D}_n$ be the set of dissimilarity measures $\mathscr{D}_n : [0,1]x[0,1]^n \longrightarrow [0,1]$ for any n, and let X be a set such that $|X| = n$ and $X \subset [0,1]$, and $\{A_n : [0,1]^n \mapsto [0,1], n \in N\}$ be a FAO. Then we will say that:

- $\{A_n\}$ is an *R-Stable family with respect to* $\mathscr{D}_n$ if: $\forall \varepsilon > 0, \exists \delta > 0 / \mathscr{D}_{|X|}(x_a, X) \le \delta \rightarrow |A_{n+1}(X, x_a) - A_n(X)| \le \varepsilon$
- $\{A_n\}$ is an *L-stable family with respect to* $\mathscr{D}_n$ if: $\forall \varepsilon > 0, \exists \delta > 0 / \mathscr{D}_{|X|}(x_a, X) \le \delta \rightarrow |A_{n+1}(x_a, X) - A_n(X)| \le \varepsilon$
- $\{A_n\}$ is an *LR-stable family with respect to* $\mathscr{D}_n$ if both previous properties hold.

Let us remark that, for a symmetric FAO $\{A_n : [0,1]^n \mapsto [0,1], n \in N\}$, $A_{n+1}(X, x) = A_{n+1}(x_1, \ldots, x_n, x) = A_{n+1}(x, x_1, \ldots, x_n) = A_{n+1}(x_1, \ldots, x_k, x, x_{k+1}, \ldots, x_n)$ and, hence, it is enough to prove, e.g., *R-stability*, to conclude that *L-stability* and *LR-stability* also hold.

Given this new definition of stability, it is of interest studying what is the relation with the strict stability defined for proper $FAOs$ in [18].

We have not defined so far the dissimilarity measures $\mathscr{D}$ in Definition 2, all we know is that they depend eventually on a family of aggregation operators. They could actually be defined in several ways, let us distinguish the following two cases:

- Let $\{\mathscr{D}A_n\}$ be the "distance" from x_a to the aggregation of the elements of X, i.e. $\mathscr{D}A_n(x_a, X) = |x_a - A_n(X)|$
- Let $\{A\mathscr{D}_n\}$ be the aggregation of the "distances" from x_a to the each of the elements of $X = \{x_1, \ldots, x_n\}$, i.e. $A\mathscr{D}_n(x_a, X) = A_n(|xa - x_1|, \ldots, |xa - x_n|)$

It is clear that, for the mean, this both definitions are equal, while in general they are unequal for any given FAO $\{A_n\}_n$. We will see too that the first one can be applied to the study of proper $FAOs$ and it is actually the concept used in the previous studies, while the second one is of applicability for proper and improper $FAOs$ and is actually more intuitive and widely used in real clustering techniques. For instance let us recall the known single linkage clustering, complete linkage clustering and group average clustering.

Now, let us see that the definition presented above implies, in some circumstances, the previous concept of strict stability:

Proposition 1. *Let the dissimilarity measure given by* $\mathscr{D}A_n(x_a, X) = |x_a - A_n(X)|$:

- *If* $\{A_n\}$ *is a family* R-stable *with respect to* $\mathscr{D}A_n$, *then* $\{A_n\}$ *is* R-strictly stable.
- *If* $\{A_n\}$ *is a family* L-stable *with respect to* $\mathscr{D}A_n$, *then* $\{A_n\}$ *is* L-strictly stable.
- *If* $\{A_n\}$ *is a family* LR-stable *with respect to* $\mathscr{D}A_n$, *then* $\{A_n\}$ *is* LR-strictly stable.

Proof. Let be $\{A_n\}$ *R-stable with respect to* $\mathscr{D}A_n$, then $\forall X = \{x_1, \ldots, x_n\}$ and $\forall \varepsilon > 0, \exists \delta > 0$:

$$\text{If } \mathscr{D}A_n x_a, X = |x_a - A_n(X)| \le \delta \text{ then } |A_{(n+1)}(X, x_a) - A_n(X)| \le \varepsilon$$

In particular, for $x_a = A_n(X)$:

$\forall \delta, |x_a - A_n(X)| = 0 < \delta$, therefore $\forall \varepsilon > 0 |A_{(n+1)}(X, x_a) - A_n(X)| \leq \varepsilon$, and since this holds $\forall \varepsilon > 0$ no matter how close to zero it is, then necessarily

$$|A_{(n+1)}(X, x_a) - A_n(X)| = 0$$

Therefore, $\{An\}$ is R-strictly stable.   □

The proof is similar for L-strict stability and hence for LR-strict stability.

Let us insist that this proposition relates the new definition of stability with the previous of strict stability, i.e. a concept that can only be applied to proper $FAOs$ as already has been said, so notice that it is not surprising that the dissimilarity measure used in the proposition is, among both defined above, $\mathscr{D}A_n$, that is, the one that can be applied to such $FPAOs$. Indeed, if $A_n(X)$ cannot be regarded as an element comparable to those belonging to X, then it has no sense to compare the new element x_a with it, as in $\mathscr{D}A_n(x_a, X) = |x_a - A_n(X)|$.

In [18] the strict stability properties were studied for the most known and used $FAOs$ (minimum, maximum, median, arithmetic mean, geometric mean, harmonic mean, owa, weighted mean and product).

Such $FAOs$ classification with respect to their strict stability level is based on the dissimilarity definition given by $\mathscr{D}A_n(x_a, X) = |x_a - A_n(X)|$ as has been shown above. It is possible to use the dissimilarity measure $A\mathscr{D}_n$ instead, referred to the aggregation of the distances using A_n, that is, $A\mathscr{D}_n(x, X) =_n$ $(|x - x_1|, |x - x_2|, \ldots, |x - x_n|)$. This distinction makes the difference between the study of stability among proper $FAOs$ (since as has been said the former dissimilarity measure can only be applied to such $FAOs$) and improper $FAOs$. However, what follows can be applied to both proper and improper $FAOs$. Moreover, since the analysis cannot be made here for any kind of measure as it would be unmanageable, three types have been chosen to do this, based on three well known and widely used clustering techniques, namely:

- single linkage clustering (where $_n = \min_n$ and hence $A\mathscr{D}_n(x, X) = \mathscr{D}^{\min_n}(x, X) = \min_n(|x - x_1|, |x - x_2|, \ldots, |x - x_n|)$),
- complete linkage clustering (where $A_n = \max_n$ and hence $A\mathscr{D}_n(x, X) = \mathscr{D}^{\max_n}(x, X) = \max_n(|x - x_1|, |x - x_2|, \ldots, |x - x_n|)$),
- group average clustering (where $A_n = mean_n(X) = \mu_n$ and hence $A\mathscr{D}_n(x_k, X) = \mathscr{D}^{\mu_n}(x, X) = \frac{\sum_{i=1}^{n} |x - x_i|}{n}$)

Using these three dissimilarity measures, let us study stability for the FAO of the minimum. Notice that, since $\{\min_n\}_{n \in N}$ is symmetric, it is enough to prove R-stability.

Proposition 2. The FAO $\{\min_n\}_{n \in N}$ is LR-stable wrt $\mathscr{D}^{\min_n}(x, X) = \min_n(|x - x_1|, |x - x_2|, \ldots, |x - x_n|)$.

Proof. By definition, $\{\min_n\}_{n \in N}$ is R-stable wrt $\mathscr{D}^{\min_n}$ iff $\forall X = x_1, \ldots, x_n$ and $\forall \varepsilon > 0, \delta > 0$:

$\mathscr{D}^{\min_n}(x, X) \leq \delta \Rightarrow |\min_{n+1}(x, X) - \min_n(X)| \leq \varepsilon$.

Hence, let $X = \{x_1, \ldots, x_n\}$ and $\varepsilon > 0$, and let $\delta = \varepsilon$, then $\forall x / \mathscr{D}^{\min_n}(x, X) \leq \delta$ there are two possible situations:

- If $x \geq \min(X) = x_{(1)}$ then $|\min_{n+1}(x, X) - \min_n(X)| = |x_{(1)} - x_{(1)}| = 0 \leq \varepsilon$;

- If $x < x_{(1)}$ then $|\min_{n+1}(x, X) - \min_n(X)| = |x - x_{(1)}|$ and since $|x - x_{(1)}| = \min_n(|x - x_1|, |x - x_2|, \ldots, |x - x_n|) = \mathscr{D}^{\min_n}(x, X) = \leq \delta = \varepsilon$. $\qquad \square$

Proposition 3. *The FAO* $\{\min_n\}_{n \in N}$ *is LR-stable wrt* $\mathscr{D}^{\max_n}(x, X) = \max_n(|x - x_1|, |x - x_2|, \ldots, |x - x_n|)$.

Proof. By definition, $\{\min_n\}_{n \in N}$ is *R-stable wrt* $\mathscr{D}^{\max_n}$ iff $\forall X = x_1, \ldots, x_n$ and $\forall \varepsilon > 0, \delta > 0$:
$\mathscr{D}^{\max_n}(x, X) \leq \delta \Rightarrow |\min_{n+1}(x, X) - \min_n(X)| \leq \varepsilon$.

Hence, let $X = \{x_1, \ldots, x_n\}$ and $\varepsilon > 0$, and let $\delta = |\max(X) - \min(x)| + \varepsilon = |x_{(n)} - x_{(1)}| + \varepsilon$, then $\forall x \, / \, \mathscr{D}^{\max_n}(x, X) \leq \delta$ there are two possible situations:

- If $x \geq \min(X) = x_{(1)}$ then $|\min_{n+1}(x, X) - \min_n(X)| = |x_{(1)} - x_{(1)}| = 0 \leq \varepsilon$
- If $x < x_{(1)}$ then $|\min_{n+1}(x, X) - \min_n(X)| = |x - x_{(1)}|$ and since $|x - x_{(1)}| = \max_n(|x - x_1|, |x - x_2|, \ldots, |x - x_n|) - |x_{(n)} - x_{(1)}| = \mathscr{D}^{\max_n}(x, X) - |x_{(n)} - x_{(1)}| \leq \delta - |x_{(n)} - x_{(1)}| = |x_{(n)} - x_{(1)}| + \varepsilon - |x_{(n)} - x_{(1)}| = \varepsilon$, i.e. $|x - x_{(1)}| \leq \varepsilon$

$\qquad \square$

Proposition 4. *The FAO* $\{\min_n\}_{n \in N}$ *is LR-stable wrt* $\mathscr{D}^{\mu_n}(x, X) =_n (|x - x_1|, |x - x_2|, \ldots, |x - x_n|) = \frac{\sum_{i=1}^n |x - x_i|}{n}$.

Proof. By definition, $\{\min_n\}_{n \in N}$ is *R-stable wrt* $\mathscr{D}^{\mu_n}$ iff $\forall X = x_1, \ldots, x_n$ and $\forall \varepsilon > 0, \delta > 0$:
$\mathscr{D}^{\mu_n}(x, X) \leq \delta \Rightarrow |\min_{n+1}(x, X) - \min_n(X)| \leq \varepsilon$.

Hence, let $X = \{x_1, \ldots, x_n\}$ and $\varepsilon > 0$, and let $\delta = \varepsilon + |\mu_n - x_{(1)}|$, then $\forall x \, / \, \mathscr{D}^{\mu_n}(x, X) = |x - \mu_n| \leq \delta$ there are two possible situations:

- If $x \geq \min(X) = x_{(1)}$ then $|\min_{n+1}(x, X) - \min_n(X)| = |x_{(1)} - x_{(1)}| = 0 \leq \varepsilon$
- If $x < x_{(1)}$ then $|\min_{n+1}(x, X) - \min_n(X)| = |x - x_{(1)}| = x_{(1)} - x$, notice too that, in this case, $\mathscr{D}^{\mu_n}(x, X) = \frac{\sum_{i=1}^n |x - x_i|}{n} = |x - \frac{\sum_{i=1}^n x_i}{n}| = |x - \mu_n| = \mathscr{D}A_n(x, X)$, and since $\mu_n \geq x_{(1)}$, then $|x - x_{(1)}| = |x - \mu_n| - |\mu_n - x_{(1)}| \leq \delta + |\mu_n - x_{(1)}| = \varepsilon + |\mu_n - x_{(1)}| - |\mu_n - x_{(1)}| = \varepsilon$ $\qquad \square$

4   Final Remarks

The new stability property defined in this work for $FAOs$ is more flexible than the strict stability defined in previous works. It can be applied to proper and improper $FAOs$ and different similarity measures can be used, widening the field of research and applications. As we have seen here, for instance, even the FAO $\{\min_n\}_{n \in N}$, that was already regarded as an stable family, can be considered as a stable family for several dissimilarity measures, while other families may not. Our definition is closer to the principle of continuity and meanwhile it covers the previous concept of strict stability. Moreover, the new concept introduced can include other Families of aggregation functions that the previous concept did not reach, and different lines of research and applications remain open.

Acknowledgment. This research has been partially supported by the Government of Spain, grant TIN2015-66471.

References

1. Amo, A., Montero, J., Molina, E.: Representation of consistent recursive rules. Eur. J. Oper. Res. **130**, 29–53 (2001)
2. Calvo, T., Mayor, G., Torrens, J., Suner, J., Mas, M., Carbonell, M.: Generation of weighting triangles associated with aggregation fuctions. Int. J. Uncertain. Fuzziness Knowl.-Based Syst. **8**(4), 417–451 (2000)
3. Calvo, T., Kolesarova, A., Komornikova, M., Mesiar, R.: Aggregation operators, properties, classes and construction methods. In: Calvo, T., Mayor, G., Mesiar, R. (eds.) Aggregation Operators New trends ans Aplications, pp. 3–104. Physica-Verlag, Heidelberg (2002)
4. Beliakov, G., Pradera, A., Calvo, T.: Aggregation Functions: A Guide to Practitioners. Springer, Berlin (2007)
5. Beliakov, G., Gomez, D., Rodriguez, J.T., Montero, J.: Learning stable weights for data of varying dimension. In: Baczynski, M., De Baets, B., Mesiar, R. (eds.) Proceedings of 8th International Summer School on Aggregation Operators AGOP. University of Silesia, Katowice (2015)
6. Beliakov, G., Gómez, D., James, S., Montero, J., Rodríguez, J.T.: Approaches to learning strictly-stable weights for data with missing values. Fuzzy Sets Syst. (2017)
7. Gmez, D., Rojas, K., Montero, J., Rodrguez, J.T., Beliakov, G.: Consistency and stability in aggregation operators an application to missing data problems. Int. J. Comput. Intell. Syst. **7**, 595–604 (2017). Ed. Taylor Francis
8. Cutello, V., Montero, J.: Hierarchical aggregation of OWA operators: basic measures and related computational problems. Uncertain. Fuzzinesss Knowl.-Based Syst. **3**, 17–26 (1995)
9. Cutello, V., Montero, J.: Recursive families of OWA operators. In: Proceedings of IEEE 3rd International Fuzzy Systems Conference on IEEE World Congress on Computational Intelligence, Piscataway (1994)
10. Dujmovic, J.: Aggregation operators and observable properties of human reasoning. In: Bustince, H., et al. (eds.) Aggregation Functions in Theory and in Practise: Proceedings of the 7th International Summer School on Aggregation Operators at the Public University of Navarra. Advances in Intelligent Systems and Computing, Pamplona, Spain, vol. 228, pp. 5–16, Springer, Heidelberg (2013)
11. Dujmovic, J., Tre, G.D., Singh, N., Tomasevich, D., Yokoohji, R.: Soft computing models in online real estate. In: Jamshidi, M., et al. (eds.) Advance Trends in Soft Computing, WCSC 2013. Studies in Fuzziness and Soft Computing, vol. 312, pp. 77–91. Springer, Cham (2013)
12. Dujmovic, J.J.: The problem of missing data in LSP aggregation. In: Greco, S., et al. (eds.) IPMU 2012, Part III. CCIS, vol. 299, pp. 336–346. Springer, Heidelberg (2012)
13. Gomez, D., Montero, J.: A discussion of aggregation functions. Kybernetika **40**, 107–120 (2004)
14. Gomez, D., Rojas, K., Rodriguez, J.T., Montero, J.: Stability in aggregation operators. In: Greco, S., et al. (eds.) IPMU 2012, Part III. CCIS, vol. 299, pp. 317–325. Springer, Heidelberg (2012)

15. Gomez, D., Rojas, K., Montero, J., Rodriguez, J.T.: Consistency and stability in aggregation operators: an application to missing data problems. In: Bustince, H., et al. (eds.) Aggregation Functions in Theory and in Practise. Advances in Intelligent Systems and Computing, pp. 507–518. Springer, Heidelberg (2012). doi:10.1007/978-3-642-39165-1_48

16. Grabisch, M., Marichal, J., Mesiar, R., Pap, E.: Aggregation functions. In: Encyclopedia of Mathematics and its Applications (2009)

17. Rojas, K., Gomez, D., Rodriguez, J.T., Montero, J.: Some properties of consistency in the families of aggregation functions. In: Melo-Pinto, P., Couto, P., Serodio, C., Fodor, J., De Baets, B. (eds.) Advances in Intelligent and Soft Computing, vol. 107, pp. 169–176. Springer, Heidelberg (2011)

18. Rojas, K., Gomez, D., Rodriguez, J.T., Montero, J.: Strict stability in aggregation operators. Fuzzy Sets Syst. **228**, 44–63 (2013)

19. Rojas, K., Gomez, D., Rodriguez, J.T., Montero, J., Valdivia, A., Paiva, F.: Development of child's home environment indexes based on consistent families of aggregation operators with prioritized hierarchical information. Fuzzy Sets Syst. **241**, 41–60 (2013)

20. Wojtowicz, A., Zywica, P., Stachowiak, A., Dyczkowski, K.: Interval- valued aggregation as a tool to improve medical diagnosis. In: Proceedings of the AGOP 2015, pp. 239–244. Katowice, Poland (2015)

21. Yager, R.R., Rybalov, A.: Nonconmutative self-identity aggregation. Fuzzy Sets Syst. **85**, 73–82 (1997)

Sizes, Super Level Measures and Integrals

Lenka Halčinová[(✉)]

Faculty of Science, Institute of Mathematics,
Pavol Jozef Šafárik University in Košice, Jesenná 5, 040 01 Košice, Slovakia
lenka.halcinova@upjs.sk

Abstract. The concept of super level measures as a generalization of classical level measures is discussed and studied in detail. Following the developing of the theory of L_p-spaces introduced by non-additive integrals based on super level measures we discuss the integration theory modified by super level measures and we compare it with the classical approach.

1 Basic Notions and Preliminaries

A common feature of various classes of non-additive integrals (see [6,7,9,10,12]) is the so-called level measure occurring in their definition. Modifying the level measure using the concept of a "size" we get the so-called super level measure recently introduced in [5]. As authors say, the concept of super level measures and integrals with respect to them turns out to be a very suitable component connecting the theory of Carleson measures and the time-frequency analysis. Therefore it is both interesting and valuable to study this new theory much deeper.

In order to make this paper as self-contained as possible, we recall here all the basic notations and definitions together with a few examples of important terms. To avoid too abstract setting, we shall assume that X is a topological space. We denote by $\mathbf{E_B}$ the σ-algebra of Borel sets of X. Then, the pair $(X, \mathbf{E_B})$ will be called a *Borel space* associated with X. Further, for several reasons that we mention later, a subcollection $\mathbf{E} \subseteq \mathbf{E_B}$ will be important in this theory. Similarly, the term measure will be understood in its most general sense: a *measure* is a set function $m : \mathscr{S} \to [0, +\infty]$ on $(X, \mathscr{S})$, where $\mathscr{S}$ is a non-empty class of subsets of X with the only (natural) condition $m(\emptyset) = 0$ whenever $\emptyset \in \mathscr{S}$.

In the following we present a slightly modified definition of sizes originally introduced in [5]. This modified concept is established in our paper [8]. The modification lies in the fact that we suppose the size is defined on all Borel subsets of X instead of a subcollection $\mathbf{E}$ only. And, what is more important, we leave the original approach where the subcollection $\mathbf{E}$ is linked to the outer measure generated by the so-called pre-measure on a subcollection $\mathbf{E}$ (see [5, Definition 2.1]). The generating procedure (deeply described in [5]) cannot produce an arbitrary monotone set function (in fact, it produces only a sub-additive measure μ in the original setting). Thus, the original concept of super level measure will not allow a generalization of many classical (non-additive) integrals.

V. Torra et al. (eds.), *Aggregation Functions in Theory and in Practice*,
Advances in Intelligent Systems and Computing 581, DOI 10.1007/978-3-319-59306-7_19

Definition 1. Let $(X, \mathbf{E_B})$ be a Borel space and $\mathscr{B}(X)$ be the set of all complex-valued Borel-measurable functions on X. A *size* is a map

$$\mathsf{s} : \mathscr{B}(X) \to [0, +\infty]^{\mathbf{E_B}}$$

such that for any $f, g \in \mathscr{B}(X)$ and $E \in \mathbf{E_B}$ it holds

(i) if $|f| \leq |g|$, then $\mathsf{s}(f)(E) \leq \mathsf{s}(g)(E)$;   (monotonicity)
(ii) $\mathsf{s}(\lambda f)(E) = |\lambda| \, \mathsf{s}(f)(E)$ for each $\lambda \in \mathbb{C}$;   (scaling property)
(iii) $\mathsf{s}(f + g)(E) \leq C_\mathsf{s} \, \mathsf{s}(f)(E) + C_\mathsf{s} \, \mathsf{s}(g)(E)$ for some fixed $C_\mathsf{s} \geq 1$
depending only on s.   (quasi-sublinearity)

The concept of sizes can be viewed as a form of averaging the positive functions from the class $\mathscr{B}(X)$ over the subcollection $\mathbf{E}$. This approach involves averaging such as the classical arithmetic mean, generalized arithmetic mean, weighting the integrals, as well as the supremum of a function over a set. The last mentioned example does not reflect averaging in its original sense, but it is a size and it plays an important role in the theory of super level measures.

Example 1. In the following we list some examples of sizes.

(a) *Supremum.* The mapping $\mathsf{s}_\infty : \mathscr{B}(X) \to [0, +\infty]^{\mathbf{E_B}}$ of the form

$$\mathsf{s}_\infty(f)(E) = \sup_{x \in E} |f(x)| = \sup |f|[E]$$

is a size. It is the classical supremum (or an L_∞-based average). This average takes only the function f and the set E into account and does not depend on any further external input (e.g. on a measure of basic sets as it is in the following example).

(b) *Standard (discrete) p-mean.* For a non-empty finite set X with discrete topology we define the mapping $\tilde{\mathsf{s}}_{\nu,p} : \mathscr{B}(X) \to [0, +\infty]^{\mathbf{E_B}}$ by

$$\tilde{\mathsf{s}}_{\nu,p}(f)(E) = \begin{cases} \left(\frac{1}{\nu(E)} \sum_{x \in E} |f(x)|^p \right)^{\frac{1}{p}}, & \text{if } \nu(E) \neq 0, \\ 0, & \text{if } \nu(E) \in \{0, +\infty\}, \end{cases}$$

with $p \in \mathbb{R}$, $p > 0$. Here ν is an arbitrary measure defined on $\mathbf{E_B}$.

(c) *Choquet integral.* The mapping $\mathsf{s}_{\text{int},m}^{(\text{Ch})} : \mathscr{B}(X) \to [0, +\infty]^{\mathbf{E_B}}$ of a Borel-measurable function $f : X \to \mathbb{C}$ over a Borel set $E \subseteq X$ given by

$$\mathsf{s}_{\text{int},m}^{(\text{Ch})}(f)(E) := (\text{Ch}) \int_E |f| \, dm = \int_0^\infty m\left(\{x \in E \cap X; |f(x)| \geq \alpha\}\right) d\alpha,$$

where m is a monotone measure, is the well-known Choquet integral and it is a size. The integral on the right-hand side is the improper Riemann integral. The monotonicity and the scaling property follows from [12, Theorem 7.2] and, although, the Choquet integral is not sublinear in general, the following inequality is always true, see [1, p. 14]

$$s_{int,m}^{(Ch)}(f+g)(E) \leq 2\left(s_{int,m}^{(Ch)}(f)(E) + s_{int,m}^{(Ch)}(g)(E)\right).$$

Especially, if monotone measure m is *submodular*, i.e.,

$$m(E \cup F) + m(E \cap F) \leq m(E) + m(F) \text{ for all } E, F \in \mathbf{E_B},$$

the Choquet integral is sublinear, cf. [10, Theorem 7.7]. Consequently, the mapping $s_{int,m}^{(Ch)}$ is a sublinear size $(C_s = 1)$ whenever m is a submodular measure.

(d) *Shilkret integral.* Under the additional conditions on a measure to be monotone the Shilkret integral defined for a Borel-measurable functions $f : X \to \mathbb{C}$ over a Borel set $E \subseteq X$ as

$$s_{int,m}^{(Sh)}(f)(E) := (Sh)\int_E |f|\,dm = \sup_{\alpha > 0}\{\alpha \cdot m\left(\{x \in A \cap X; |f(x)| \geq \alpha\}\right)\}$$

is a size. Especially, if monotone measure is *maxitive*, i.e.,

$$m(E \cup F) = \max\{m(E), m(F)\} \text{ for all disjoint sets } E, F \in \mathbf{E_B},$$

the Shilkret integral is sublinear, cf [11, pp. 112–113] and the mapping $s_{int,m}^{(Sh)}$ is a sublinear size.

New sizes can be generated by an appropriate multiplication. Indeed, the mapping $\frac{1}{\nu} \boxdot s_{int,m}^{(Ch)} : \mathscr{B}(X) \to [0, +\infty]^{\mathbf{E_B}}$ defined by[1]

$$\left(\frac{1}{\nu} \boxdot s_{int,m}^{(Ch)}\right)(f)(E) := \frac{1}{\nu}(E)(Ch)\int_E |f|\,dm$$

with ν being an arbitrary measure defined on $\mathbf{E_B}$ and with the usual convention "$0 \cdot \infty = 0$", is again a size. Numerous examples of sizes can be found in the paper [8].

2   Super Level Measures

In the definition of super level measure the so-called outer essential supremum plays a crucial role[2].

---

[1] The mapping $\frac{1}{\nu} : \mathbf{E_B} \to [0, \infty]$ assigns 0 if $\nu(E) = +\infty$, assigns $+\infty$ if $\nu(E) = 0$ and the value $\frac{1}{\nu(E)}$ otherwise.

[2] By $\mathbf{1}_F$ we denote the characteristic function of the set F, i.e., $\mathbf{1}_F(x) = 1$ if $x \in F$ and $\mathbf{1}_F(x) = 0$ otherwise.

Definition 2. Let X be a topological space. The *outer essential supremum* of a function $f \in \mathscr{B}(X)$ over a set $F \in \mathbf{E_B}$ with respect to a size s and a subcollection $\mathbf{E} \subseteq \mathbf{E_B}$ is defined by

$$\operatorname*{outsup}_{F} \mathsf{s}(f)\langle \mathbf{E} \rangle := \sup_{E \in \mathbf{E}} \mathsf{s}(f \mathbf{1}_F)(E).$$

In the previous definition one can consider different subcollections, for example, the subcollection of all open balls in $\mathbb{R}^m$ ($\mathbf{E}_{\text{ball}}$), the subcollection of tents in the open upper half-plane $\mathbb{R} \times (0, \infty)$ ($\mathbf{E}_{\text{tent}}$), etc. For more details see Subsect. 2.2 in [5]. It is obvious that different subcollections can lead to different outcomes. Usually we omit the indication of subcollection $\mathbf{E}$ when it is identified with $\mathbf{E_B}$ or when there is no possible confusion. Let us remark that the original definition of the outer essential supremum, see [5, Definition 2.4] differs from ours. Indeed, the authors in [5] consider the subcollection $\mathbf{E}$ to be the same as the domain of size which has some limitations. And, as it was mentioned, the subcollection $\mathbf{E}$ is not linked to the outer measure μ generated by a pre-measure.

Example 2. Let $f : X \to \mathbb{R}$, $X = \{x_1, x_2\}$ be a function such that $f(x_1) = 1$ and $f(x_2) = 2$. Let us consider the size $\tilde{\mathsf{s}}_{\nu,p}$ with ν being the counting measure, $p = 1$ and two subcollections, the subcollection $\mathbf{E}_1 = \{\{x_1\}, \{x_2\}\}$ and the subcollection $\mathbf{E}_2 = \{\{x_1, x_2\}\}$. Then the outer essential supremum for each Borel set $F \in \mathbf{E_B}$, $\mathbf{E_B} = 2^X$ is summarized in the following table. In the first column of the table are all possible sets $F \in \mathbf{E_B}$, in the first raw are sets E necessary for computing the outer essential supremum on the corresponding subcollections. The table is then filled by values $\tilde{\mathsf{s}}_{\nu,1}(f \mathbf{1}_F)(E)$.

E \ F	$\{x_1\}$	$\{x_2\}$	$\{x_1,x_2\}$	$\operatorname{outsup}_F \tilde{\mathsf{s}}_{\nu,1}(f)\langle \mathbf{E}_1 \rangle$	$\operatorname{outsup}_F \tilde{\mathsf{s}}_{\nu,1}(f)\langle \mathbf{E}_2 \rangle$
$\emptyset$	0	0	0	0	0
$\{x_1\}$	1	0	0.5	1	0.5
$\{x_2\}$	0	2	1	2	1
$\{x_1,x_2\}$	1	2	1.5	2	1.5

As can be seen, different subcollections lead to different outcomes of the outer essential supremum.

In the following text let us consider a topological space X and a subcollection $\mathbf{E} \subseteq \mathbf{E_B}$.

Definition 3. The quantity

$$\mu(\mathsf{s}(f)\langle \mathbf{E} \rangle > \alpha) := \inf \left\{ \mu(F) : \ F \in \mathbf{E_B}, \ \operatorname*{outsup}_{X \backslash F} \mathsf{s}(f)\langle \mathbf{E} \rangle \leq \alpha \right\}, \quad \alpha > 0,$$

is called a *super level measure* of $f \in \mathscr{B}(X)$ with respect to monotone measure μ, size s and subcollection $\mathbf{E}$.

In the following we shall use the notation

$$h_{\mu,\mathsf{s},\mathbf{E},f}(\alpha) := \mu(\mathsf{s}(f)\langle \mathbf{E} \rangle > \alpha).$$

Example 3. Following the same inputs as in the Example 2 and considering monotone measure μ being the counting measure, the value of super level measure with respect to different subcollections is the following

$$\begin{aligned}
h_{\mu,\tilde{\mathsf{s}}_{\nu,1},\mathbf{E}_1,f}(\alpha) &= \mu(\tilde{\mathsf{s}}_{\nu,1}(f)\langle \mathbf{E}_1 \rangle > \alpha) = 2 \cdot \mathbf{1}_{[0,1[}(\alpha) + 1 \cdot \mathbf{1}_{[1,2[}(\alpha), \\
h_{\mu,\tilde{\mathsf{s}}_{\nu,1},\mathbf{E}_2,f}(\alpha) &= \mu(\tilde{\mathsf{s}}_{\nu,1}(f)\langle \mathbf{E}_2 \rangle > \alpha) = 2 \cdot \mathbf{1}_{[0,0.5[}(\alpha) + 1 \cdot \mathbf{1}_{[0.5,1.5[}(\alpha).
\end{aligned}$$

The concept of super level measure is a generalization of the classical level measure usually denoted by $h_{m,f}$. Indeed, for each $\alpha > 0$ we may write

$$h_{m,f}(\alpha) = m(\{x \in X; |f(x)| > \alpha\}) = \inf\{m(F) : F \in \mathbf{E}_B, (\forall x \in X \setminus F)\, |f(x)| \le \alpha\}.$$

Modifying the previous formula by outer essential supremum we get the definition of the super level measure. Now, the connection with the standard concept is more evident. Also, a natural question arises: When does the concept of super level measure coincide with the standard measure? In what follows, under μ-ess sup we understand the standard essential supremum associated with a monotone measure μ.

Proposition 1. Let $\mathbf{E} \subseteq \mathbf{E}_B$ and let s be a size such that

$$\operatorname*{outsup}_{F} \mathsf{s}(f) = \mu\text{-ess sup} |f\mathbf{1}_F|, \quad F \in \mathbf{E}_B.$$

Then $\mu(\mathsf{s}(f)\langle \mathbf{E} \rangle > \alpha) = \mu(\{x \in X; |f(x)| > \alpha\})$. In particular, if $\operatorname{outsup}_F \mathsf{s}(f)$ $\langle \mathbf{E} \rangle = \sup |f|[F]$ with $F \in \mathbf{E}_B$, then $\mu(\mathsf{s}(f)\langle \mathbf{E} \rangle > \alpha) = \mu(\{x \in X; |f(x)| > \alpha\})$.

Proof. Let $F \in \mathbf{E}_B$. If μ-ess sup $|f\mathbf{1}_{X \setminus F}| \le \alpha$, then there is $B \in \mathbf{E}_B$ such that $\mu(B) = 0$ and $(X \setminus F) \setminus B \subseteq \{x \in X; |f(x)| \le \alpha\} \setminus B$, and thus $\{x \in X; |f(x)| > \alpha\} \setminus B \subseteq F \setminus B$. From monotonicity of μ we then have

$$\mu(\{x \in X; |f(x)| > \alpha\}) = \mu(\{x \in X; |f(x)| > \alpha\} \setminus B) \le \mu(F \setminus B) = \mu(F),$$

and finally we obtain

$$\begin{aligned}
\mu(\mathsf{s}(f)\langle \mathbf{E} \rangle > \alpha) &= \inf\left\{\mu(F) : F \in \mathbf{E}_B, \operatorname*{outsup}_{X \setminus F} \mathsf{s}(f)\langle \mathbf{E} \rangle \le \alpha\right\} \\
&= \inf\{\mu(F) : F \in \mathbf{E}_B, \mu\text{-ess sup} |f\mathbf{1}_{X \setminus F}| \le \alpha\} \\
&= \mu(\{x \in X; |f(x)| > \alpha\}).
\end{aligned}$$

To prove the particular case, one can consider a measure μ which assigns 1 to each non-empty set. $\qquad\square$

The equality outsup$_F$ $\mathsf{s}(f)\langle \mathbf{E}\rangle = \sup |f|[F]$ is true for example when one takes s_∞ size in the formula of outer essential supremum under the condition that the set F is covered by sets from $\mathbf{E} \subseteq \mathbf{E}_\mathrm{B}$. If the set F is not covered by sets from the subcollection, the value of outer essential supremum is 0, while the supremum of function f on F can be totally different. So, in this particular case the outer essential supremum coincides with the classical supremum and the super level measure is just the outer measure of the upper level set.

3   Integration with Respect to Super Level Measures

Naturally, together with the theory of (super level) measure, the concept of integrals can be introduced. For a subcollection $\mathbf{E} \subseteq \mathbf{E}_\mathrm{B}$, a size s and functions $f \in \mathscr{B}(X)$ the analogues of the Choquet and the Shilkret integral, respectively, are as follows

$$\mathbf{I}_{\mathrm{Ch}}(\mu, \mathsf{s}, \mathbf{E}, f) := \int_0^\infty \mu(\mathsf{s}(f)\langle \mathbf{E}\rangle > \alpha)\, d\alpha,$$

$$\mathbf{I}_{\mathrm{Sh}}(\mu, \mathsf{s}, \mathbf{E}, f) := \sup_{\alpha>0} \big\{\alpha \cdot \mu(\mathsf{s}(f)\langle \mathbf{E}\rangle > \alpha)\big\}.$$

From the definition of super level measure immediately follows that the non-negative real-valued function $\mu(\mathsf{s}(f)\langle \mathbf{E}\rangle > \alpha)$ is monotone in α (the same is true for the standard level measure), so the generalized Choquet integral is well defined and it is a number from $[0, +\infty]$. According to Proposition 1 the classical Choquet as well as Shilkret integral with respect to a monotone measures are included in the definitions above. It is enough to take s_∞ size or the others for which the condition of Proposition 1 is true. Immediately, from the properties of sizes and corresponding super level measures we get the following result.

Proposition 2. Let $\mathbf{E} \subseteq \mathbf{E}_\mathrm{B}$, s be a size and $f, g \in \mathscr{B}(X)$. Then for each $\mathrm{N} \in \{\mathrm{Ch}, \mathrm{Sh}\}$ it holds

(i) if $|f| \leq |g|$, then $\mathbf{I}_\mathrm{N}(\mu, \mathsf{s}, \mathbf{E}, f) \leq \mathbf{I}_\mathrm{N}(\mu, \mathsf{s}, \mathbf{E}, g)$;
(ii) for each $\lambda \in \mathbb{C}$ we have $\mathbf{I}_\mathrm{N}(\mu, \mathsf{s}, \mathbf{E}, \lambda f) = |\lambda|\, \mathbf{I}_\mathrm{N}(\mu, \mathsf{s}, \mathbf{E}, f)$;
(iii) there is a constant $C_{\mathsf{s},\mathrm{N}}$ independent of f, g such that

$$\mathbf{I}_\mathrm{N}(\mu, \mathsf{s}, \mathbf{E}, f + g) \leq C_{\mathsf{s},\mathrm{N}} \Big(\mathbf{I}_\mathrm{N}(\mu, \mathsf{s}, \mathbf{E}, f) + \mathbf{I}_\mathrm{N}(\mu, \mathsf{s}, \mathbf{E}, g)\Big).$$

As can be seen, we have introduced only two types of non-additive integrals. One reason is that they provide a background for defining outer L_p-spaces in the original paper [5]. A deeper investigation of their properties could be useful for developing of the theory of L_p-spaces. However, following this pattern we can also generalize the another types of non-additive integrals. For example, the well-known Sugeno integral does not satisfy the scalling property, therefore it is not a size, but there is no barrier to define a generalized form of Sugeno integral via super level measures as follows

$$\mathbf{I}_{\mathrm{Su}}(\mu, \mathsf{s}, \mathbf{E}, f) := \sup_{\alpha>0} \min\{\alpha, \mu(\mathsf{s}(f)\langle \mathbf{E}\rangle > \alpha)\}.$$

Or, we can generalize the other non-additive integrals, the generalized upper Sugeno integral [2], the seminormed integral [3,4] (considering restriction to $[0,1]$), etc.

In the following we compare the standard Choquet integral (based on the level measure) with its modified version defined via super level measure considering $s_{int,m}^{(Ch)}$ size (see Example 1c).

Example 4. Let $f : X \to \mathbb{R}$, $X = \{x_1, x_2\}$ be a function such that $f(x_1) = 1$ and $f(x_2) = 2$. Let $m : 2^X \to [0, +\infty)$ be a monotone measure (capacity) defined by $m(\emptyset) = 0$, $m(\{x_1\}) = 0.4$, $m(\{x_2\}) = 0.6$ and $m(\{x_1, x_2\}) = 1$.

Then the classical level measure takes the form

$$h_{m,f}(\alpha) = 1 \cdot \mathbf{1}_{[0,1[}(\alpha) + 0.6 \cdot \mathbf{1}_{[1,2[}(\alpha).$$

The standard Choquet integral defined via level measure takes the value 1.6, i.e. $(Ch) \int f \, dm = \int_0^\infty h_{m,f}(\alpha) \, d\alpha = 1.6$.

On the other hand, the modified Choquet integral takes different values. Let ν_1 be the counting measure, ν_2 be the measure defined as $\nu_2(\emptyset) = 0$, $\nu_2(\{x_1\}) = 0.5$, $\nu_2(\{x_2\}) = 0.5$ and $\nu_2(\{x_1, x_2\}) = 1$. For the subcollection $\mathbf{E} = 2^X$ we get the following values of outer essential supremum on the corresponding set F

F	$\emptyset$	$\{x_1\}$	$\{x_2\}$	$\{x_1, x_2\}$
$\underset{F}{\text{outsup}} \frac{1}{\nu_1} \boxdot s_{int,m}^{(Ch)}(f)$	0	0.4	1.2	1.6
$\underset{F}{\text{outsup}} \frac{1}{\nu_2} \boxdot s_{int,m}^{(Ch)}(f)$	0	0.8	2.4	2.4

and the corresponding super level measure takes the form

$$h_{m, \frac{1}{\nu_1} \boxdot s_{int,m}^{(Ch)}, f}(\alpha) = 1 \cdot \mathbf{1}_{[0,0.4[}(\alpha) + 0.6 \cdot \mathbf{1}_{[0.4,1.2[}(\alpha) + 0.4 \cdot \mathbf{1}_{[1.2,1.6[}(\alpha),$$

$$h_{m, \frac{1}{\nu_2} \boxdot s_{int,m}^{(Ch)}, f}(\alpha) = 1 \cdot \mathbf{1}_{[0,0.8[}(\alpha) + 0.6 \cdot \mathbf{1}_{[0.8,2.4[}(\alpha).$$

Consequently,

$$\mathbf{I}_{Ch}\left(m, \frac{1}{\nu_1} \boxdot s_{int,m}^{(Ch)}, 2^X, f\right) = 1.04, \quad \mathbf{I}_{Ch}\left(m, \frac{1}{\nu_2} \boxdot s_{int,m}^{(Ch)}, 2^X, f\right) = 2.24.$$

The previous Example demonstrates that the approach via super level measures provides results different from the standard approach. Moreover, using different sizes in the definition of super level measure we can get higher as well as lower values of modified integral in comparison with the standard approach.

Acknowledgements. The author kindly acknowledges the support of the grant VVGS-2016-255, and thanks the co-authors of the paper [8] who collaborated on this research.

References

1. Adams, D.R.: Choquet integrals in potential theory. Publ. Mat. **42**, 3–66 (1998)
2. Boczek, M., Kaluszka, M.: On the Minkowski-Hölder type inequalities for generalized Sugeno integrals with an application. Kybernetika **52**(3), 329–347 (2016)
3. Borzová-Molnárová, J., Halčinová, L., Hutník, O.: The smallest semicopula-based universal integrals I: properties and characterizations. Fuzzy Sets Syst. **271**, 1–17 (2015)
4. Borzová-Molnárová, J., Halčinová, L., Hutník, O.: The smallest semicopula-based universal integrals II: convergence theorems. Fuzzy Sets Syst. **271**, 18–30 (2015)
5. Do, Y., Thiele, C.: L^p theory for outer measures and two themes of Lennart Carleson united. Bull. Amer. Math. Sci. **52**(2), 249–296 (2015)
6. Grabisch, M.: Set Functions, Games and Capacities in Decision Making. Theory and Decision Library C, vol. 46. Springer, Switzerland (2016)
7. Greco, S., Matarazzo, B., Giove, S.: The Choquet integral with respect to a level dependent capacity. Fuzzy Sets Syst. **175**(1), 1–35 (2011)
8. Halčinová, L., Hutník, O., Kiseľák, J., Šupina, J.: Beyond the scope of super level measures, submitted
9. Klement, E.P., Mesiar, R., Spizzichino, F., Stupňanová, A.: Universal integrals based on copulas. Fuzzy Optim. Decis. Making **13**(3), 273–286 (2014)
10. Pap, E.: Null-Additive Set Functions. Kluwer - Ister Science, Dordrecht - Bratislava (1995)
11. Shilkret, N.: Maxitive measure and integration. Indag. Math. **33**, 109–116 (1971)
12. Wang, Z., Klir, G.: Generalized Measure Theory. Springer, Heidelberg (2009)

Monotonicity in the Construction of Ordinal Sums of Fuzzy Implications

Michał Baczyński[1]([⊠]), Paweł Drygaś[2], and Radko Mesiar[3]

[1] Institute of Mathematics, University of Silesia in Katowice, ul. Bankowa 14,
40-007 Katowice, Poland
michal.baczynski@us.edu.pl
[2] Faculty of Mathematics and Natural Sciences, University of Rzeszów,
Rzeszów, Poland
paweldrs@ur.edu.pl
[3] Faculty of Civil Engineering, Slovak University of Technology, Radlinského 11,
810 05 Bratislava, Slovakia
mesiar@math.sk

Abstract. In this contribution we discus the problem of monotonicity of intervals in the ordinal sums of fuzzy implication constructions. As a result, new ways of constructing of ordinal sums of fuzzy implications are obtained. These methods allow to adapt the value of fuzzy implication to specific requirements. For our new methods of construction, several sufficient properties for obtaining a fuzzy implication as a result are presented. Moreover, preservation of some properties of the ordinal sums are examined. Among others neutrality property, identity property, and ordering property are considered.

1 Introduction

Fuzzy implications find applications in many fields such as fuzzy control, approximate reasoning, and decision support systems. This is why new families of these connectives are the subject of investigation. One of the directions of such research is considering an ordinal sum of fuzzy implications on the pattern of the ordinal sum of t-norms. Some interesting results connected to representation of the residual implication corresponding to a fuzzy conjunction (for example continuous or at least left-continuous t-norms) given by an ordinal sum were obtained in [2,9,14] (see also [13]). In [15] Su et al. introduced a concept of ordinal sum of fuzzy implications similar to the construction of the ordinal sum of t-norms. In [6–8] other constructions of ordinal sums of fuzzy implications were described.

In this paper, some problems connected with the monotonicity are discussed and consequently new possibilities of defining ordinal sums of fuzzy implications are proposed. In previously known constructions, if arguments belong to the interval $[a_k, b_k]$, then values also belong to this interval. This means, that values are increasing with respect to index set. Now we change this situation. Moreover, we can change the length of the set of values of implication on given interval. The first construction generates a fuzzy implication without any additional

© Springer International Publishing AG 2018
V. Torra et al. (eds.), *Aggregation Functions in Theory and in Practice*,
Advances in Intelligent Systems and Computing 581, DOI 10.1007/978-3-319-59306-7_20

assumptions on summands. In the other one sufficient properties for obtaining a fuzzy implication are presented.

In Sect. 2 some basic information about fuzzy connectives, in particular triangular norms and fuzzy implications, including their ordinal sum are presented. In Sect. 3 the construction of ordinal sums of fuzzy implications and properties of these methods are examined.

2   Preliminaries

Here we recall the notions of a t-norm and a fuzzy implication, as well as some of the constructions of ordinal sums of these fuzzy connectives.

2.1   Triangular Norms

Firstly, we put the definition of a t-norm and some important class of t-norms with several examples of these operations.

Definition 1 [12]. A t-norm is an increasing, commutative and associative operation $T\colon [0,1]^2 \to [0,1]$ with a neutral element 1.

Example 1 [12, p. 4], [10, p. 7]. Here, we list well-known t-norms.

$$T_{\mathrm{M}}(x,y) = \min(x,y), \qquad \text{minimum t-norm,}$$

$$T_{\mathrm{P}}(x,y) = xy, \qquad \text{product t-norm,}$$

$$T_{\mathrm{LK}}(x,y) = \max(x+y-1,0), \qquad \text{Łukasiewicz t-norm,}$$

$$T_{\mathrm{D}}(x,y) = \begin{cases} x, & \text{if } y = 1, \\ y, & \text{if } x = 1, \\ 0, & \text{otherwise,} \end{cases} \qquad \text{drastic t-norm,}$$

$$T_{\mathrm{nM}}(x,y) = \begin{cases} 0, & \text{if } x+y \leq 1, \\ \min(x,y), & \text{otherwise,} \end{cases} \qquad \text{nilpotent minimum t-norm.}$$

Next, let us recall the generalized Ordinal Sum Theorem for t-norms [11, Corollary 2].

Theorem 1 (cf. [5,11,12]). *Let $\{[a_k, b_k]\}_{k \in \mathscr{A}}$ be a countable family of nonoverlapping, closed, proper subintervals of $[0,1]$, where $\mathscr{A}$ is a finite or infinite index set. Let T be an operation in $[0,1]$ defined by*

$$T(x,y) = \begin{cases} a_k + (b_k - a_k)T_k \left(\frac{x-a_k}{b_k-a_k}, \frac{y-a_k}{b_k-a_k} \right), & \text{if } (x,y) \in (a_k, b_k]^2, \\ \min(x,y), & \text{otherwise,} \end{cases} \qquad (1)$$

where for each k the binary operation $T_k\colon [0,1]^2 \to [0,1]$ is associative, commutative increasing such that $T_k \leq \min$, i.e., T_k is a t-subnorm. Moreover, if $b_k = a_l$ for some l, k and T_l is with a zero divisor, then T_k has a neutral element $e = 1$. We also assume that if $b_k = 1$ for some k, then the operation T_k has a neutral element $e = 1$. Then the operation T is a t-norm.

Definition 2. T-norm T defined as in Theorem 1 in Eq. (1) is called an ordinal sum of $\{([a_i, b_i], T_i)\}_{i \in \mathscr{A}}$ and each T_i is called a summand.

For the general structure of such ordinal sum of t-norms see Fig. 1.

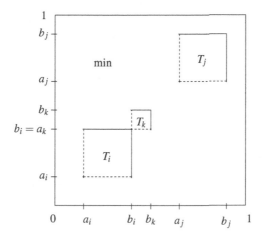

Fig. 1. The structure of an ordinal sum of t-norms given by (1).

2.2   Fuzzy Implications

Now, we focus on the class of fuzzy implications.

Definition 3 [1,10]. A function $I \colon [0,1]^2 \to [0,1]$ is called a fuzzy implication if it satisfies the following conditions

(I1) I is non-increasing with respect to the first variable,
(I2) I is non-decreasing with respect to the second variable,
(I3) $I(0,0) = 1$,
(I4) $I(1,1) = 1$,
(I5) $I(1,0) = 0$.

Directly from the above definition we obtain as follows.

Corollary 1. *A fuzzy implication has a right zero element 1 and fulfils the condition*

$$I(0,y) = 1, \qquad x,y \in [0,1].$$

There are other properties the fuzzy implication may also have. Some of them are listed below. The property (CB), i.e., the boundary condition plays a spacial role in the sequel.

Definition 4 (cf. [1, p. 9], [3,4]). We say that a fuzzy implication I fulfils:

- the neutrality property (NP), if

$$I(1,y) = y, \quad y \in [0,1], \tag{NP}$$

- the identity principle (IP), if

$$I(x, x) = 1, \quad x \in [0, 1], \tag{IP}$$

- the ordering property (OP), if

$$I(x, y) = 1 \Leftrightarrow x \leq y, \quad x, y \in [0, 1], \tag{OP}$$

- the property (CB), if

$$I(x, y) \geq y, \quad x, y \in [0, 1], \tag{CB}$$

- the left ordering property (LOP), if

$$x \leq y \Rightarrow I(x, y) = 1, \quad x, y \in [0, 1], \tag{LOP}$$

- the right ordering property (ROP), if

$$I(x, y) = 1 \Rightarrow x \leq y, \quad x, y \in [0, 1], \tag{ROP}$$

- the strong boundary condition (SBC), if

$$x \neq 0 \Rightarrow I(x, 0) = 0, \quad x, y \in [0, 1], \tag{SBC}$$

- the strong corner condition for 0 (SCC0), if

$$I(x, y) = 0 \Rightarrow x = 1 \wedge y = 0, \quad x, y \in [0, 1], \tag{SCC0}$$

- the strong corner condition for 1 (SCC1), if

$$I(x, y) = 1 \Rightarrow x = 0 \vee y = 1, \quad x, y \in [0, 1]. \tag{SCC1}$$

Remark 1. Let us notice that the property (CB) is equivalent to the following one

$$I(1, y) \geq y, \quad x, y \in [0, 1]. \tag{CB'}$$

Moreover, if a fuzzy implication satisfies (NP), then it satisfies (CB).

Example 2 (see [1]*).* Let us present the following family of fuzzy implications for $\alpha \in [0, 1]$

$$I_\alpha(x, y) = \begin{cases} 0, & \text{if } x = 1, y = 0, \\ 1, & \text{if } x = 0 \text{ or } y = 1, \\ \alpha & \text{otherwise.} \end{cases}$$

The operations I_0 and I_1 are the least and the greatest fuzzy implication, respectively, where

$$I_0(x, y) = \begin{cases} 1, & \text{if } x = 0 \text{ or } y = 1, \\ 0, & \text{otherwise,} \end{cases}$$

$$I_1(x, y) = \begin{cases} 0, & \text{if } x = 1, y = 0, \\ 1, & \text{otherwise.} \end{cases}$$

The following are other examples of classical fuzzy implications.

$$I_{\text{LK}}(x,y) = \min(1-x+y,1), \qquad \text{Łukasiewicz implication},$$

$$I_{\text{GD}}(x,y) = \begin{cases} 1, & \text{if } x \leq y, \\ y, & \text{otherwise}, \end{cases} \qquad \text{Gödel implication},$$

$$I_{\text{RC}}(x,y) = 1-x+xy, \qquad \text{Reichenbach implication},$$

$$I_{\text{DN}}(x,y) = \max(1-x,y), \qquad \text{Dienes implication},$$

$$I_{\text{GG}}(x,y) = \begin{cases} 1, & \text{if } x \leq y, \\ \frac{y}{x}, & \text{otherwise}, \end{cases} \qquad \text{Goguen implication},$$

$$I_{\text{RS}}(x,y) = \begin{cases} 1, & \text{if } x \leq y, \\ 0, & \text{otherwise}, \end{cases} \qquad \text{Rescher implication},$$

$$I_{\text{YG}}(x,y) = \begin{cases} 1, & \text{if } x = y = 0, \\ y^x, & \text{otherwise}, \end{cases} \qquad \text{Yager implication},$$

$$I_{\text{FD}}(x,y) = \begin{cases} 1, & \text{if } x \leq y, \\ \max(1-x,y), & \text{otherwise}, \end{cases} \qquad \text{Fodor implication},$$

$$I_{\text{WB}}(x,y) = \begin{cases} 1, & \text{if } x < 1, \\ y, & \text{otherwise}, \end{cases} \qquad \text{Weber implication}$$

$$I_{\text{DP}}(x,y) = \begin{cases} y, & \text{if } x = 1, \\ 1-x, & \text{if } y = 0, \\ 1 & \text{otherwise}, \end{cases} \qquad \text{Dubois-Prade implication}.$$

Except for I_α for $\alpha \in [0,1)$ and I_{RS}, the fuzzy implications from this example fulfil property (CB).

3   Main Results

At the beginning we recall one of the construction of ordinal sum of fuzzy implication.

Definition 5 [15]. Let $\{I_k\}_{k \in \mathscr{A}}$ be a family of implications and $\{[a_k, b_k]\}_{k \in \mathscr{A}}$ be a family of pairwise disjoint closed subintervals of $[0,1]$ with $0 < a_k < b_k$ for all $k \in \mathscr{A}$, where $\mathscr{A}$ is a finite or infinite index set. The mapping $I : [0,1]^2 \to [0,1]$ given by

$$I(x,y) = \begin{cases} a_k + (b_k - a_k) I_k \left(\frac{x-a_k}{b_k-a_k}, \frac{y-a_k}{b_k-a_k} \right), & \text{if } x,y \in [a_k, b_k], \\ I_{RS}(x,y), & \text{otherwise}, \end{cases} \tag{2}$$

we call ordinal sum of fuzzy implications $\{I_k\}_{k \in \mathscr{A}}$.

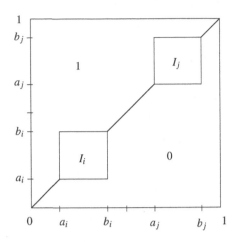

Fig. 2. The structure of an ordinal sum of fuzzy implications given by (2).

For the general structure of the above ordinal sum of fuzzy implications see Fig. 2.

Theorem 2 [15]. *Let* $\{I_k\}_{k\in\mathscr{A}}$ *be a family of implications. Then ordinal sum of implication given by* (2) *is a fuzzy implication.*

In this construction order in the set $\mathscr{A}$ is closely related to order between members of the family of the intervals $\{[a_k, b_k]\}_{k\in\mathscr{A}}$. Moreover, values for these intervals are ordered in the same manner. The following example shows that it is not necessary.

Example 3. Let us consider an operation given by

$$I(x,y) = \begin{cases} 1, & \text{if } x \leq y, \\ 0.5, & \text{if } x > y \text{ and } (x,y \in [0.2, 0.4] \text{ or } x,y \in [0.6, 0.8]), \\ 0, & \text{otherwise.} \end{cases}$$

For the plot of this function see Fig. 3. To show relationships of the above implication of the construction of ordinal sum of fuzzy implication we will write our implication in another form

$$I(x,y) = \begin{cases} 0.5 + 0.5 I_{RS}\left(\frac{x-0.2}{0.2}, \frac{y-0.2}{0.2}\right), & \text{if } x,y \in [0.2, 0.4], \\ 0.5 + 0.5 I_{RS}\left(\frac{x-0.6}{0.2}, \frac{y-0.6}{0.2}\right), & \text{if } x,y \in [0.6, 0.8], \\ I_{RS}, & \text{otherwise.} \end{cases}$$

As a generalization of the example above, we present the general construction.

Definition 6. Let $\{I_k\}_{k\in\mathscr{A}}$ be a family of fuzzy implications and $\{[a_k, b_k]\}_{k\in\mathscr{A}}$ be a family of pairwise disjoint subintervals of $(0,1)$, with $a_k < b_k$ for all $k \in \mathscr{A}$,

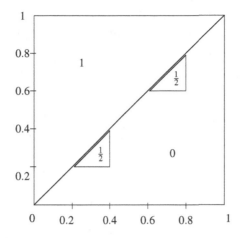

Fig. 3. The structure of fuzzy implication given in Example 3.

where $\mathscr{A}$ is a finite or countably infinite index set. Moreover, let $\{[c_k, d_k]\}_{k \in \mathscr{A}}$ be a family of subintervals of $[0, 1]$, with $c_k \leq d_k$ for all $k \in \mathscr{A}$. Let us define an operation $I \colon [0, 1]^2 \to [0, 1]$ by the following formula:

$$I(x, y) = \begin{cases} c_k + (d_k - c_k) I_k \left(\frac{x - a_k}{b_k - a_k}, \frac{y - a_k}{b_k - a_k} \right), & \text{if } x, y \in [a_k, b_k], \\ I_{RS}(x, y), & \text{otherwise.} \end{cases} \tag{3}$$

For the general structure of the above ordinal sum of fuzzy implications see Fig. 4.

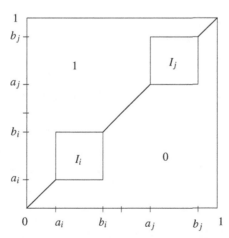

Fig. 4. The structure of a fuzzy implication given by (3).

Theorem 3. *Let* $\{I_k\}_{k \in \mathscr{A}}$ *be a family of implications. Then the operation* I *given by* (3) *is a fuzzy implication.*

Proof. Firstly, we consider the condition (I1). Let $x_1, x_2, y \in [0, 1]$ such that $x_1 < x_2$. If $y \in [a_k, b_k]$ for some $k \in \mathscr{A}$, then we consider the following cases:

1. $b_k < x_2$. Then $I(x_1, y) \geq 0 = I_{RS}(x_2, y) = I(x_2, y)$.
2. $x_1 < a_k$. Then $I(x_1, y) = I_{RS}(x_1, y) = 1 \geq I(x_2, y)$.
3. $a_k \leq x_1 < x_2 \leq b_k$. Then using monotonicity of I_k we have

$$I(x_1, y) = c_k + (d_k - c_k)I_k \left(\frac{x_1 - a_k}{b_k - a_k}, \frac{y - a_k}{b_k - a_k} \right)$$

$$\geq c_k + (d_k - c_k)I_k \left(\frac{x_2 - a_k}{b_k - a_k}, \frac{y - a_k}{b_k - a_k} \right) = I(x_2, y).$$

If $y \notin [a_k, b_k]$ for all $k \in \mathscr{A}$, then $I(x_1, y) = I_{RS}(x_1, y) \geq I_{RS}(x_2, y) = I(x_2, y)$. So, I satisfies (I1).

Next, let us consider the condition (I2). Let $x, y_1, y_2 \in [0, 1]$, $y_1 < y_2$. If $x \in [a_k, b_k]$ for some $k \in \mathscr{A}$, then we consider the following cases:

1. $b_k < y_2$. Then $I(x, y_1) \leq 1 = I_{RS}(x, y_2) = I(x, y_2)$.
2. $y_1 < a_k$. Then $I(x, y_1) = I_{RS}(x, y_1) = 0 \leq I(x, y_2)$.
3. $a_k \leq y_1 < y_2 \leq b_k$. Then using monotonicity of I_k we have

$$I(x, y_1) = c_k + (d_k - c_k)I_k \left(\frac{x - a_k}{b_k - a_k}, \frac{y_1 - a_k}{b_k - a_k} \right)$$

$$\leq c_k + (d_k - c_k)I_k \left(\frac{x - a_k}{b_k - a_k}, \frac{y_2 - a_k}{b_k - a_k} \right) = I(x, y_2).$$

If $x \notin [a_k, b_k]$ for all $k \in \mathscr{A}$, then $I(x, y_1) = I_{RS}(x, y_1) \leq I_{RS}(x, y_2) = I(x, y_2)$. So, I satisfies also (I2).

Directly from (3) we have $I(0, 0) = I_{RS}(0, 0) = 1$ and $I(1, 1) = I_{RS}(1, 1) = 1$, $I(1, 0) = I_{RS}(1, 0) = 0$. So I fulfils (I3), (I4) and (I5). $\qquad \blacksquare$

We have also the following result for which the proof we omit.

Theorem 4. *Let $\{I_k\}_{k \in \mathscr{A}}$ be a family of fuzzy implications and operation I be given by (3).*

(i) I does not satisfy (NP).
(ii) I satisfies (IP) if and only if for all $k \in \mathscr{A}$ $d_k = 1$ and I_k satisfies (IP).
(iii) I satisfies (LOP) ((ROP), (OP), respectively) if and only if for all $k \in \mathscr{A}$ $d_k = 1$ and I_k satisfies (LOP) ((ROP), (OP), respectively).
(iv) I does not satisfy (CB).
(v) I satisfies (SBC).
(vi) I does not satisfy (SCC0) and (SCC1).

In this paper we propose yet another method of generating fuzzy implications by the use of an ordinal sum of fuzzy implications.

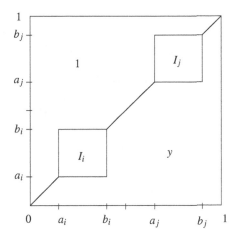

Fig. 5. The structure of a fuzzy implication given by (4).

Definition 7. Let $\{I_k\}_{k \in \mathscr{A}}$ be a family of fuzzy implications and $\{[a_k, b_k]\}_{k \in \mathscr{A}}$ be a family of pairwise disjoint subintervals of $(0, 1)$, with $a_k < b_k$ for all $k \in \mathscr{A}$, where $\mathscr{A}$ is a finite or countably infinite index set. Moreover, let $\{[c_k, d_k]\}_{k \in \mathscr{A}}$ be a family of subintervals of $[0, 1]$, with $c_k \leq d_k$ for all $k \in \mathscr{A}$. Let us define an operation $I : [0, 1]^2 \rightarrow [0, 1]$ by the following formula:

$$I(x, y) = \begin{cases} c_k + (d_k - c_k)I_k\left(\frac{x-a_k}{b_k-a_k}, \frac{y-a_k}{b_k-a_k}\right), & \text{if } x, y \in [a_k, b_k], \\ I_{GD}(x, y), & \text{otherwise.} \end{cases} \quad (4)$$

For the general structure of the above ordinal sum of fuzzy implications see Fig. 5.

Directly from (4) we have the following fact.

Theorem 5. Let $\{I_k\}_{k \in \mathscr{A}}$ be a family of implications. The operation I given by (4) fulfils (13), (14) and (15).

Theorem 6. Let $\{I_k\}_{k \in \mathscr{A}}$ be a family of implications. If for all $k \in \mathscr{A}$ we have $c_k \geq b_k$, then the operation I given by (4) is a fuzzy implication.

Theorem 7. Let $\{I_k\}_{k \in \mathscr{A}}$ be a family of fuzzy implications and the operation I given by (4) be a fuzzy implication.

(i) I satisfies (NP).
(ii) I satisfies (IP) if and only if for all $k \in \mathscr{A}$ $d_k = 1$ and I_k satisfies IP.
(iii) I satisfies (LOP) ((ROP), (OP), respectively) if and only if for all $k \in \mathscr{A}$ $d_k = 1$ and I_k satisfies (LOP) ((ROP), (OP), respectively).
(iv) I satisfies (CB).
(v) I satisfies (SBC).
(vi) I does not satisfy (SCC0) and (SCC1).

4 Conclusions

In this paper two methods of constructing ordinal sums of fuzzy implications are presented. Sufficient properties of summands for obtaining a fuzzy implication as a result are examined. Basic properties of thus obtained implications have been examined. It seems useful to examine other properties of the component of introduced ordinal sums which can be preserved by the ordinal sums.

Acknowledgment. The work on this paper for Michał Baczyński was partially supported by the National Science Centre, Poland, under Grant No. 2015/19/B/ST6/03259. The work on this paper for Paweł Drygaś was partially supported by the Centre for Innovation and Transfer of Natural Sciences and Engineering Knowledge in Rzeszów, through Project Number RPPK.01.03.00-18-001/10. The work on this paper for Radko Mesiar was supported by the Grant APVV-14-0013.

References

1. Baczyński, M., Jayaram, B.: Fuzzy Implications. Springer, Berlin (2008)
2. De Baets, B., Mesiar, R.: Residual implicators of continuous t-norms. In: Zimmermann, H.J. (ed.) Proceedings of the 4th European Congress on Intelligent Techniques and Soft Computing EUFIT 1996, Aachen, Germany, 26 September 1996. ELITE, Aachen, pp. 27-31 (1996)
3. Dimuro, G.P., Bedregal, B., Santiago, R.H.N.: On (G, N)-implications derived from grouping functions. Inform. Sci. **279**, 1–17 (2014)
4. Dimuro, G.P., Bedregal, B.: On residual implications derived from overlap functions. Inform. Sci. **312**, 78–88 (2015)
5. Drewniak, J., Drygaś, P.: Ordered semigroups in constructions of uninorms and nullnorms. In: Issues in Soft Computing Theory and Applications, pp. 147–158. EXIT, Warszawa (2005)
6. Drygaś, P., Król, A.: Various kinds of ordinal sums of fuzzy implications. In: Atanassov, K., ct al. (eds.) Novel Developments in Uncertainty Representation and Processing. Advances in Intelligent Systems and Computing, vol. 401, pp. 37–49. Springer, Cham (2016)
7. Drygaś, P., Król, A.: Generating fuzzy implications by ordinal sums. Tatra Mt. Math. Publ. **66**, 39–50 (2016)
8. Drygaś, P., Król, A.: Two constructions of ordinal sums of fuzzy implications. In: Proceedings of the Fifteenth International Workshop on Intuitionistic Fuzzy Sets and Generalized Nets, Warsaw, Poland, 12–14 October 2016
9. Durante, F., Klement, E.P., Mesiar, R., Sempi, C.: Conjunctors and their residual implicators: characterizations and construction methods. Mediterr. J. Math. **4**, 343–356 (2007)
10. Fodor, J., Roubens, M.: Fuzzy Preference Modelling and Multicriteria Decision Support. Kluwer Academic Publishers, Dordrecht (1994)
11. Jenei, S.: A note on the ordinal sum theorem and its consequence for the construction of triangular norm. Fuzzy Sets Syst. **126**, 199–205 (2002)
12. Klement, E.P., Mesiar, R., Pap, E.: Triangular Norms. Kluwer Academic Publishers, Dordrecht (2000)

13. Król, A.: Generating of fuzzy implications. In: Montero, J., Pasi, G., Ciucci, D. (eds.) Proceedings of the 8th Conference of the European Society for Fuzzy Logic and Technology, EUSFLAT-13, Milano, Italy, 11–13 September 2013, pp. 758–763. Atlantis Press (2013)
14. Mesiar, R., Mesiarová, A.: Residual implications and left-continuous t-norms which are ordinal sums of semigroups. Fuzzy Sets Syst. **143**, 47–57 (2004)
15. Su, Y., Xie, A., Liu, H.: On ordinal sum implications. Inform. Sci. **293**, 251–262 (2015)

On the Visualization of Discrete Non-additive Measures

Juhee Bae, Elio Ventocilla, Maria Riveiro, and Vicenç Torra[✉]

School of Informatics, University of Skövde, Skövde, Sweden
{juhee.bae,elio.ventocilla,maria.riveiro,vtorra}@his.se

Abstract. Non-additive measures generalize additive measures, and have been utilized in several applications. They are used to represent different types of uncertainty and also to represent importance in data aggregation. As non-additive measures are set functions, the number of values to be considered grows exponentially. This makes difficult their definition but also their interpretation and understanding. In order to support understability, this paper explores the topic of visualizing discrete non-additive measures using node-link diagram representations.

1 Introduction

Non-additive measures are monotonic set functions. They generalize additive measures as e.g. probabilities and the Lebesgue measure. Several names are used to represent this concept; they are also called fuzzy measures (name introduced by Sugeno in 1972 [18,19]), capacities (see e.g. Choquet's seminal work [7]) and monotonic games (see e.g. [24]).

Non-additive measures can be used for representing uncertainty. In this case, several families of measures have been defined, see e.g. probabilities, belief and plausibility, as well as possibility and necessity. It is usual to use functions to combine and aggregate these uncertainty measures. For instance, the Demspter-Shafer rule of combination is used for belief measures.

Non-additive measures are also used to represent importance or relevance of information sources in data aggregation [5,11,23]. This is the case when we use the Choquet [7] and the Sugeno integral [19]. These integrals aggregate a set of values proceeding from a set of information sources taking into account the relevance of the sources. Non-additive measures are used to represent our background knowledge on this relevance of the sources. The measures permit us to have more flexibility than the one offered by additive measures. They do not longer require that the measure of a set is the addition of the measure of its components. This permits to represent positive and negative interaction of the elements. That is, we can have for two disjoint sets A and B (i.e., $A \cap B = \emptyset$) that either $\mu(A \cup B) > \mu(A) + \mu(B)$, $\mu(A \cup B) < \mu(A) + \mu(B)$ or just $\mu(A \cup B) = \mu(A) + \mu(B)$ as it is the case for probabilities.

This additional flexibility is at the cost of a more complex definition. As non-additive measures do not satisfy the additivity axiom, we need to supply

© Springer International Publishing AG 2018
V. Torra et al. (eds.), *Aggregation Functions in Theory and in Practice*,
Advances in Intelligent Systems and Computing 581, DOI 10.1007/978-3-319-59306-7_21

values for each subset of the reference set. Being a set function, this means that we need to supply $O(2^n)$ where n is the number of elements of the reference set.

In order to help in the definition of these measures, a few families of measures have been defined with reduced complexity. This is the case of Sugeno λ-measures [19], $\perp$-decomposable fuzzy measures, hierarchically decomposable fuzzy measures [22], distorted and m-dimensional distorted probabilities [14], k-additive measures [10]. There have also been approaches to learn these measures from data. This is the case of e.g. [1,16].

Due to the number of parameters needed to define these measures, it is also difficult to understand what exactly represents a fuzzy measure. For this purpose, several (mathematical) indices can be used. The Shapley [17] and Banzhaf [3] indices are two of them.

In this paper we propose and explore an alternative way to understand these type of measures using graphical representations of the measures. As we will discuss later, our proposal is based on graph visualizations, in particular, node-link diagram representations.

Node-link diagrams [4,13,20] (see e.g. Fig. 1) are widely used to draw relationships between elements in a model. They are used in social networks, process models, and on hierarchical structures [6]. This type of graphs depict a collection of elements (vertices or nodes) and a set of relations between them (edges). Edges may indicate a weight (such as the strength of the relationship), as well as the direction of the relationship between the nodes. It is easier to read and understand node-link diagrams when the underlying relations are simple and sparse [8], however, they are less preferred with many overlapping links, that can generate occlusion problems [4]. The interpretation of the nodes' and links' depends on the application. In fact, one prior user study depicting multivariate data sets [2] gave weights to the links with selected visual cues to better

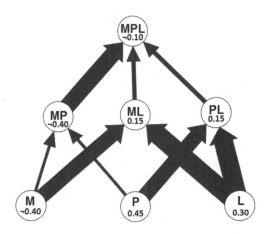

Fig. 1. Visualization of Example 1 where the difference between measures are represented by means of thickness.

understand the relationships' strength. Using similar principles, in this paper we propose the use of brightness and width to better understand and emphasize the relationships of discrete non-additive measures.

The problem of visualizing non-additive measures have also been considered by Murofushi's lab [15,21,25]. They have also used graphs to represent fuzzy measures (Hasse diagrams). As we do here, nodes represent subsets $A \subseteq X$. Then, [15] locates the nodes in the picture taking into account the measure of the sets. In [15] they use a combinatorial optimization problem with exhaustive search to determine the position of the nodes in the picture. In [25] they use a branch and bound algorithm for the same purpose. Their approach is different to our approach here where measures are represented by brightness and width of the edges.

The structure of the paper is as follows. In Sect. 2 we review some basic definitions that we need in this paper. Section 3 introduces our approach for visualizing the measures and Sect. 4 provides visualization examples. The paper finishes with a summary and lines for future work.

2   Preliminaries

In this section we review the definition of non-additive measures. We also give the definition of the Choquet integral, one of the tools used to aggregate data from a set of information sources with respect to the non-additive measure.

Definition 1. A non-additive (or fuzzy) measure μ on a set X is a set function $\mu : \wp(X) \to [0, 1]$ satisfying the following axioms:

(i) $\mu(\emptyset) = 0$, $\mu(X) = 1$ (boundary conditions)
(ii) $A \subseteq B$ implies $\mu(A) \leq \mu(B)$ (monotonicity)

Here, $\wp(X)$ represents the power set of X.

Note that in this definition the additivity axiom $\mu(A \cup B) = \mu(A) + \mu(B)$ for $A \cap B = \emptyset$ is replaced by the monotonicity condition.

Given a set of information sources X (e.g., sensors or experts) we can represent the value supplied by each information source x in X by $f(x)$. Then, μ represents the importance of the sets $A \subseteq X$. That is, μ represents the importance of a set A of information sources.

When the additivity takes place, we have that the importance of a set corresponds to the addition of the importance of its terms. That is $\mu(A) = \sum_{x \in A} \mu(\{x\})$. As this is no longer a requirement we may represent positive interactions between elements and negative interactions. Note that we have a positive interaction between A and B (with $A \cap B = \emptyset$) when

$$\mu(A \cup B) > \mu(A) + \mu(B)$$

and that we have a negative interaction when

$$\mu(A \cup B) < \mu(A) + \mu(B).$$

A well known example of a non-additive measure is the one introduced in [9]. This example is about the evaluation of students of a high school in terms of their ratings in three subjects: mathematics, physics, and literature. The importance of these subjects is expressed by means of a measure. We revise this example below as we will use it for illustration in this paper. The formulation follows [23].

Example 1. The director of a high school has to evaluate the students according to their level in mathematics (M), physics (P), and literature (L). The evaluation consists of obtaining a final rating as an average of the ratings of the three subjects. For each student, the final rating depends on the importance given to the subjects. To settle these importances, a non-additive measure is used. Here, X is the set of all subjects (i.e., $X = \{M, P, L\}$), and $\mu(A)$ is the importance of a particular set of subjects A. The definition of the measure considers the following elements.

1. Boundary conditions:
   $\mu(\emptyset) = 0$, $\mu(\{M, P, L\}) = 1$
   The importance of the empty set is 0. The set consisting of all objects has maximum importance.
2. Relative importance of scientific versus literary subjects:
   $\mu(\{M\}) = \mu(\{P\}) = 0.45$, $\mu(\{L\}) = 0.3$
   The importance of mathematics and physics is greater than the importance of literature.
3. *Redundancy* between mathematics and physics:
   $\mu(\{M, P\}) = 0.5 < \mu(\{M\}) + \mu(\{P\})$
   Mathematics and physics are similar subjects. The importance of the set containing both should not be larger than their addition.
4. Support between literature and scientific subjects:
   $\mu(\{M, L\}) = \mu(\{P, L\}) = 0.9 > \mu(\{P\}) + \mu(\{L\}) = 0.45 + 0.3 = 0.75$
   $\mu(\{M, L\}) = \mu(\{P, L\}) = 0.9 > \mu(\{M\}) + \mu(\{L\}) = 0.45 + 0.3 = 0.75$
   Mathematics and literature are complementary subjects.

An outline of this fuzzy measure is given in Table 1.

In this example we have seen that mathematics and literature have positive interaction while mathematics and physics have negative interaction. One of the ways to observe the positive interaction is by means of the Möbius transform.

The Möbius transform of a non-additive measure on X is a set function that assigns to each subset of X a value (either positive or negative). For each

Table 1. Non-additive measure of Example 1 based on [9].

$\mu(\emptyset) = 0$	$\mu(\{M, L\}) = 0.9$
$\mu(\{M\}) = 0.45$	$\mu(\{P, L\}) = 0.9$
$\mu(\{P\}) = 0.45$	$\mu(\{M, P\}) = 0.5$
$\mu(\{L\}) = 0.3$	$\mu(\{M, P, L\}) = 1$

Table 2. Möbius transform of the measure given in Example 1 and summarized in Table 1.

$m(\emptyset) = 0$	$m(\{M, L\}) = 0.15$
$m(\{M\}) = 0.45$	$m(\{P, L\}) = 0.15$
$m(\{P\}) = 0.45$	$m(\{M, P\}) = -0.4$
$m(\{L\}) = 0.3$	$m(\{M, P, L\}) = -0.1$

non-additive measure there is a unique Möbius transform, and for each Möbius transform there is a unique measure. Formally, a Möbius transform is a function $m : \wp(X) \to \mathbb{R}$ such that $m(\emptyset) = 0$, $\sum_{A \subseteq X} m(A) = 1$, and, if $A \subset B$, then $\sum_{C \subseteq A} m(C) \leq \sum_{C \subseteq B} m(C)$. The following definition explains how to build the Möbius transform from a measure.

Definition 2. Let μ be a fuzzy measure; then, its Möbius transform m is defined as

$$m_\mu(A) := \sum_{B \subseteq A} (-1)^{|A|-|B|} \mu(B) \tag{1}$$

for all $A \subset X$.

Note that the function m is not restricted to the $[0, 1]$ interval.

Given a function m that is a Möbius transform, we can reconstruct the original measure as follows:

$$\mu(A) = \sum_{B \subseteq A} m(B)$$

for all $A \subseteq X$.

Table 2 gives the Möbius transform of the measure in Example 1 and outlined in Table 1.

Given an assignment $f : X \to \mathbb{R}$ (that assigns a value to each information source), and a non-additive measure μ we can aggregate the values $f(x)$ for $x \in X$ by means of a Choquet integral. In Example 1 this means that given a student and three marks one for mathematics, another for physics and a third for literature, we can average them and obtain an aggregated value taking into account the importances of these subjects according to the measure μ. For illustration, we give the definition of the Choquet integral below.

Definition 3. Let μ be a non-additive measure on $X = \{x_1, \ldots, x_N\}$; then, the *Choquet integral* of a function $f : X \to \mathbb{R}^+$ with respect to the fuzzy measure μ is defined by

$$(C) \int f d\mu = \sum_{i=1}^{N} [f(x_{s(i)}) - f(x_{s(i-1)})]\mu(A_{s(i)}), \tag{2}$$

where $f(x_{s(i)})$ indicates that the indices have been permuted so that $0 \leq f(x_{s(1)}) \leq \cdots \leq f(x_{s(N)}) \leq 1$, and where $f(x_{s(0)}) = 0$ and $A_{s(i)} = \{x_{s(i)}, \ldots, x_{s(N)}\}$.

An important property of the Choquet integral is that when the measure is additive it corresponds to the Lebesgue integral. In other words, when the measure is a probability, the Choquet integral corresponds to the weighted mean of the values (where the weights corresponds to the probabilities).

3   Our Approach

In order to visualize graphically a non-additive measure, we first build a graph from the measure, and then we use node-link diagrams to depict the graph. It is well known that a graph consists of nodes or vertices – basic elements–, and edges – relationships between these elements–. That is, a graph G is defined by the pair $G = (V, E)$ where V is the set of vertices and $E \subset V \times V$ is the set of edges. In our case, we consider labeled graphs where both vertices and edges have a label. So, in addition to V and E we have also two label functions l_V and l_E.

The construction of a graph for a non-additive measure μ on the reference set X is as follows.

- Define the set of vertices as the subsets of X excluding the empty set. That is, $V = \wp(X) \setminus \emptyset$.
- Define the set of edges in terms of set inclusion on $\wp(X)$ between sets that only differ in one element. That is,

$$E = \cup_{a \subset X, c \notin a} \{(a, a \cup c)\}.$$

- Assign to each vertex the Möbius transform of the corresponding set. That is, $l_V(A) = m(A)$.
- Assign to each edge (a, b) the difference between the measure on the largest set and the measure on the smallest set. That is, for (a, b) with $a \subset b$ define $l_E((a, b)) = \mu(b) - \mu(a)$.

Then, we depict this graph using a node-link diagram, that is, we represent each vertex (i.e., the corresponding subset of X and its Möbius transform) and the edges (i.e., the difference between the values of the non-additive measures $l_E((a, b))$). We have considered two graphical representations for l_E. In one case this information is depicted by brightness. The values of brightness range from 0.0 to 0.9, where the value 0.0 represents the biggest difference (darker blue), and 0.9 the smallest difference (brighter blue). Then, we transform the difference between values (say d) into brightness using $1 - d$. In the other case, we use the thickness of the link between the nodes for depicting l_E.

To illustrate this construction, we consider the non-additive measure in Example 1. The graph contains 7 nodes corresponding to the subsets of $X = \{M, L, P\}$. That is,

$$V = \{\{M\}, \{P\}, \{L\}, \{M, L\}, \{M, P\}, \{P, L\}, \{M, P, L\}\}.$$

Table 3. Labelling function for the graph constructed for Example 1 and summarized in Table 1.

$(M, ML) = 0.45$	$(L, PL) = 0.60$
$(M, MP) = 0.05$	$(ML, MPL) = 0.1$
$(P, PL) = 0.45$	$(PL, MPL) = 0.1$
$(P, MP) = 0.05$	$(MP, MPL) = 0.5$
$(L, ML) = 0.60$	

Edges will be defined for $(M, ML), (M, MP), (P, PL), (P, MP), (L, PL),$ $(L, MPL), (MP, MPL), (PL, MPL), (ML, MPL)$. Then, l_V is defined for each node according to Table 2. Finally, l_E is defined according to Table 3. As an example, we give the computation of $l_E((M, ML))$ and $l_E((M, MP))$. We use MP to represent the set $\{M, L\}$.

$$l_E((M, ML)) = \mu(ML) - \mu(M) = 0.9 - 0.45 = 0.45$$

$$l_E((M, MP)) = \mu(MP) - \mu(M) = 0.5 - 0.45 = 0.05$$

Figures 1 and 2 represent this graph. Figure 2 corresponds to the case of using brightness. For instance, l_E of the edge (L, PL) is 0.60 and thus a *high* value (therefore, it is depicted by a dark edge), while the l_E of the edge (PL, MPL) is 0.1 (therefore, it is shown with a brighter edge). The default value of the edge's width was 0.43px and hue valued 240 from the HSB model. So, the visualization shows with dark arrows when the measure increase is significant. We can also see that the measure of $\{M, P\}$ is not changed much with respect to the one of $\{M\}$ and $\{P\}$ (all inputs have arrows with light colours) and this causes that the Möbius transformation is negative. In contrast, $\{M, L\}$ and $\{P, L\}$ receive two dark arrows and the Möbius is positive.

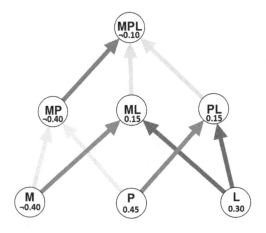

Fig. 2. Visualization of Example 1 where the difference between measures are represented by means of brightness. (Color figure online)

Figure 1 corresponds to the use of thickness to represent the difference between measures. However, it may be perceptually challenging to differentiate between edges with similar thickness values. Thus, we suggest to utilize brightness to encode differences between measures in the next section. Brightness, as well as hue and width, has been used previously for encoding correlation degree in graphs, see e.g. [12].

4   Examples of Visualizations

In this section we present the visualization of another measure that contains five elements, and thus, more relationships. It is a hierarchically decomposable fuzzy measure (see [22] for details) that is based on the structure represented in Fig. 3.

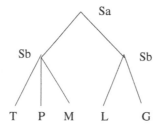

Fig. 3. Graphical representation of a hierarchical decomposable fuzzy measure on the reference set $\{T, M, P, L, G\}$. The reference set contains two subjects for humanities (literature and classical greek) and three scientific subjects (topology, mathematics and physics).

The measure is similar to the one of Example 1, but the reference set includes five subjects instead of three. There are three scientific subjects: mathematics (M), physics (P) and topology $(T$ – in fact, in the original example this is mathematical logics but we use T here for convenience), and two humanistic subjects: literature (L) and greek (G). The measure has some similarities to Example 1 as scientific subjects have more weight than humanistic ones, and interactions between scientific and humanistic are positive while interactions between scientific subjects, and interactions between humanistic are negative.

In this sense note that the Möbius transform can be misleading as $m(\{T, P, M\}) = 0.35$ but $\mu(\{T, P, M\}) = 0.50$ with $\mu(\{T, P\}) = \mu(\{T, M\}) = \mu(\{P, M\}) = 0.47$.

Two visualizations of this measure are given in Fig. 4. Both describe the information by means of the brightness of the colour. One uses standard arrows and the other uses tapered arrows. In this case, the nodes contain the value of the measure for the set (instead of the Möbius transform). Again, we can see the most significant changes.

Figure 5 gives another representation of the measure. In this case, only the edges with a significant difference between measures are shown (i.e., a difference larger than 0.1). The nodes include the Möbius transform. The brightness of

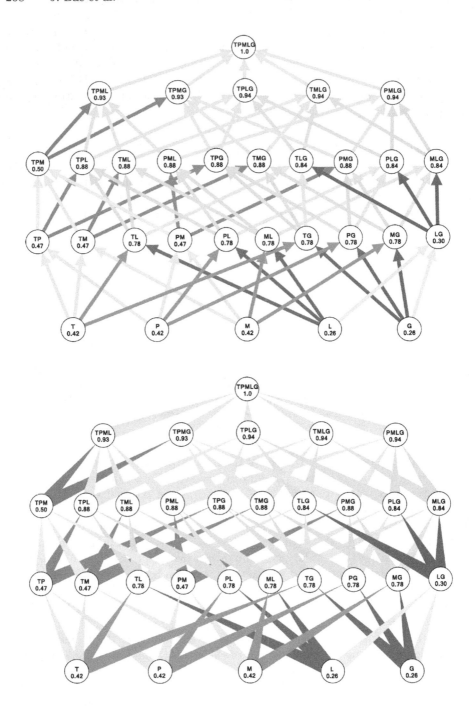

Fig. 4. Visualization of the hierarchical decomposable fuzzy measure. (Color figure online)

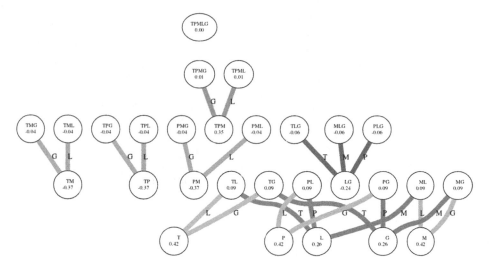

Fig. 5. Visualization of the hierarchical decomposable fuzzy measure. (Color figure online)

the edges is in relation to the difference. Finally, for each edge we have a label corresponding to the element that we are adding. That is, in the edge between $\{T, P, M\}$ and $\{T, P, M, G\}$ we have the label G.

5 Future Work

In this paper we have suggested and explored the visualization of non-additive measures using node-link diagrams. Future work will focus on evaluating the initiatives presented here with users, as well as developing alternative ways for visualizing discrete non-additive measures including, e.g., other indices. For instance, the size of the nodes could encode the Möbius transformation. Moreover, we plan to look at measures on larger sets, and use other tools to build the visualizations, exploring, for example, the use of hypergraphs. We will also explore visualization tools for measures on larger sets, and see if our approach scales well or alternative tools are needed.

Acknowledgements. This research has been conducted within BIDAF 2014/32 and NOVA 20140294 projects, supported by the Swedish Knowledge Foundation.

References

1. Abril, D., Navarro-Arribas, G., Torra, V.: Choquet integral for record linkage. Ann. Oper. Res. **195**, 97–110 (2012)
2. Bae, J., Ventocilla, E., Riveiro, M., Helldin, T., Falkman, G.: Evaluating multi-attributes on cause and effect relationship visualization. In: Proceedings of the International Conference on Information Visualization Theory and Applications (IVAPP), Porto, Portugal (2017)

3. Banzhaf, J.F.: III Weighted voting does not work: a mathematical analysis. Rutgers Law Rev. **19**, 317–343 (1965)
4. Battista, G., Eades, P., Tamassia, R., Tollis, I.: Algorithms for the Visualisation of Graphs. Prentice Hall, Upper Saddle River (1999)
5. Beliakov, G., Pradera, A., Calvo, T.: A Guide for Practitioners. Springer, Heidelberg (2008)
6. Blythe, J., McGrath, C., Krackhardt, D.: The effect of graph layout on inference from social network data. In: Proceedings of the Symposium on Graph Drawing 1995, Passau, Germany (1995)
7. Choquet, G.: Theory of capacities. Ann. Inst. Fourier **5**, 131–295 (1953/54)
8. Ghoniem, M., Fekete, J.-D., Castagliola, P.: A comparison of the readability of graphs using node-link and matrix-based representations. In: Proceedings of the IEEE Symposium on Information Visualization, pp. 17–24 (2004)
9. Grabisch, M.: Fuzzy integral in multicriteria decision making. Fuzzy Sets Syst. **69**, 279–298 (1995)
10. Grabisch, M.: k-order additive discrete fuzzy measures and their representation. Fuzzy Sets Syst. **92**(2), 167–189 (1997)
11. Grabisch, M., Marichal, J.-L., Mesiar, R., Pap, E.: Aggregation Functions. Cambridge University Press, Cambridge (2009)
12. Guo, H., Huang, J., Laidlaw, D.: Representing uncertainty in graph edges: an evaluation of paired visual variables. IEEE Trans. Vis. Comput. Graph. **21**(10), 1173–1186 (2015)
13. Herman, I., Melancon, G., Marshall, M.: Graph visualization and navigation in information visualization: a survey. IEEE Trans. Vis. Comput. Graph. **6**, 24–43 (2003)
14. Narukawa, Y., Torra, V.: Fuzzy measure and probability distributions: distorted probabilities. IEEE Trans. Fuzzy Syst. **13**(5), 617–629 (2005)
15. Ren, W., Murofushi, T.: A visualization system of fuzzy measures using Hasse diagrams. In: 24th Fuzzy Workshop Proceedings, pp. 51–54 (2005)
16. Soria-Frisch, A.: Unsupervised construction of fuzzy measures through self-organizing feature maps and its application in color image segmentation. Int. J. Approx. Reason. **41**, 23–42 (2006)
17. Shapley, L.: A value for n-person games. Ann. Math. Stud. **28**, 307–317 (1953)
18. Sugeno, M.: Fuzzy measures and fuzzy integrals (in Japanese). Trans. Soc. Instrum. Control Eng. **8**, 2 (1972)
19. Sugeno, M.: Theory of fuzzy integrals and its applications. Ph.D. Dissertation, Tokyo Institute of Technology, Tokyo, Japan (1974)
20. Sugiyama, K., Tagawa, S., Toda, M.: Methods for visual understanding of hierarchical system structures. IEEE Trans. Syst. Man Cybern. SMC **11**(2), 109–125 (1981)
21. Takatsuki, R., Murofushi, T.: Interface design for monotonic set function Hasse diagram display system. In: 41st Fuzzy Workshop Proceedings, pp. 21–24 (2015)
22. Torra, V.: On hierarchically S-decomposable fuzzy measures. Int. J. Intel. Syst. **14**(9), 923–934 (1999)
23. Torra, V., Narukawa, Y.: Information Fusion and Aggregation Operators. Springer, Heidelberg (2007)
24. von Neumann, J., Morgenstern, O.: Theory of Games and Economic Behavior. Princeton University Press, Princeton (1944)
25. Watanabe, R., Murofushi, T.: Automated visualization of monotone set functions using the Hasse diagrams. In: 39th Fuzzy Workshop Proceedings, pp. 93–96 (2013)

Generating Recommendations in GDM
with an Allocation of Information Granularity

Francisco Javier Cabrerizo[1]($\boxtimes$), Juan Antonio Morente-Molinera[2],
Sergio Alonso[3], Ignacio Javier Pérez[4], Raquel Ureña[4],
and Enrique Herrera-Viedma[1]

[1] Department of Computer Science and Artificial Intelligence,
University of Granada, Granada, Spain
{cabrerizo,viedma}@decsai.ugr.es
[2] Universidad Internacional de La Rioja (UNIR), Logroño, Spain
juan.morente@unir.net
[3] Department of Software Engineering, University of Granada, Granada, Spain
zerjioi@ugr.es
[4] Department of Computer Science and Engineering,
University of Cádiz, Cádiz, Spain
{ignaciojavier.perez,raquel.urena}@uca.es

Abstract. A Group decision making process is carried out when human beings jointly make an election from a possible collection of alternatives. Here, a question of importance is to avoid winners and losers, in the sense that the choice is not any more attributable to any single individual, but all group members contribute to the decision. For this reason, the agreement or consensus achieved among all the individuals should be as high as possible. In this contribution, a feedback mechanism is presented in order to increase the consensus achieved among the decision makers involved in this kind of problems. It is based on granular computing, which is utilized here to provide the necessary flexibility to increase the consensus. The feedback mechanism is able to deal with heterogeneous contexts, that is, contexts in which the decision makers have importance degrees considering their capacity or talent to handle the problem.

Keywords: Group decision making · Consensus · Feedback mechanism · Granular computing · Heterogeneous contexts

1 Introduction

A group decision making problem (GDM) consists of more than one decision maker interacting to make a decision. In particular, this situation is defined as that one in which a group of decision makers, $E = \{e_1, e_2, \ldots, e_m\}$ ($m \geq 2$), give their assessments about a collection of possible alternatives, $X = \{x_1, x_2, \ldots, x_n\}$ ($n \geq 2$), by means of a particular preference structure. The decision makers aim to reach a common solution by ranking the alternatives from best to worst [7,11,20]. If the decision process is defined in a fuzzy

© Springer International Publishing AG 2018
V. Torra et al. (eds.), *Aggregation Functions in Theory and in Practice*,
Advances in Intelligent Systems and Computing 581, DOI 10.1007/978-3-319-59306-7_22

environment, some degrees of preferences given in the unit interval are associated with the alternatives.

To solve this kind of decision problems, different questions have to be considered:

- To model the evaluations expressed by the decision makers, various representation structures have been used in this research field [5,6]. Among them, the most common one, caused by its efficiency in modeling decision process, is the preference relation because the attempt to perform pairwise evaluations is more reasonable contrasted with any experimental overhead required when giving membership grades to all possible alternatives in an only one stage. This means the decision maker must assess each alternative against all the others as a whole, and it is not an easy job. In particular, researchers have widely used the fuzzy preference relations [27,33] as they have useful properties to operate with them without trouble and also provide an expressive preference representation [5,16].
- In most of the approaches, the opinions provided by the decision makers are considered equally important. However, decision makers usually possess distinct understanding levels and background about the problem at hand. To tackle the different importance levels among the decision makers, a weight value is assigned to each decision maker. This weight value is usually utilized in the aggregation phase to model this heterogeneity [28,38].
- In order to obtain an adequate level of agreement among the assessments conveyed by the decision makers involved in a GDM situation, it is advisable that the decision makers talk about their reasons for providing their preferences. If this process is not effectuated, some decision makers could reject the solutions achieved as they might not accept them [2,32]. Therefore, a consensus process is usually carried out before obtaining the solution in a GDM situation [8,10,20,26,37].

In this contribution, a feedback mechanism is presented to improve the consensus among the decision makers implicated in a GDM problem defined in a heterogeneous context, i.e., in a situation in which it is considered that the decision makers possess distinct level of understanding about the alternatives and, therefore, importance degrees are given to them in order to reflect their importance to solve the problem. The goal of the feedback mechanism is to give advice about how the decision makers should adjust their opinions for the purpose of increasing the level of agreement. Concretely, the feedback mechanism proposed here uses, as a key component, an allocation of information granularity [29,31,35] to generate the advice. That is, because decision makers should be ready to modify their initial assessments on the alternatives, information granularity could provide the required level of flexibility by utilizing the initial assessments which are adjusted in order to increase the consensus. Assuming that the opinions communicated by the decision makers are represented by fuzzy preference relations, this flexibility is achieved by considering granular fuzzy preference relations in which each value is considered as an information granule in place of a single numeric one.

The rest of this manuscript is set up as follows. In Sect. 2, a classical consensus reaching process is described. The proposed feedback mechanism is presented in Sect. 3. An application example of the proposed approach is illustrated in Sect. 4. Lastly, some conclusions are pointed out in Sect. 5.

2   Consensus Reaching Process

A consensus reaching process is a discussion process carried out repeatedly and composed of a number of negotiation rounds. Here, it is assumed that the decision makers want to cooperate and consent to change their evaluations following the recommendations provided by a moderator, who is aware of the agreement degree in each round of the consensus reaching process, for the purpose of increasing the level of consensus. Concretely, a typical consensus process is composed of the following steps:

1. The problem and the collection of possible alternatives to solve it are shown to the decision makers.
2. Decision makers discuss and share their knowledge about the problem with the aim of facilitating the process of providing their preferences.
3. Decision makers express their evaluations about the alternatives in a particular structure of preference representation (in this contribution, fuzzy preference relations).
4. The moderator uses the decision makers' opinions to compute consensus measures allowing him/her to identify if an adequate level of consensus has been obtained.
5. The consensus reaching process stops if an enough level of consensus has been reached and, then, the solution can be obtained by applying a selection process [4, 19]. If not, a feedback mechanism is applied in which the moderator gives advice to the decision makers.
6. The advice is provided to the decision makers and the round is completed. Then, decision makers discuss their evaluations again in order to increase the level of consensus (Step 2).

The moderator knows the consensus level achieved among the decision makers by calculating some consensus measures. To compute them, coincidence among the opinions is calculated [3, 14]. As aforementioned, in this contribution we assume that fuzzy preference relations are used to model the evaluations conveyed by the decision makers [21, 27, 33].

Definition 1. *A fuzzy preference relation PR on a set of alternatives X is a fuzzy set on the Cartesian product $X \times X$, i.e., it is characterized by a membership function $\mu_{PR} : X \times X \rightarrow [0,1]$.*

The $n \times n$ matrix $PR = (pr_{ij})$ is usually utilized to represent a fuzzy preference relation PR. Here, $pr_{ij} = \mu_{PR}(x_i, x_j)$ $(\forall i, j \in \{1, \ldots, n\})$ is the preference degree of the alternative x_i over x_j: $pr_{ij} = 0.5$ means indifference between x_i and

x_j ($x_i \sim x_j$), $pr_{ij} = 1$ means that x_i is entirely preferred to x_j, and $pr_{ij} > 0.5$ means that x_i is preferred to x_j ($x_i \succ x_j$). According to this interpretation, $pr_{ii} = 0.5 \ \forall i \in \{1, \ldots, n\}$ ($x_i \sim x_i$). Since the entries of the main diagonal (pr_{ii}) do not matter here, they will be written as '—' instead of 0.5 [21].

When fuzzy preference relations are used, consensus degrees are given at three different levels: pairs of alternatives, alternatives, and relation. In such a way, consensus degrees are computed as follows [3,36]:

1. A similarity matrix, $SM^{kl} = (sm_{ij}^{kl})$, is determined for each pair of decision makers (e_k, e_l) ($k = 1, \ldots, m-1$, $l = k+1, \ldots, m$) as follows:

$$sm_{ij}^{kl} = 1 - D(pr_{ij}^{k}, pr_{ij}^{l}) \tag{1}$$

where $D : [0,1] \times [0,1] \rightarrow [0,1]$ is a distance function [9]. The closer sm_{ij}^{kl} to 1, the more similar pr_{ij}^{k} and pr_{ij}^{l}.

2. All the $(m-1) \times (m-2)$ similarity matrices are aggregated using an aggregation function ϕ to obtain a consensus matrix $CM = (cm_{ij})$.

$$cm_{ij} = \phi(sm_{ij}^{kl}), \ k = 1, \ldots, m-1, \ l = k+1, \ldots, m \tag{2}$$

3. Then, the consensus degrees are obtained at the three different levels of a fuzzy preference relation:

(a) *Consensus degree on the pairs of alternatives*, cp_{ij}. It is defined to estimate the consensus degree on a pair of alternatives (x_i, x_j), among all the decision makers. It is expressed by the entry of CM:

$$cp_{ij} = cm_{ij} \tag{3}$$

(b) *Consensus degree on the alternatives*, ca_i. This measure is defined to estimate the consensus degree on an alternative (x_i), among all the decision makers. This value is estimated by aggregating the consensus degrees of all the pair of alternatives involving it:

$$ca_i = \phi(cp_{ij}), \ j = 1, \ldots, n \wedge j \neq i \tag{4}$$

(c) *Consensus degree on the relation*, cr. It expresses the global consensus degree among the evaluations given by all the decision makers. It is computed by aggregating all the consensus degrees at the level of alternatives:

$$cr = \phi(ca_i), \ i = 1, \ldots, n \tag{5}$$

The value used to check the consensus state is cr. Concretely, the closer cr to 1, the greater the consensus among all the decision makers' judgments.

3 Generating Advice with an Allocation of Information Granularity

The generation of advice to the decision makers is an essential point for the purpose of increasing the consensus in the next discussion rounds. Therefore, many consensus approaches incorporate a feedback mechanism helping decision makers to discover the alterations they need to make in their assessments to improve the consensus.

At the beginning, the advice was given by the moderator [1, 12, 13, 21–23]. Nevertheless, because of the moderator can introduce some subjectivity in the discussion process, consensus models were developed by the researchers in which the moderator figure was substituted by automatic tools [15, 17, 18, 36].

In GDM situations, a collaborative and cooperative atmosphere is recommendable and, therefore, it is supposed that decision makers are ready to modify their evaluations in order to get better consensus solutions. In such a way, the decision makers have to allow a certain flexibility in their initial assessments. Based on this assumption, we propose a feedback mechanism in which the required flexibility is brought by acknowledging the entries of the fuzzy preference relations as information granules instead of numbers. To emphasize that the feedback mechanism uses granular fuzzy preference relations, the notation $G(PR)$ is employed, being $G(.)$ a specific granular formalism being utilized (fuzzy sets, rough sets, probability density functions, intervals, and so on). Concretely, the feedback mechanism exploits the role of information granularity as a way to increase the accord by treating the granularity level as synonymous of flexibility, which is utilized to optimize an optimization criterion related to the level of consensus. Furthermore, the feedback mechanism is carried out automatically, without a moderator, making more efficient and effective the discussion process.

This optimization problem is not an easy task and, therefore, it requires the use of a global optimization technique. In particular, among different optimization techniques, the Particle Swarm Optimization (PSO) framework is utilized here [24], which is based on the social behavior metaphor. Here, first, a population of particles (candidate solutions) is initialized randomly. Second, a randomized velocity is allocated to each particle, which is iteratively moved though the search-space according to simple mathematical formulae over the particle's velocity and position. The movement of each particle is guided in the direction of the position of the best fitness reached up to now by the particle itself and by the position of the best fitness reached so far across the entire population [24, 25, 34]. This optimization environment has been chosen because it does not present a high level of computational overhead and because it provides an important flexibility level in the optimization.

In what follows, the details of this optimization framework are introduced. In particular, we describe the fitness function, the particle's representation and the algorithm used here.

3.1 Fitness Function

In the PSO environment, during the particle's movement, its performance is evaluated by means of a fitness function or performance criterion. In this contribution, the PSO aims to maximize the consensus achieved among the decision makers. Concretely, the consensus degree cr is used to compute it. Therefore, the optimization problem reads as follows:

$$\text{Max}_{PR^1, PR^2, \ldots, PR^m \in P(PR)} cr \tag{6}$$

Here, we assume intervals to articulate the granularity of information and, therefore, the length of the intervals can be sought as a level of granularity α. Hence, this maximization problem is performed by the feedback mechanism for all interval-valued fuzzy preference relations that are possible according to the fixed level of information granularity α. In addition, because interval-valued fuzzy preference relations are used, $G(PR) = P(PR)$, where $P(.)$ signifies a family of intervals.

In summary, the following fitness function f will be used by the PSO environment:

$$f = cr \tag{7}$$

3.2 Particle

In the PSO framework, how to find a proper association between the problem solution and the particle's representation is very important. In the GDM context assumed here, each particle is represented by a vector whose values belong to the unit interval, i.e., if we assume a group of m decision makers and a set of n possible alternatives, the particle is made of $m \cdot n(n-1)$ entries.

Let us suppose an element pr_{ij} and a granularity level α placed in the $[0, 1]$ interval. If we start with an initial fuzzy preference relation expressed by a decision maker, the interval of acceptable values of this element of $P(PR)$ is calculated as follows:

$$[a, b] = [\max(0, pr_{ij} - \alpha/2), \min(1, pr_{ij} + \alpha/2)] \tag{8}$$

For instance, if we have $pr_{ij} = 0.5$, being the level of granularity $\alpha = 0.2$, and the related element of the particle x equal to 0.2, then, the related interval of the interval-valued fuzzy preference relation computed using Eq. (8) is $[a, b] = [0.40, 0.60]$. Then, using the expression $z = a + (b - a)x$, the modified value of pr_{ij} changes into 0.44.

3.3 Algorithm

The common configuration of the PSO environment is applied in this contribution. Then, the particle's velocity is updated as follows:

$$\mathbf{v}(t+1) = w \times \mathbf{v}(t) + c_1 \mathbf{a} \cdot (\mathbf{z}_p - \mathbf{z}) + c_2 \mathbf{b} \cdot (\mathbf{z}_g - \mathbf{z}) \tag{9}$$

being "t" an index of the generation or iteration, · a vector multiplication carried out coordinate-wise, z_g the best position overall and developed up to this point across the swarm, and z_p the best position obtained up to this point for the particle under study.

The inertia component (w) scales the actual velocity $v(t)$ and stresses some effect of resistance to modify the actual velocity. Its value is usually 0.2 and it is kept constant through all the optimization process [30]. a and b represent vectors of random numbers obtained from the uniform distribution over the unit interval. These vectors help form an appropriate combination of the components of the velocity. Particularly interesting is the second expression controlling the modification in the particle's velocity because it captures the connections between the particle's history and the overall population's history in terms of their performance obtained up to this point.

Finally, in generation "t + 1", the particle's position is calculated as:

$$z(t + 1) = z(t) + v(t + 1) \tag{10}$$

Once the fuzzy preference relations have been optimized by the PSO environment, they are presented to the decision makers in order to improve the consensus.

4   Practical Example

In this section, we illustrate an application example in order to quantify the increase of the consensus when our approach is used.

Let us suppose a GDM problem in which a group of four decision makers, $\{e_1, e_2, e_3, e_4\}$, have to express their opinions about a collection of four alternatives, $\{x_1, x_2, x_3, x_4\}$. Furthermore, because the decision makers possess different levels of knowledge about the problem, the following weight values are assigned to them: $w_1 = 0.35$, $w_2 = 0.25$, $w_3 = 0.20$, and $w_4 = 0.20$.

At the beginning of the discussion process, the following fuzzy preference relations are provided by the group of four decision makers:

$$PR^1 = \begin{pmatrix} - & 0.40 & 0.70 & 0.20 \\ 0.60 & - & 0.60 & 0.70 \\ 0.30 & 0.10 & - & 0.20 \\ 0.80 & 0.20 & 0.90 & - \end{pmatrix} \quad PR^2 = \begin{pmatrix} - & 0.20 & 0.60 & 0.70 \\ 0.90 & - & 0.40 & 0.20 \\ 0.20 & 0.70 & - & 0.70 \\ 0.10 & 0.70 & 0.20 & - \end{pmatrix}$$

$$PR^3 = \begin{pmatrix} - & 0.90 & 0.10 & 0.10 \\ 0.10 & - & 0.80 & 0.80 \\ 0.80 & 0.10 & - & 0.80 \\ 0.10 & 0.20 & 0.20 & - \end{pmatrix} \quad PR^4 = \begin{pmatrix} - & 0.80 & 0.30 & 0.60 \\ 0.20 & - & 0.70 & 0.30 \\ 0.80 & 0.40 & - & 0.40 \\ 0.40 & 0.90 & 0.60 & - \end{pmatrix}$$

Once the preferences have been provided, the consensus measures are computed as presented in Sect. 2. In this example, the arithmetic mean is used as aggregation function ϕ, and the following distance measure is used:

$$D(pr_{ij}^k, pr_{ij}^l) = |pr_{ij}^k - pr_{ij}^l| \tag{11}$$

The consensus matrix is equal to:

$$CM = \begin{pmatrix} - & 0.58 & 0.65 & 0.63 \\ 0.53 & - & 0.78 & 0.63 \\ 0.62 & 0.65 & - & 0.65 \\ 0.58 & 0.57 & 0.58 & - \end{pmatrix}$$

The element (i, j) of the consensus matrix represents the consensus degrees on the pair of alternatives (x_i, x_j).

The consensus degrees on the alternatives are:

$$ca_1 = 0.60$$
$$ca_2 = 0.63$$
$$ca_3 = 0.66$$
$$ca_4 = 0.61$$

And the consensus on the relation is:

$$cr = 0.62$$

Here, we suppose the consensus achieved is not sufficient and, therefore, the decision makers must adjust their evaluations in order to increase the level of consensus before applying the selection process. To help them, our feedback mechanism is applied.

First, as a consequence of an exhaustive experimentation, in the PSO environment, the following values of the parameters were selected:

- 50 particles formed the swarm, as this value was found to obtain stable results.
- 200 iterations were performed as it was observed that, after this number of iterations, were no further modifications of the values of the fitness function.
- c_1 and c_2 were set as 2 because these values are usually found in the current literature.

Given a level of granularity $\alpha = 0.6$, the feedback mechanism recommends the following fuzzy preference relations to the decision makers:

$$PR^1 = \begin{pmatrix} - & 0.24 & 0.57 & 0.19 \\ 0.53 & - & 0.56 & 0.66 \\ 0.08 & 0.20 & - & 0.23 \\ 0.80 & 0.33 & 0.60 & - \end{pmatrix} \quad PR^2 = \begin{pmatrix} - & 0.26 & 0.65 & 0.67 \\ 0.80 & - & 0.36 & 0.24 \\ 0.33 & 0.57 & - & 0.60 \\ 0.20 & 0.62 & 0.05 & - \end{pmatrix}$$

$$PR^3 = \begin{pmatrix} - & 0.83 & 0.19 & 0.24 \\ 0.11 & - & 0.65 & 0.76 \\ 0.62 & 0.17 & - & 0.65 \\ 0.77 & 0.19 & 0.05 & - \end{pmatrix} \quad PR^4 = \begin{pmatrix} - & 0.70 & 0.16 & 0.55 \\ 0.32 & - & 0.60 & 0.14 \\ 0.67 & 0.43 & - & 0.29 \\ 0.22 & 0.80 & 0.30 & - \end{pmatrix}$$

In the next discussion round, it is assumed that the decision makers accept the advice generated by the feedback mechanism and, therefore, the consensus measures are calculated again.

The consensus matrix is equal to:

$$CM = \begin{pmatrix} - & 0.63 & 0.69 & 0.71 \\ 0.62 & - & 0.85 & 0.62 \\ 0.66 & 0.76 & - & 0.74 \\ 0.61 & 0.65 & 0.68 & - \end{pmatrix}$$

The element (i, j) of the consensus matrix represents the consensus degrees on the pair of alternatives (x_i, x_j).

The consensus degrees on the alternatives are:

$$ca_1 = 0.66$$
$$ca_2 = 0.69$$
$$ca_3 = 0.73$$
$$ca_4 = 0.67$$

And the consensus on the relation is:

$$cr = 0.69$$

At this step, if we consider that the level of consensus is enough, the selection process would be applied in order to obtain the solution set of alternatives. In this process, the weights of the decision makers should be considered when aggregating their opinions.

To conclude this practical example, it should be noted that a granularity level equal to 0.6 has been utilized here. Using this granularity level, we have observed that the consensus among the decision makers has been improved. In such a way, the higher the level of granularity, the higher the level of flexibility and, therefore, the possibility of increasing the consensus. For instance, if a granularity level of 0.3 is used, the consensus achieved is equal to 0.65, and if a granularity level of 0.8 is used, the consensus reached is equal to 0.75.

5   Conclusions

An allocation of information granularity has been used in this contribution to give advice that helps the decision makers involved in a GDM situation to increase the level of consensus. Concretely, a feedback mechanism generating recommendations has been presented, in which the level of granularity has been used as synonymous of flexibility. This granularity level has been utilized to optimize the consensus level (optimization criterion): the higher the level of granularity, the higher the level of adaptability and, in such a way, the level of consensus achieved.

Finally, we would like to clarify the following aspects of the feedback mechanism presented here:

– The higher the level of granularity α, the higher the changes advised by the feedback mechanism and, in such a way, the possibility of obtaining the required level of consensus is higher. However, if the granularity level α is very high, the changes advised could be very different in comparison with the first opinions provided by the decision makers and, hence, the decision makers could not accept them.

- The PSO framework has been shown as a suitable optimization technique for this problem. Nevertheless, the fitness function is optimized by the PSO framework and the result obtained is the best one being formed by the PSO, but there is no promise that it is an optimal result.

Acknowlededement. The authors would like to acknowledge FEDER financial support from the Projects TIN2013-40658-P and TIN2016-75850-P.

References

1. Bordogna, G., Fedrizzi, M., Pasi, A.: A linguistic modeling of consensus in group decision making based on OWA operators. IEEE Trans. Syst. Man Cybern.-Part A: Syst. Hum. **27**(1), 126–133 (1997)
2. Butler, C.T., Rothstein, A.: On Conflict and Consensus: A Handbook on Formal Consensus Decision Making. Tahoma Park (2006)
3. Cabrerizo, F.J., Moreno, J.M., Pérez, I.J., Herrera-Viedma, E.: Analyzing consensus approaches in fuzzy group decision making: advantages and drawbacks. Soft Comput. **14**(5), 451–463 (2010)
4. Cabrerizo, F.J., Heradio, R., Pérez, I.J., Herrera-Viedma, E.: A selection process based on additive consistency to deal with incomplete fuzzy linguistic information. J. Univ. Comput. Sci. **16**(1), 62–81 (2010)
5. Chiclana, F., Herrera, F., Herrera-Viedma, E.: Integrating three representation models in fuzzy multipurpose decision making based on fuzzy preference relations. Fuzzy Sets Syst. **97**(1), 33–48 (1998)
6. Chiclana, F., Herrera, F., Herrera-Viedma, E.: A note on the internal consistency of various preference representations. Fuzzy Sets Syst. **131**(1), 75–78 (2002)
7. Chen, S.J., Hwang, C.L.: Fuzzy Multiple Attributive Decision Making: Theory and its Applications. Springer, Berlin (1992)
8. Chu, J., Liu, X., Wang, Y., Chin, K.-S.: A group decision making model considering both the additive consistency and group consensus of intuitionistic fuzzy preference relations. Comput. Ind. Eng. **101**, 227–242 (2016)
9. Deza, M.M., Deza, E.: Encyclopedia of Distances. Springer, Berlin (2009)
10. Dong, Y., Xiao, J., Zhang, H., Wang, T.: Managing consensus and weights in iterative multiple-attribute group decision making. Appl. Soft Comput. **48**, 80–90 (2016)
11. Fodor, J., Roubens, M.: Fuzzy preference modelling and multicriteria decision support. Kluwer, Dordrecht (1994)
12. Herrera, F., Herrera-Viedma, E., Verdegay, J.L.: A model of consensus in group decision making under linguistic assessments. Fuzzy Sets Syst. **7**(1), 73–87 (1996)
13. Herrera, F., Herrera-Viedma, E., Verdegay, J.L.: A rational consensus model in group decision making using linguistic assessments. Fuzzy Sets Syst. **88**(1), 31–49 (1997)
14. Herrera, F., Herrera-Viedma, E., Verdegay, J.L.: Linguistic measures based on fuzzy coincidence for reaching consensus in group decision making. Int. J. Approx. Reason. **16**(3–4), 309–334 (1997)

15. Herrera-Viedma, E., Herrera, F., Chiclana, F.: A consensus model for multiperson decision making with different preference structures. IEEE Trans. Syst. Man Cybern.-Part A: Syst. Hum. **32**(3), 394–402 (2002)
16. Herrera-Viedma, E., Herrera, F., Chiclana, F., Luque, M.: Some issues on consistency of fuzzy preference relations. Eur. J. Oper. Res. **154**(1), 98–109 (2004)
17. Herrera-Viedma, E., Martínez, L., Mata, F., Chiclana, F.: A consensus support system model for group decision-making problems with multigranular linguistic preference relations. IEEE Trans. Fuzzy Syst. **3**(5), 644–658 (2005)
18. Herrera-Viedma, E., Alonso, S., Chiclana, F., Herrera, F.: A consensus model for group decision making with incomplete fuzzy preference relations. IEEE Trans. Fuzzy Syst. **15**(5), 863–877 (2007)
19. Herrera-Viedma, E., Herrera, F., Alonso, S.: Group decision-making model with incomplete fuzzy preference relations based on additive consistency. IEEE Trans. Syst. Man Cybern.-Part B: Cybern. **37**(1), 176–189 (2007)
20. Herrera-Viedma, E., Cabrerizo, F.J., Kacprzyk, J., Pedrycz, W.: A review of soft consensus models in a fuzzy environment. Inf. Fusion **17**, 4–13 (2014)
21. Kacprzyk, J., Fedrizzi, M.: 'Soft' consensus measures for monitoring real consensus reaching processes under fuzzy preferences. Control Cybern. **15**(3–4), 309–323 (1986)
22. Kacprzyk, J., Fedrizzi, M.: A 'soft' measure of consensus in the setting of partial (fuzzy) preferences. Eur. J. Oper. Res. **34**(3), 316–325 (1988)
23. Kacprzyk, J., Fedrizzi, M., Nurmi, H.: Group decision making and consensus under fuzzy preferences and fuzzy majority. Fuzzy Sets Syst. **49**(1), 21–31 (1992)
24. Kennedy, J., Eberhart, R.C.: Particle swarm optimization. In: Proceedings of the IEEE International Conference on Neural Networks, pp. 1942–1948 (1995)
25. Kennedy, J., Eberhart, R.C., Shi, Y.: Swarm Intelligence. Morgan Kaufmann Publishers, San Francisco (2001)
26. Ma, L.-C.: A new group ranking approach for ordinal preferences based on group maximum consensus sequences. Eur. J. Oper. Res. **251**(1), 171–181 (2016)
27. Orlovski, S.A.: Decision-making with a fuzzy preference relation. Fuzzy Sets Syst. **1**(3), 155–167 (1978)
28. Pérez, I.J., Cabrerizo, F.J., Alonso, S., Herrera-Viedma, E.: A new consensus model for group decision making problems with non-homogeneous experts. IEEE Trans. Syst. Man Cybern.: Hum. **44**(4), 494–498 (2014)
29. Pedrycz, W.: The principle of justifiable granularity and an optimization of information granularity allocation as fundamentals of granular computing. J. Inf. Process. Syst. **7**(3), 397–412 (2011)
30. Pedrycz, A., Hirota, K., Pedrycz, W., Dong, F.: Granular representation and granular computing with fuzzy sets. Fuzzy Sets Syst. **203**, 17–32 (2012)
31. Pedrycz, W.: Knowledge management and semantic modeling: a role of information granularity. Int. J. Softw. Eng. Knowl. **23**(1), 5–12 (2013)
32. Saint, S., Lawson, J.R.: Rules for Reaching Consensus: A Moderm Approach to Decision Making. Jossey-Bass, San Francisco (1994)
33. Tanino, T.: Fuzzy preference orderings in group decision making. Fuzzy Sets Syst. **12**(2), 117–131 (1984)
34. Trelea, I.C.: The particle swarm optimization algorithm: convergence analysis and parameter selection. Inf. Process. Lett. **85**, 317–325 (2003)
35. Wang, X., Pedrycz, W., Gacek, A., Liu, X.: From numeric data to information granules: a design through clustering and the principle of justifiable granularity. Knowl.-Based Syst. **101**, 100–113 (2016)

36. Ureña, M.R., Cabrerizo, F.J., Morente-Molinera, J.A., Herrera-Viedma, E.: GDM-R: a new framework in R to support fuzzy group decision making processes. Inf. Sci. **357**, 161–181 (2016)
37. Wu, Z., Xu, J.: Managing consistency and consensus in group decision making with hesitant fuzzy linguistic preference relations. Omega **65**, 28–40 (2016)
38. Yager, R.R.: Weighted maximum entropy owa aggregation with applications to decision making under risk. IEEE Trans. Syst. Man Cybern.-Part A: Syst. Hum. **39**(3), 555–564 (2009)

Aggregation Functions, Similarity and Fuzzy Measures

Surajit Borkotokey[1], Magdaléna Komorníková[2], Jun Li[3(⊠)],
and Radko Mesiar[2]

[1] Dibrugarh University, Dibrugarh 786004, Assam, India
surajitbor@yahoo.com
[2] Faculty of Civil Engineering, Slovak University of Technology,
Radlinského 11, 810 05 Bratislava, Slovakia
{magdalena.komornikova,radko.mesiar}@stuba.sk
[3] School of Science, Communication University of China, Beijing 100024, China
lijun@cuc.edu.cn

Abstract. We propose a new method for constructing fuzzy measures. This method is based on a fixed aggregation function A, similarity measure S and a vector $\mathbf{x} \in [0,1]^n$. Some illustrative examples yielding parametric families of fuzzy measures are given, and some properties of our method are studied.

Keywords: Aggregation function · Similarity · Fuzzy measures

1 Introduction

When studying links between aggregation functions and fuzzy measures, mostly several kinds of integrals are considered, assigning to a fixed fuzzy measure m the related aggregation function. Recall, as a distinguished example, the Choquet integral (i.e., construction of comonotone additive aggregation functions) and the Sugeno integral (i.e., the construction of comonotone maxitive and minitive aggregation functions). For more details we recommend [1,4,6,7]. An alternative view, when, based on a fixed aggregation function, some related fuzzy measures are derived, was proposed by Benvenuti et al. [2]. In this paper, we introduce a more sophisticated approach how to derive fuzzy measure based on a given fixed aggregation function A. In our approach, we deal with a similarity measure S, too, and with an arbitrary n-ary vector $\mathbf{x} \in [0,1]^n$. Note that Benvenuti's approach was based on constant vector $\mathbf{c} \in [0,1]^n$ only, requiring $A(\mathbf{c}) > 0$.

The paper is organized as follows. In the next section, we recall necessary preliminaries. Section 3 brings our new method for constructing fuzzy measures. This method is exemplified in Sect. 4. Some properties and particular instances of our method are contained in Sect. 5. Finally, some concluding remarks are added.

© Springer International Publishing AG 2018
V. Torra et al. (eds.), *Aggregation Functions in Theory and in Practice*,
Advances in Intelligent Systems and Computing 581, DOI 10.1007/978-3-319-59306-7_23

2   Preliminaries

We will deal with universe $\mathcal{N} = \{1, \ldots, n\}$, where $n \geq 2$ can be seen, for example, as a number of considered criteria. An aggregation function $A : [0,1]^n \to [0,1]$ is a monotone function preserving the bounds, $A(\mathbf{0}) = 0$ and $A(\mathbf{1}) = 1$ (i.e., it is an order-preserving homomorphism of lattices $[0,1]^n$ and $[0,1]$). Similarly, a fuzzy measure $m : 2^{\mathcal{N}} \to [0,1]$ is an order preserving homomorphism, i.e., m is a monotone set function satisfying $m(\emptyset) = 0$ and $m(\mathcal{N}) = 1$. In what follows, we will work with the characteristic function 1_E of the set $E \subseteq \mathcal{N}$, i.e.,

$$1_E(i) = \begin{cases} 1 & \text{if } i \in E, \\ 0 & \text{otherwise.} \end{cases}$$

It is not difficult to see that, for any aggregation function $A : [0,1]^n \to [0,1]$, the mapping $m_A : 2^{\mathcal{N}} \to [0,1]$ given by

$$m_A(E) = A(1_E), \ E \subseteq \mathcal{N},$$

is a fuzzy measure. Similarly, if $A(\mathbf{c}) > 0$ for some constant vector $\mathbf{c} \in]0,1]^n$, $c \in]0,1]$, then $m_{A,c} : 2^{\mathcal{N}} \to [0,1]$ given by

$$m_{A,c}(E) = \frac{A(c \cdot 1_E)}{A(\mathbf{c})}, E \subseteq \mathcal{N},$$

is a fuzzy measure, see [2].

We will work also with similarity measures [3,5]. Recall that a function $S : [0,1]^2 \to [0,1]$ is called a similarity measure whenever it satisfies the next axioms:

(S1) $S(x,x) = 1$ for each $x \in [0,1]$;
(S2) $S(0,1) = 0$;
(S3) $S(x,y) = S(y,x)$ for each $x,y \in [0,1]$;
(S4) for each $0 \leq x \leq y \leq z \leq 1$, $S(x,z) \leq S(x,y) \wedge S(y,z)$.

3   A New Construction Method for Fuzzy Measures

Consider a fixed aggregation function $A : [0,1]^n \to [0,1]$, a fixed similarity measure $S : [0,1]^2 \to [0,1]$, and a vector $\mathbf{x} \in [0,1]^n$. For $E \subseteq \mathcal{N}$, let

$$E_{A,\mathbf{x}}^+ = A(\mathbf{x} \vee 1_{E^c})$$

and

$$E_{A,\mathbf{x}}^- = A(\mathbf{x} \wedge 1_E),$$

where E^c is the complement of the set E, $E^c = \mathcal{N} \setminus E$.

It is not difficult to check that:

(i) $\emptyset_{A,\mathbf{x}}^+ = A(\mathbf{1}) = 1$ and $\emptyset_{A,\mathbf{x}}^- = A(\mathbf{0}) = 0$;
(ii) $\mathcal{N}_{A,\mathbf{x}}^+ = A(\mathbf{x}) = \mathcal{N}_{A,\mathbf{x}}^-$;

(iii) if $E \subseteq F \subseteq \mathcal{N}$ then $E_{A,\mathbf{x}}^- \leq F_{A,\mathbf{x}}^- \leq F_{A,\mathbf{x}}^+ \leq E_{A,\mathbf{x}}^+$.

Based on the above properties, the next claim can be verified.

Theorem 1. *Let* $A : [0,1]^n \to [0,1]$ *be an aggregation function,* $S : [0,1]^2 \to [0,1]$ *be a similarity measure and* $\mathbf{x} \in [0,1]^n$. *Then the mapping* $m_{A,S,\mathbf{x}} : 2^{\mathcal{N}} \to [0,1]$ *given by*

$$m_{A,S,\mathbf{x}}(E) = S(E_{A,\mathbf{x}}^-, E_{A,\mathbf{x}}^+)$$

is a fuzzy measure.

Proof. The monotonicity of $m_{A,S,\mathbf{x}}$ follows from the above property (iii). Concerning the boundary conditions, we have $m_{A,S,\mathbf{x}}(\emptyset) = S(1,0) = 0$, see the property (i) above, and $m_{A,S,\mathbf{x}}(\mathcal{N})) = S(A(\mathbf{x}), A(\mathbf{x})) = 1$, see property (ii) above. Summarizing, we see that $m_{A,S,\mathbf{x}}$ is a fuzzy measure.

4   Examples

In this section we give some illustrations of Theorem 1.

Example 1. Let $A : [0,1]^3 \to [0,1]$ be an OWA operator [8] given by

$$A(\mathbf{x}) = 0.2 \max(\mathbf{x}) + 0.5 \operatorname{med}(\mathbf{x}) + 0.3 \min(\mathbf{x}).$$

Consider $S = S_1$, i.e., $S(x,y) = 1 - |x - y|$, and let $\mathbf{x} = (x_1, x_2, x_3)$ be such that $x_1 = \operatorname{med}(\mathbf{x})$, $x_2 = \min(\mathbf{x})$ and $x_3 = \max(\mathbf{x})$. Then, for $m = m_{A,S,\mathbf{x}}$, it holds:

$$m(\{1\}) = 0.3 - 0.1x_1, \quad m(\{2\}) = 0.3 - 0.1x_2, \quad m(\{3\}) = 0.3 - 0.1x_3,$$

$$m(\{1,2\}) = 0.8 - 0.3x_1 + 0.2x_2, \quad m(\{2,3\}) = 0.8 - 0.3x_3 + 0.2x_2$$

and

$$m(\{1,3\}) = 0.8 - 0.3x_3 + 0.2x_1.$$

Example 2. Consider a weighted arithmetic mean $W : [0,1]^n \to [0,1]$ given by $W(\mathbf{x}) = \sum_{i=1}^n w_i x_i$, where $\mathbf{w} \in [0,1]^n$ and $\sum_{i=1}^n w_i = 1$. Then, for $E \subseteq \mathcal{N}$,

$$E_{W,\mathbf{x}}^+ = \sum_{i \notin E} w_i + \sum_{i \in E} w_i x_i$$

and

$$E_{W,\mathbf{x}}^- = \sum_{i \in E} w_i x_i.$$

For $\lambda \in]0, \infty[$, define $S_\lambda : [0,1]^2 \to [0,1]$ by

$$S_\lambda(x,y) = 1 - (x - y)^\lambda.$$

Then S_λ is a similarity measure. Based on Theorem 1, we have a parametric class $(m_{W,S_\lambda,\mathbf{x}})_{\lambda \in]0,\infty[, \, \mathbf{x} \in [0,1]^n}$ of fuzzy measures given by

$$m_{W,S_\lambda,\mathbf{x}}(E) = 1 - \left(\sum_{i \notin E} w_i \right)^\lambda.$$

In this class, the choice of $\mathbf{x} \in [0,1]^n$ is irrelevant, $m_{W,S_\lambda,\mathbf{x}} = m_{W,S_\lambda,\mathbf{y}}$ for all $\mathbf{x}, \mathbf{y} \in [0,1]^n$. In particular, if $\lambda = 1$, $m_{W,S_1,\mathbf{x}}(E) = \sum_{i \in E} w_i$, i.e., we have obtained a probability measure on $\mathcal{N}$ related to the normed weighting vector $\mathbf{w} = (w_1, \ldots, w_n)$.

Put now $S : [0,1]^2 \to [0,1]$,

$$S(x,y) = \frac{x \wedge y}{x \vee y},$$

with convention $\frac{0}{0} = 1$. Then S is a similarity measure and the fuzzy measure $m_{W,S,\mathbf{x}} : 2^\mathcal{N} \to [0,1]$ is given by

$$m_{W,S,\mathbf{x}}(E) = \frac{\sum_{i \in E} w_i x_i}{1 - \sum_{i \in E} w_i(1 - x_i)}.$$

Then $m_{W,S,1} = m_{W,S_1,1}$ is the probability measure related to $\mathbf{w} = (w_1, \ldots, w_n)$, $m_{W,S,0} = m_*$ is the smallest fuzzy measure,

$$m_*(E) = \begin{cases} 1 & \text{if } E = \mathcal{N}, \\ 0 & \text{otherwise,} \end{cases}$$

etc. We see now that $m_{W,S,\mathbf{x}}$ depends on $\mathbf{x}$.

5   Some Properties and Particular Cases

When considering dualities, the next result can be shown.

Proposition 1. *Let $A : [0,1]^n \to [0,1]$ be an aggregation function, $S : [0,1]^2 \to [0,1]$ be a similarity measure and $\mathbf{x} \in [0,1]^n$. Then $m_{A,S,\mathbf{x}} = m_{A^d,S^d,1-\mathbf{x}}$, where $A^d : [0,1]^n \to [0,1]$ is the dual aggregation function given by $A^d(\mathbf{x}) = 1 - A(1 - \mathbf{x})$, and $S^d : [0,1]^2 \to [0,1]$ is the dual similarity measure given by $S^d(x,y) = S(1 - x, 1 - y)$.*

For extremal aggregation functions $A^*, A_* : [0,1]^n \to [0,1]$ given by

$$A^*(\mathbf{x}) = \begin{cases} 0 & \text{if } \mathbf{x} = \mathbf{0}, \\ 1 & \text{otherwise} \end{cases}$$

and

$$A_*(\mathbf{x}) = \begin{cases} 1 & \text{if } \mathbf{x} = \mathbf{1}, \\ 0 & \text{otherwise,} \end{cases}$$

we have the following result, independent of the considered similarity measure S:

$$m_{A^*,S,\mathbf{x}}(E) = \begin{cases} 0 & \text{if } E \neq \mathcal{N}, \ E \cap \text{supp } \mathbf{x} = \emptyset, \\ 1 & \text{otherwise} \end{cases}$$

and

$$m_{A_*,S,\mathbf{x}}(E) = \begin{cases} 0 & \text{if } E \neq \mathcal{N}, \ E \subseteq \ker \mathbf{x}, \\ 1 & \text{otherwise}, \end{cases}$$

where supp $\mathbf{x} = \{i \in \mathcal{N} \,|\, x_i > 0\}$ and ker $\mathbf{x} = \{i \in \mathcal{N} \,|\, x_i = 1\}$.

As an interesting particular case consider $\mathbf{x} \in \{0,1\}^n$, i.e., the case when $\mathbf{x} = 1_B$ for some set $B \subseteq \mathcal{N}$. Then $\mathbf{x} \vee 1_{E^c} = 1_{B \cup E^c}$ and $\mathbf{x} \wedge 1_{E^c} = 1_{B \cap E}$, and thus for determining values $E_{A,\mathbf{x}}^+$ and $E_{A,\mathbf{x}}^-$, we need to know the fuzzy measure m_A only. Denoting, for simplification, $m_A = m$, we have then

$$E_{A,\mathbf{x}}^+ = m(B \cup E^c) \text{ and } E_{A,\mathbf{x}}^- = m(B \cap E).$$

These facts ensure the next result.

Theorem 2. *Let $m : 2^{\mathcal{N}} \to [0,1]$ be a fuzzy measure. Then, for any similarity measure $S : [0,1]^2 \to [0,1]$ and any $B \subseteq \mathcal{N}$, the mapping $m_{S,B} : 2^{\mathcal{N}} \to [0,1]$ given by*

$$m_{S,B}(E) = S(m(B \cap E), m(B \cup E^c))$$

is a fuzzy measure.

The proof follows from the above facts and Theorem 1 directly and therefore it is omitted.

Using the notation from Sect. 4, we see that

- $m_{S_\lambda,\emptyset} = 1 - (m(E^c))^\lambda = (m^\lambda)^d (E)$, and, in particular, $m_{S_1,0} = m^d$;
- $m_{S_\lambda,\mathcal{N}} = 1 - (1 - m(E))^\lambda$, and, in particular, $m_{S_1,\mathcal{N}} = m$;
- $m_{S,B}(E) = \frac{m(B \cap E)}{m(B \cup E^c)}$ (with convention $\frac{0}{0} = 1$), and, in particular,

$$m_{S,\mathcal{N} \setminus \{i\}}(E) = \begin{cases} m(E) & \text{if } i \notin E, \\ \frac{m(E \setminus \{i\})}{m(\mathcal{N} \setminus \{i\})} & \text{if } i \in E. \end{cases}$$

As a by–product, we see also a construction method based on m and S only (the first result is related to $B = \emptyset$, the second one to $B = \mathcal{N}$).

Proposition 2. *Let $m : 2^{\mathcal{N}} \to [0,1]$ be a fuzzy measure and $S : [0,1]^2 \to [0,1]$ be a similarity measure. Then the mappings $m_S^{(1)}, m_S^{(2)} : 2^{\mathcal{N}} \to [0,1]$ given by*

$$m_S^{(1)}(E) = S(0, m(E^c))$$

and

$$m_S^{(2)}(E) = S(m(E), 1)$$

are fuzzy measures.

6   Concluding Remarks

We have introduced and discussed new construction methods for fuzzy measures. In its most general form, our method is based on an aggregation function A, a similarity measure S and a vector $\mathbf{x}$. We have proposed also more specific methods, based, e.g., on a fixed fuzzy measure m, a similarity measure S and a set B. Though in some cases our approach is limited, see, e.g., $(m_{W,S_\lambda,\mathbf{x}})$ family discussed in Sect. 4, in most of the cases all considered parameters A, S and $\mathbf{x}$ have their influence on the resulting fuzzy measure $m_{A,S,\mathbf{x}}$. Note that our approach modifies the earlier approach of Benvenuti et al. [2], allowing to consider any vector $\mathbf{x} \in [0,1]^n$. We expect application of our approach in multicriteria decision support exploiting fuzzy measures, where some perturbation of an a-priori given fuzzy measure m is necessary to fit better to the considered real data.

Acknowledgement. This work was supported in part by Slovak Research and Development Agency under contract No. APVV–14–0013, the National Natural Science Foundation of China (Grants No. 11371332 and No. 11571106). Surajit Borkotokey acknowledges SAIA, Slovakia for providing him a scholarship under the National Scholarship Programme and the Department of Mathematics, STU, Bratislava for the hospitality received during January–March 2017.

References

1. Beliakov, G., Pradera, A., Calvo, T.: Aggregation Functions: A Guide for Practitioners. Springer, Berlin (2007)
2. Benvenuti, P., Vivona, D., Divari, M.: Aggregation operators and associated fuzzy measures. Int. J. Uncertain. Fuzziness Knowl. Based Syst. **9**(2), 197 (2001). doi:10. 1142/S0218488501000739
3. De Baets, B., De Meyer, H., Naessens, H.: A class of rational cardinality-based similarity measures. J. Comput. Appl. Math. **132**, 51–69 (2001)
4. Grabisch, M., Marichal, J.-L., Mesiar, R., Pap, E.: Aggregation Functions. Encyclopedia of Mathematics and Its Applications, vol. 127. Cambridge University Press, Cambridge (2009)
5. Pekala, B.: Similarity measure defined from overlap function. In: Proceedings of the AGOP 2015, Katowice, pp. 205–210 (2015)
6. Torra, V., Narukawa, Y.: Modeling Decisions: Information Fusion and Aggregation Operators. Springer, Berlin (2007)
7. Wang, Z., Klir, G.J.: Generalized Measure Theory. Springer, New York (2009)
8. Yager, R.R.: On ordered weighted averaging aggregation operators in multicriteria decision making. IEEE Trans. Syst. Man Cybern. **18**, 183–190 (1988)

On the Construction of Associative, Commutative and Increasing Operations by Paving

Wenwen Zong[1], Yong Su[1], Hua-Wen Liu[1(✉)], and Bernard De Baets[2]

[1] School of Mathematics, Shandong University, Jinan, Shandong 250100, China
zongwen198811@163.com, yongsu88@163.com, hw.liu@sdu.edu.cn
[2] KERMIT, Department of Mathematical Modelling, Statistics and Bioinformatics,
Ghent University, Coupure Links 653, Ghent 9000, Belgium
bernard.debaets@ugent.be

Abstract. Bodjanova, Kalina and Král' recently introduced a construction method, called paving, which enables to define a new associative, commutative and increasing operation from a given one and a discrete representable partial operation. As a matter of fact, not every discrete t-norm is representable, i.e. it can not always be generated by some additive generator, and this also holds for t-conorms and uninorms. Inspired by this fact and the method of paving, we construct some new associative, commutative and increasing operations on the unit interval from a t-norm on the unit interval and a discrete t-norm, t-superconorm, t-conorm or uninorm. Because of the duality between t-norms and t-conorms, we also define some operations from a t-conorm and a discrete t-norm, t-subnorm, t-conorm or uninorm.

Keywords: Associative operations · Uninorms · T-norms · T-conorms · Paving

1 Introduction

The associativity models the independence of the aggregation on the grouping of input values and it allows to investigate binary aggregation operators only (as far as their n-ary extensions are then determined uniquely). It is needless to emphasize the key role of associative operations (t-norms, t-conorms, uninorms, nullnorms, etc.) not only in fuzzy set theory, but also in many areas of application, especially in decision-making under uncertainty [5], image processing [1,6], fuzzy neural networks [7] and so on. The most important classes of associative, commutative, increasing operations in the framework of fuzzy sets is that of uninorms ([4,5,18]), which includes t-norms [10,17] and t-conorms [10] as two special classes. A large number of methods to construct uninorms (including t-norms and t-conorms) are introduced: Klement et al. [10], Schweizer and Sklar [17], Jenei [8], Ling [13], Maes and De Baets [11], Mas et al. [12], Mesiarová-Zemánková [14–16] and so on.

© Springer International Publishing AG 2018
V. Torra et al. (eds.), *Aggregation Functions in Theory and in Practice*,
Advances in Intelligent Systems and Computing 581, DOI 10.1007/978-3-319-59306-7_24

Kalina et al. [2,9] introduced a construction method called paving. The main idea is as follows: the unit interval is split into countably many disjoint sub-intervals $(I_i)_{i \in J_n}$ with J_n an index-set and with the help of an appropriate operation $*'$ on J_n and a family of increasing transformations $\varphi_i : I_i \to [0,1]$, a new operation $\oplus$ is defined by

$$x \oplus y = \varphi_{i*'j}^{-1}(\varphi_i(x) * \varphi_j(y)), \quad x \in I_i, \ y \in I_j. \tag{1}$$

Unfortunately, Kalina et al. only consider discrete representable associative operations as operation $*'$, which is rather restrictive. For instance, not every discrete t-norm can be generated by some additive generator, and this applies to t-conorms and uninorms. Moreover, the operation $*'$ in [2] is not always internal on J_n. In this paper, we will consider a general discrete associative operation as operation $*'$ on J_n, to construct some new associative, commutative and increasing operations. The graphical schema of paving is depicted in Fig. 1

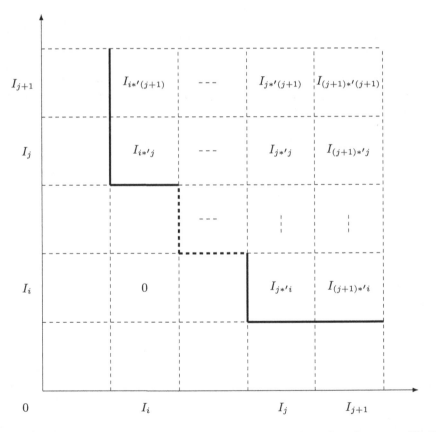

Fig. 1. The structure of $\oplus$, where the thick line is the boundary between $\{(i,j) \mid i *' j = 0\}$ and $\{(i,j) \mid i *' j > 0\}$. Inside the blocks it is shown in which sub-interval the operation $\oplus$ takes its values.

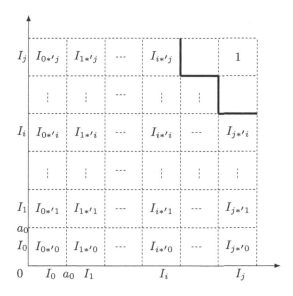

Fig. 2. The structure of $\oplus$, where the thick line is the boundary between $\{(i,j) \mid i *' j = n\}$ and $\{(i,j) \mid i *' j < n)\}$. Inside the blocks it is shown in which sub-interval the operation $\oplus$ takes its values.

(which depicts the construction of a conjunctive operation $\oplus$) and Fig. 2 (which depicts the construction of a disjunctive operation $\oplus$).

The paper is organized as follows. In Sect. 2, we present some preliminary notions and results that are necessary for the rest of the paper. Starting from Eq. (1), when $*$ is a t-norm and $*'$ is a discrete t-norm, t-superconorm, t-conorm or uninorm, we construct some new associative, commutative and increasing operations in Sect. 3. At the same time, all the dual constructions when $*$ is a t-conorm are also listed in Sect. 3.

2   Preliminaries

In this section we recall some basic notions and facts that are necessary for the understanding of what follows.

Definition 1 [10]. *A decreasing function* $N : [0,1] \rightarrow [0,1]$ *is called a fuzzy negation if* $N(0) = 1$ *and* $N(1) = 0$. *Moreover, a fuzzy negation* N *is called strong if it is involutive, i.e., if* $N(N(x)) = x$ *for all* $x \in [0,1]$.

Definition 2 [18]. *A binary operation* $U : [0,1]^2 \rightarrow [0,1]$ *is called a uninorm if it is associative, commutative, increasing and has a neutral element* $e \in [0,1]$, *i.e.,* $U(x,e) = x$ *for all* $x \in [0,1]$.

A uninorm with neutral element $e = 1$ is a *t-norm* [10,17] and a uninorm with neutral element $e = 0$ is a *t-conorm* [10]. We say that a uninorm U is *proper* if $e \in \;]0,1[$. If $U(1,0) = 0$, then U is called *conjunctive*. If $U(1,0) = 1$, then U is called *disjunctive*. Conjunctive and disjunctive uninorms are dual to each other. For an arbitrary disjunctive uninorm U and a strong negation N, its N-dual conjunctive uninorm is given by

$$U_N^d(x,y) = N(U(N(x), N(y))). \tag{2}$$

For an overview of basic properties of uninorms, we refer to [3].

Remark 1. Note that, for a strong negation N, the N-dual operation to a t-norm T defined by $S(x,y) = N(T(N(x), N(y)))$ is a t-conorm. For more information, see, e.g., [10].

Definition 3 [8]. *(i) A binary operation $\widetilde{T} : [0,1]^2 \to [0,1]$ is called a triangular subnorm (t-subnorm, for short), if it is associative, commutative, increasing and fulfills the condition $\widetilde{T}(x,y) \leq \min(x,y)$ for all $(x,y) \in [0,1]^2$.*

(ii) A binary operation $\widetilde{S} : [0,1]^2 \to [0,1]$ is called a triangular superconorm (t-superconorm, for short), if it is associative, commutative, increasing and fulfills the condition $\widetilde{S}(x,y) \geq \max(x,y)$ for all $(x,y) \in [0,1]^2$.

Definition 4. *Let $* : [0,1]^2 \to [0,1]$ be a commutative operation. Fix a value $a \in [0,1]$. We say that $x \in [0,1]$, $x \neq a$, is an a-divisor if there exists $y \in [0,1]$, $y \neq a$, such that*

$$x * y = a. \tag{3}$$

3   Construction of New Operations

The main idea of our construction method is described in Eq. (1) with the help of a discrete associative operation $*'$. For the rest of this paper, we adopt the following notations.

Let $\mathbb{N}$ be the set of all positive integers. We consider an index-set

$$J_n = \{0, 1, 2, \ldots, n\}$$

for some $n \in \mathbb{N}$.

We will split the interval $[0,1]$ into $n+1$ sub-intervals by choosing the end-points of the system of sub-intervals

$$0 = a_{-1} < a_0 < a_1 < a_2 < \ldots < a_{n-1} < a_n = 1.$$

Because of this partition, we will use half-open intervals, i.e., either left-open or right-open. We will use indexing of the chosen sub-intervals in accordance with the right end-point. For the case of left-open sub-intervals, $I_i = \;]a_{i-1}, a_i]$; for the case of right-open sub-intervals, $I_i = [a_{i-1}, a_i[$.

For a fixed system of right-open sub-intervals $(I_i)_{i=0}^n$, $\varphi_i : I_i \to [0,1[$ are increasing bijections. For a fixed system of left-open sub-intervals $(I_i)_{i=0}^n$, $\chi_i : I_i \to \;]0,1]$ are increasing bijections.

Remark 2 [2]. In order not to get out of the range of the transformations χ_i when using left-open sub-intervals, the starting operation $*$ (the basic paving stone) must be without zero-divisors. Similarly, when using right-open sub-intervals, $*$ must be without one-divisors.

Here, we consider to construct new associative, commutative and increasing operations from a given one $*$, and two certain cases of associative, commutative and increasing operations will be taken into account: the case that $*$ is a t-norm and the case that $*$ is a t-conorm.

3.1   The Case that $*$ Is a T-Norm

In this subsection, we construct some new associative, commutative and increasing operations on the unit interval from a t-norm on the unit interval and a discrete t-norm/t-superconorm/t-conorm/uninorm.

Firstly, we construct a new operation $\oplus$ from a t-norm $*$ and a discrete t-norm $*'$ in Eq. (1). Because of the partition of unit interval, we distinguish two cases: when right-open sub-intervals of $[0,1[$ and left-open sub-intervals of $]0,1]$.

Proposition 1. *Let* $* : [0,1]^2 \rightarrow [0,1]$ *be a t-norm,* $(I_i)_{i=0}^n$ *be a partition of* $[0,1[$ *consisting of right-open sub-intervals. Assume that* $*'$ *is a discrete t-norm on* $J_n = \{0, \ldots, n\}$ *such that* $*'$ *is strictly increasing on the domain* $\{(i,j) \mid i,j \in J_n, \; i *' j > 0\}$. *Then the operation* $\oplus_1$ *defined by*

$$x \oplus_1 y = \begin{cases} \varphi_{i*'j}^{-1}(\varphi_i(x) * \varphi_j(y)), & \text{if } x \in I_i, \; y \in I_j \text{ and } i *' j > 0, \\ \min(x,y), & \text{if } \max(x,y) = 1, \\ 0, & \text{otherwise,} \end{cases} \qquad (4)$$

is a t-norm.

In fact, $\oplus_1$ is not always increasing without the condition that $*'$ is strictly increasing on the domain $\{(i,j) \mid i,j \in J, \; i *' j > 0\}$.

Example 1. Assume that $J_7 = \{0,1,2,\ldots,7\}$, $(I_i = [i/8, (i+1)/8[)_{i=0}^7$ is a partition of $[0,1[$. Let $*$ be the t-norm $T_M(x,y) = \min(x,y)$ on $[0,1]$, $*'$ be the discrete t-norm $T_M(i,j) = \min(i,j)$ on J_7, $\varphi_i(x) = \frac{x - a_{i-1}}{a_i - a_{i-1}}$. Define $x \oplus y$ as follows:

$$x \oplus y = \begin{cases} \varphi_{\min(i,j)}^{-1}(\min(\frac{x-a_{i-1}}{a_i-a_{i-1}}, \frac{y-a_{j-1}}{a_j-a_{j-1}})), & \text{if } x \in I_i, \; y \in I_j, \text{ and } \min(i,j) > 0, \\ \min(x,y), & \text{if } \max(x,y) = 1, \\ 0, & \text{otherwise.} \end{cases}$$

Consider that $x = \frac{3}{16}$, $y = \frac{3}{16}$ and $z = \frac{1}{4}$, then we have that

$$x \oplus y = \varphi_1^{-1}\left(\frac{1}{2}\right) = \frac{3}{16} > \frac{1}{8} = \varphi_1^{-1}(0) = x \oplus z. \qquad (5)$$

That is, $\oplus$ is not increasing.

By (4), we can see that for any t-norm $*$, its values on the upper right boundary of the unit square $[0,1]^2$ have no impact on the properties of $\oplus_1$. Moreover, It is obvious that associativity, commutativity and monotonicity of $\oplus_1$ are determined by the corresponding properties of $*$, respectively. Thus, we can easily obtain that Proposition 1 holds for t-subnorm instead of t-norm.

Example 2. Assume that $J_n = \{0,1,2,\ldots,n\}$, $(I_i)_{i=0}^n$ is a partition of $[0,1[$ consisting of right-open sub-intervals. Let $*$ be the t-subnorm $\widetilde{T} = \max(\min(x,\frac{1}{2}) + \min(y,\frac{1}{2}) - \frac{3}{4},0)$ on $[0,1]$, $*$ be the discrete t-norm $T_L(i,j) = \max(0, i+j-n)$ on J_n, $\varphi_i(x) = \frac{x - a_{i-1}}{a_i - a_{i-1}}$. Define $x \oplus y$ as follows:

$$
x \oplus y = \begin{cases} \varphi_{i+j-n}^{-1}(\widetilde{T}(\frac{x-a_{i-1}}{a_i-a_{i-1}}, \frac{y-a_{j-1}}{a_j-a_{j-1}})), & \text{if } x \in I_i, \ y \in I_j \text{ and } i+j > n, \\ \min(x,y), & \text{if } \max(x,y) = 1, \\ 0, & \text{otherwise,} \end{cases} \tag{6}
$$

is a t-norm.

As stated earlier, $*$ must be a t-norm without zero-divisors when left-open sub-intervals are taken into account. Similar to Proposition 1, the following proposition can be obtained:

Proposition 2. *Let* $* : [0,1]^2 \to [0,1]$ *be a t-norm without zero-divisors,* $(I_i)_{i=0}^n$ *be a partition of* $]0,1]$ *consisting of left-open sub-intervals. Assume that* $*'$ *is a discrete t-norm on* J_n *such that* $*'$ *is strictly increasing on the domain* $\{(i,j) \mid i,j \in J_n, \ i *' j > 0\}$. *Then the operation* $\oplus_2$ *defined by*

$$
x \oplus_2 y = \begin{cases} \min(x,y), & \text{if } \max(x,y) = 1, \\ \chi_{i*'j}^{-1}(\chi_i(x) * \chi_j(y)), & \text{if } x \in I_i \setminus \{1\}, \ y \in I_j \setminus \{1\} \text{ and } i *' j > 0, \\ 0, & \text{otherwise,} \end{cases} \tag{7}
$$

is a t-norm.

Next, we discuss the construction when $*$ is a t-norm and $*'$ is a discrete t-superconorm. Analogously, two cases of right-open sub-intervals of $[0,1[$ and left-open sub-intervals of $]0,1]$ are taken into account. We start with the case of the right-open sub-intervals.

Proposition 3. *Let* $* : [0,1]^2 \to [0,1]$ *be a t-norm,* $(I_i)_{i=0}^n$ *be a partition of* $[0,1[$ *consisting of right-open sub-intervals. Assume that* $*'$ *is a discrete t-superconorm on* J_n *such that* $*'$ *is strictly increasing and* $i *' j > \max(i,j)$ *on the domain* $\{(i,j) \mid i,j \in J_n, \ i *' j < n\}$. *Then the operation* $\oplus_3$ *defined by*

$$
x \oplus_3 y = \begin{cases} \varphi_{i*'j}^{-1}(\varphi_i(x) * \varphi_j(y)), & \text{if } x \in I_i, \ y \in I_j \text{ and } i *' j < n, \\ 1, & \text{otherwise,} \end{cases} \tag{8}
$$

is a t-superconorm.

Without the condition that $*'$ is strictly increasing on the domain $\{(i,j) \mid i,j \in J_n, \ i *' j < n\}$, $\oplus_3$ is not always increasing. We have the following counterexample.

Example 3. Assume that $J_7 = \{0, 1, 2, \ldots, 7\}$, $(I_i = [i/8, (i+1)/8[)_{i=0}^7$ is a partition of $[0, 1[$. Let $*$ be the t-norm $T_M(x, y) = \min(x, y)$ on $[0, 1]$, $*'$ be the discrete t-superconorm $\widetilde{S} = \min(n, \max(i, j) + 4)$ on J_7, $\varphi_i(x) = \frac{x - a_{i-1}}{a_i - a_{i-1}}$. Define $x \oplus y$ as follows:

$$x \oplus y = \begin{cases} \varphi_{\widetilde{S}(i,j)}^{-1}(\min(\frac{x - a_{i-1}}{a_i - a_{i-1}}, \frac{y - a_{j-1}}{a_j - a_{j-1}})), & \text{if } x \in I_i,\ y \in I_j \text{ and } i *' j < n, \\ 1, & \text{otherwise.} \end{cases}$$

Consider that $x = \frac{1}{16}$, $y = \frac{1}{8}$ and $z = \frac{3}{16}$, then we have that

$$x \oplus z = \varphi_5^{-1}\left(\frac{1}{2}\right) = \frac{11}{16} > \frac{5}{8} = \varphi_5^{-1}(0) = y \oplus z. \tag{9}$$

Obviously, $\oplus$ is not increasing.

In Eq. (8), let $x \oplus_3 y = \max(x, y)$ on the domain $\{(x, y) \mid x, y \in [0, 1], \min (x, y) = 0\}$. We can easily prove that the operation $\oplus_3$ is a t-conorm by simple calculations.

Similarly, when left-open sub-intervals are taken into account, $*$ must be a t-norm without zero-divisors. Then, the following proposition can be obtained:

Proposition 4. *Let* $* : [0, 1]^2 \to [0, 1]$ *be a t-norm without zero-divisors,* $(I_i)_{i=0}^n$ *be a partition of* $]0, 1]$ *consisting of left-open sub-intervals. Assume that* $*'$ *is a discrete t-superconorm on* J_n *such that* $*'$ *is strictly increasing and* $i *' j > \max(i, j)$ *on the domain* $\{(i, j) \mid i, j \in J_n,\ i *' j < n\}$. *Then the operation* $\oplus_4$ *defined by*

$$x \oplus_4 y = \begin{cases} \chi_{i*'j}^{-1}(\chi_i(x) * \chi_j(y)), & \text{if } x \in I_i,\ y \in I_j \text{ and } i *' j < n, \\ \max(x, y), & \text{if } \min(x, y) = 0, \\ 1, & \text{otherwise,} \end{cases} \tag{10}$$

is a t-conorm.

In what follows, we construct a new operation from a t-norm $*$ and a discrete uninorm $*'$.

Proposition 5. *Let* $* : [0, 1]^2 \to [0, 1]$ *be a t-norm,* $(I_i)_{i=0}^n$ *be a partition of* $[0, 1[$ *consisting of right-open sub-intervals. Assume that* $*'$ *is a discrete uninorm on* J_n *with neutral element* h *such that* $*'$ *is strictly increasing on the domain* $\{(i, j) \mid i, j \in J_n,\ \max(i, j) \le h, i *' j > 0\}$ *and* $\{(i, j) \mid i, j \in J_n,\ \min(i, j) \ge h, i *' j < n\}$. *Then the operation* $\oplus_5$ *defined by*

$$x \oplus_5 y = \begin{cases} a_i, & \text{if } \min(x, y) < a_h \text{ and } a_h \le a_i \le \max(x, y) < a_{i+1}, \\ \varphi_{i*'j}^{-1}(\varphi_i(x) * \varphi_j(y)), & \text{if } x \in I_i, y \in I_j, \max(i, j) \le h \text{ and } i *' j > 0, \\ & \text{or } h < \min(i, j) \text{ and } i *' j < n, \\ 1, & \text{if } x \in I_i, y \in I_j,\ h < \min(i, j) \text{ and } i *' j = n, \\ & \text{or } \max(x, y) = 1, \\ 0, & \text{otherwise,} \end{cases} \tag{11}$$

is associative, commutative and increasing.

In fact, the similar proposition does not hold when $(I_i)_{i=0}^n$ is a partition of $]0, 1]$ consisting of left-open sub-intervals. A counterexample is as follows:

Example 4. Assume that $J_4 = \{0, 1, 2, 3, 4\}$, $(I_i =]i/5, (i+1)/5])_{i=0}^4$ is a partition of $]0, 1]$. Let $*$ be the t-norm $T_M(x, y) = \min(x, y)$ on $[0, 1]$, $*'$ be the discrete uninorm U with neutral element 2:

$$U(i, j) = \begin{cases} T_L(i, j), & \text{if } 0 \le i, j \le 2, \\ S_L(i, j), & \text{if } 2 \le i, j \le 4, \\ \min(i, j), & \text{otherwise,} \end{cases}$$

where $T_L(i, j) = \max(0, i + j - 2)$, $S_L(i, j) = \min(4, i + j - 2)$.

Besides, $\varphi_i(x) = \frac{x - a_{i-1}}{a_i - a_{i-1}}$. Define $x \oplus y$ as follows:

$$x \oplus y = \begin{cases} a_{i+1}, & \text{if } \frac{3}{5} < \max(x, y) \text{ and } a_i < \min(x, y) \le a_{i+1} \le \frac{3}{5}, \\ \varphi_{U(i,j)}^{-1}(\varphi_i(x) * \varphi_j(y)), & \text{if } x \in I_i, y \in I_j, \max(i, j) \le 2 \text{ and } i *' j > 0, \\ & \text{or } 2 < \min(i, j) \text{ and } i *' j < 4, \\ 0, & \text{if } x \in I_i, y \in I_j, \ \max(i, j) \le 2 \text{ and } i *' j = 0, \\ & \text{or } \min(x, y) = 0, \\ 1, & \text{otherwise.} \end{cases}$$

Consider that $x = \frac{1}{2}$, $y = \frac{1}{2}$ and $z = \frac{4}{5}$, then we have that

$$(x \oplus y) \oplus z = a_{U(2,2)} = a_2 = \frac{3}{5} \ne \frac{1}{2} = x \oplus a_2 = x \oplus (y \oplus z). \tag{12}$$

Obviously, $\oplus$ is not associative.

When $*$ is a t-norm and $*'$ is a discrete t-conorm, we can construct some proper uninorms.

Proposition 6. *Let $* : [0, 1]^2 \to [0, 1]$ be a t-norm, $(I_i)_{i=0}^n$ be a partition of $[0, 1[$ consisting of right-open sub-intervals. Assume that $*'$ is a discrete t-conorm on J_n such that $*'$ is strictly increasing on the domain $\{(i, j) \mid i, j \in J_n, \ i *' j < n\}$. Then the operation $\oplus_6$ defined by*

$$x \oplus_6 y = \begin{cases} \varphi_{i*'j}^{-1}(\varphi_i(x) * \varphi_j(y)), & \text{if } x \in I_i \setminus \{a_0\}, \ y \in I_j \setminus \{a_0\} \text{ and } i *' j < n, \\ & \text{or } \min(x, y) \in I_0, \ \max(x, y) \in I_n, \\ y, & \text{if } x = a_0, \\ x, & \text{if } y = a_0, \\ 1, & \text{otherwise,} \end{cases} \tag{13}$$

is a proper disjunctive uninorm with neutral element a_0 if and only if $*$ has no zero-divisors.

In what follows, we give an example to illustrate that $*'$ must be strictly increasing on the domain $\{(i, j) \mid i, j \in J_n, \ i *' j < n\}$.

Example 5. Assume that $J_4 = \{0, 1, 2, 3, 4\}$, $(I_i = [i/5, (i+1)/5[)_{i=0}^4$ is a partition of $[0, 1[$. Let $*$ be the t-norm $T_M(x, y) = \min(x, y)$ on $[0, 1]$, $*'$ be the discrete t-conorm $S_M = \max(i, j)$ on J_4, $\varphi_i(x) = \frac{x - a_{i-1}}{a_i - a_{i-1}}$. Define $x \oplus y$ as follows:

$$
x \oplus y = \begin{cases}
\varphi_{\max(i,j)}^{-1}(\min(\frac{x-a_{i-1}}{a_i-a_{i-1}}, \frac{y-a_{j-1}}{a_j-a_{j-1}})), & \text{if } x \in I_i \setminus \{\frac{1}{5}\},\ y \in I_j \setminus \{\frac{1}{5}\} \text{ and } \max(i,j) < 4, \\
& \quad \text{or } \min(x, y) \in [0, \frac{1}{5}[,\ \max(x, y) \in [\frac{4}{5}, 1[, \\
y, & \text{if } x = \frac{1}{5}, \\
x, & \text{if } y = \frac{1}{5}, \\
1, & \text{otherwise.}
\end{cases}
$$

Consider that $x = \frac{3}{10}$, $y = \frac{2}{5}$ and $z = \frac{1}{2}$, then we have that

$$
x \oplus z = \varphi_2^{-1}\left(\frac{1}{2}\right) = \frac{1}{2} > \frac{2}{5} = \varphi_2^{-1}(0) = y \oplus z. \tag{14}
$$

Obviously, $\oplus$ is not increasing.

Similar to Proposition 6, when the left-open sub-intervals are taken into account, we have the following result:

Proposition 7. *Let $* : [0,1]^2 \to [0,1]$ be a t-norm without zero-divisors, $(I_i)_{i=0}^n$ be a partition of $]0, 1]$ consisting of left-open sub-intervals. Assume that $*'$ is a discrete t-conorm on J_n such that $*'$ is strictly increasing on the domain $\{(i, j) \mid i, j \in J_n,\ i *' j < n\}$. Then the operation $\oplus_7$ defined by*

$$
x \oplus_7 y = \begin{cases}
\chi_{i*'j}^{-1}(\chi_i(x) * \chi_j(y)), & \text{if } x \in I_i,\ y \in I_j \text{ and } i *' j < n, \\
& \quad \text{or } \min(x, y) \in I_0,\ \max(x, y) \in I_n, \\
0, & \text{if } \min(x, y) = 0, \\
1, & \text{otherwise,}
\end{cases} \tag{15}
$$

is a proper conjunctive uninorm with neutral element a_0.

3.2   The Case that $*$ Is a T-Conorm

Taking into account the duality between t-norms and t-conorms, the results in the case that $*$ is a t-conorm are easily obtained.

Proposition 8. *Let $* : [0,1]^2 \to [0,1]$ be a t-conorm without one-divisors, $(I_i)_{i=0}^n$ be a partition of $[0, 1[$ consisting of right-open sub-intervals. Assume that $*'$ is a discrete t-conorm on J_n such that $*'$ is strictly increasing on the domain $\{(i, j) \mid i, j \in J_n,\ i *' j < n\}$. Then the operation $\oplus^1$ defined by*

$$
x \oplus^1 y = \begin{cases}
\varphi_{i*'j}^{-1}(\varphi_i(x) * \varphi_j(y)), & \text{if } x \in I_i \setminus \{0\},\ y \in I_j \setminus \{0\} \text{ and } i *' j < n, \\
\max(x, y), & \text{if } \min(x, y) = 0, \\
1, & \text{otherwise,}
\end{cases}
$$

is a t-conorm.

Similar to the case that $*$ is a t-norm, Proposition 8 holds for t-superconorm instead of t-conorm.

Proposition 9. *Let* $* : [0,1]^2 \to [0,1]$ *be a t-conorm,* $(I_i)_{i=0}^n$ *be a partition of* $]0,1]$ *consisting of left-open sub-intervals. Assume that* $*'$ *is a discrete t-conorm on* J_n *such that* $*'$ *is strictly increasing on the domain* $\{(i,j) \mid i,j \in J_n, \ i *' j < n\}$. *Then the operation* $\oplus^2$ *defined by*

$$x \oplus^2 y = \begin{cases} \max(x,y), & \text{if } \min(x,y) = 0, \\ \chi_{i*'j}^{-1}(\chi_i(x) * \chi_j(y)), & \text{if } x \in I_i, \ y \in I_j \text{ and } i *' j < n, \\ 1, & \text{otherwise}, \end{cases}$$

is a t-conorm.

Proposition 10. *Let* $* : [0,1]^2 \to [0,1]$ *be a t-conorm without one-divisors,* $(I_i)_{i=0}^n$ *be a partition of* $[0,1[$ *consisting of right-open sub-intervals. Assume that* $*'$ *is a discrete t-subnorm on* J_n *such that* $*'$ *is strict increasing and* $i *' j < \min(i,j)$ *on the domain* $\{(i,j) \mid i,j \in J_n, \ i *' j > 0\}$. *Then operation* $\oplus^3$ *defined by*

$$x \oplus^3 y = \begin{cases} \varphi_{i*'j}^{-1}(\varphi_i(x) * \varphi_j(y)), & \text{if } x \in I_i, \ y \in I_j \text{ and } i *' j > 0, \\ \min(x,y), & \text{if } \max(x,y) = 1, \\ 0, & \text{otherwise}, \end{cases}$$

is a t-norm.

Proposition 11. *Let* $* : [0,1]^2 \to [0,1]$ *be a t-conorm,* $(I_i)_{i=0}^n$ *be a partition of* $]0,1]$ *consisting of left-open sub-intervals. Assume that* $*'$ *is a discrete t-subnorm on* J_n *such that* $*'$ *is strictly increasing and* $i *' j < \min(i,j)$ *on the domain* $\{(i,j) \mid i,j \in J_n, \ i *' j > 0\}$. *Then the operation* $\oplus^4$ *defined by*

$$x \oplus^4 y = \begin{cases} \chi_{i*'j}^{-1}(\chi_i(x) * \chi_j(y)), & \text{if } x \in I_i, \ y \in I_j \text{ and } i *' j > 0, \\ 0, & \text{otherwise}, \end{cases}$$

is a t-subnorm.

Proposition 12. *Let* $* : [0,1]^2 \to [0,1]$ *be a t-conorm without one-divisors,* $(I_i)_{i=0}^n$ *be a partition of* $[0,1[$ *consisting of right-open sub-intervals. Assume that* $*'$ *is a discrete t-norm on* J_n *such that* $*'$ *is strictly increasing on the domain* $\{(i,j) \mid i,j \in J_n, \ i *' j > 0\}$. *Then the operation* $\oplus^5$ *defined by*

$$x \oplus^5 y = \begin{cases} \varphi_{i*'j}^{-1}(\varphi_i(x) * \varphi_j(y)), & \text{if } x \in I_i, \ y \in I_j \text{ and } i *' j > 0, \\ & \quad \text{or } \min(x,y) \in I_0, \ \max(x,y) \in I_n, \\ 1, & \text{if } \max(x,y) = 1, \\ 0, & \text{otherwise}, \end{cases}$$

is a proper disjunctive uninorm with neutral element a_{n-1}.

Proposition 13. *Let* $* : [0,1]^2 \to [0,1]$ *be a t-conorm,* $(I_i)_{i=0}^n$ *be a partition of* $]0,1]$ *consisting of left-open sub-intervals. Assume that* $*'$ *is a discrete uninorm on* J_n *with neutral element* h *such that* $*'$ *is strictly increasing on the domain* $\{(i,j) \mid i,j \in J_n, \ \max(i,j) \leq h, i *' j > 0\}$ *and* $\{(i,j) \mid i,j \in J_n, \ \min(i,j) \geq h, i *' j < n\}$. *Then the operation* $\oplus^6$ *defined by*

$$
x \oplus^6 y = \begin{cases}
a_{i+1}, & \text{if } \max(x,y) > a_{h-1} \text{ and } a_i < \min(x,y) \leq a_{i+1} \leq a_{h-1}, \\
\chi_{i*'j}^{-1}(\chi_i(x) * \chi_j(y)), & \text{if } x \in I_i, y \in I_j, \max(i,j) \leq h-1 \text{ and } i *' j > 0, \\
& \text{or } h-1 < \min(i,j) \text{ and } i *' j < n, \\
0, & \text{if } x \in I_i, y \in I_j, \max(i,j) \leq h-1 \text{ and } i *' j = 0, \\
& \text{or } \min(x,y) = 0, \\
1, & \text{otherwise,}
\end{cases}
$$

is associative, commutative and increasing.

Proposition 14. *Let* $* : [0,1]^2 \to [0,1]$ *be a t-conorm,* $(I_i)_{i=0}^n$ *be a partition of* $]0,1]$ *consisting of left-open sub-intervals. Assume that* $*'$ *is a discrete t-norm on* J_n *such that* $*'$ *is strictly increasing on the domain* $\{(i,j) \mid i,j \in J_n, \ i *' j > 0\}$. *Then the operation* $\oplus^7$ *defined by*

$$
x \oplus^7 y = \begin{cases}
\chi_{i*'j}^{-1}(\chi_i(x) * \chi_j(y)), & \text{if } x \in I_i \setminus \{a_{n-1}\}, \ y \in I_j \setminus \{a_{n-1}\} \text{ and } i *' j > 0, \\
& \text{or } \min(x,y) \in I_0, \ \max(x,y) \in I_n, \\
y, & \text{if } x = a_{n-1}, \\
x, & \text{if } y = a_{n-1}, \\
0, & \text{otherwise,}
\end{cases}
$$

is a proper conjunctive uninorm with neutral element a_{n-1} *if and only if* $*$ *has no one-divisors.*

Results

Inspired by the construction method of paving, we construct some new associative, commutative and increasing operations on the unit interval from a t-norm on the unit interval and a discrete t-norm/t-superconorm/t-conorm/uninorm. Similarly, we present the dual constructions from a t-conorm and a discrete t-norm/t-subnorm/t-conorm/uninorm.

Acknowledgements. This work is supported by the National Natural Science Foundation of China No. 61573211. The author Wenwen Zong is supported by the China Scholarship Council under Grant No. 201606220121. The author Yong Su is supported by the China Scholarship Council under Grant No. 201506220039.

References

1. Barrenechea, E., Bustince, H., De Baets, B., Lopez-Molina, C.: Construction of interval-valued fuzzy relations with application to the generation of fuzzy edge images. IEEE Trans. Fuzzy Syst. **19**(5), 819–830 (2011)
2. Bodjanova, S., Kalina, M.: Block-wise construction of commutative increasing monoids. Fuzzy Sets Syst. http://dx.doi.org/10.1016/j.fss.2016.10.002

3. Calvo, T., Mayor, G., Mesiar, R.: Aggregation Operators: New Trends and Applications. Studies in Fuzziness and Soft Computing, vol. 97. Springer, Berlin (2002)
4. Czogała, E., Drewniak, J.: Associative monotonic operations in fuzzy set theory. Fuzzy Sets Syst. **12**, 249–269 (1984)
5. Dombi, J.: Basic concepts for a theory of evaluation: the aggregative operator. Eur. J. Oper. Res. **10**, 282–293 (1982)
6. González-Hidalgo, M., Massanet, S., Mir, A., Ruiz-Aguilera, D.: On the choice of the pair conjunction-implication into the fuzzy morphological edge detector. IEEE Trans. Fuzzy Syst. **23**(4), 872–884 (2015)
7. Leite, D., Costa Jr., P., Gomide, F.: Evolving granular neural network for semi-supervised data stream classification, In: IEEE World Congress on Computational Intelligence-Part IJCNN 2010, pp. 1877–1884 (2010)
8. Jenei, S.: Structure of left-continuous triangular norms with strong induced negations. (II) rotation-annihilation construction. J. Appl. Non-Classical Log. **11**, 351–366 (2001)
9. Kalina, M., Král', P.: Construction of commutative and associative operations by paving. In: Alonso, J.M., Bustince, H., Reformat, M. (eds.) IFSA-EUSFLAT 2015, pp. 1201–1207. Atlantis Press, Gijn (2015)
10. Klement, E.P., Mesiar, R., Pap, E.: Triangular Norms. Springer, Berlin (2000)
11. Maes, K.C., De Baets, B.: The triple rotation method for constructing t-norms. Fuzzy Sets Syst. **158**, 1652–1674 (2007)
12. Mas, M., Massanet, S., Ruiz-Aguilera, D., Torrens, J.: A survey on the existing classes of uninorms. J. Intell. Fuzzy Syst. **29**(3), 1021–1037 (2015)
13. Ling, C.H.: Representation of associative functions. Publ. Math. Debr. **12**, 189–212 (1965)
14. Mesiarová-Zemánková, A.: A note on decomposition of idempotent uninorms into an ordinal sum of singleton semigroups. Fuzzy Sets Syst. **299**, 140–145 (2016)
15. Mesiarová-Zemánková, A.: Ordinal sums of representable uninorms. Fuzzy Sets Syst. **308**, 42–53 (2017)
16. Mesiarová-Zemánková, A.: Ordinal sum construction for uninorms and generalized uninorms. Int. J. Approx. Reason. **76**, 1–17 (2016)
17. Schweizer, B., Sklar, A.: Probabilistic Metric Spaces. North Holland, New York (1983)
18. Yager, R.R., Rybalov, A.: Uninorm aggregation operators. Fuzzy Sets Syst. **80**, 111–120 (1996)

On Implication Operators

József Dombi$^{(\boxtimes)}$

Institute of Informatics, University of Szeged, Szeged, Hungary
`dombi@inf.u-szeged.hu`

Abstract. Distributivity properties play an important role in fuzzy research. Based on the solution of the autodistributivity functional equations, we give a characterisation of two types of distributivity of fuzzy implication. Based on the mean disjunctive operator, the mean implication operator is introduced. Using the Pliant operators -where all operators have a common generator function- we show that some weakened properties of the fuzzy mean implications are valid. In the propositions we use the fixed point of the negation as a threshold. Finally, the generalized modus ponens is examined in this framework.

1 Introduction

The implication operator ($\rightarrow$) plays a significant role in classical two-valued logic. From implication, we can obtain all other basic logical connectives of the binary logic,viz., the binary operators- and ($\wedge$), or ($\vee$)-and the unary negation operator ($\neg$). The implication operator holds the center stage in the inference mechanisms of any logic, like modus ponens and modus tollens. The truth table for the classical implication operator is given in Table 1.

In multivalued logic, several problems arise with the definition of the implication operator. Even if we find good connectives (negation and disjunction) for the definition of an implication, these do not preserve the good properties of the classical implication operator.

Now we will give some basic definition.

Definition 1. The negation operator is compatible with two-valued logic and it is continuous, strictly decreasing function and involutive, i.e. it is a strong negation. The negation operator can be characterized by its fixed point ν ($\eta(\nu)) = \nu$). The negation operator with fixed point ν will be denoted by $\eta_\nu(x)$.

Definition 2. Let $c(x, y)$ and $d(x, y)$ denote the strictly monotonously increasing Archimedian t-norm and t-conorm. The representation of these operators is

$$c(x, y) = f_c^{-1}(f_c(x) + f_c(y)) \qquad d(x, y) = f_d^{-1}(f_d(x) + f_d(y)),$$

where $f_c(x)$ and $f_d(x)$ are the generator functions of the operators and $f_c(x) : (0, 1] \rightarrow [0, \infty]$ is a strictly decreasing continuous function and $f_d(x) : [0, 1) \rightarrow [0, \infty]$ is a strictly monotonously increasing continuous function and we suppose

© Springer International Publishing AG 2018
V. Torra et al. (eds.), *Aggregation Functions in Theory and in Practice*,
Advances in Intelligent Systems and Computing 581, DOI 10.1007/978-3-319-59306-7_25

that the conjunctive and the disjuctive operator build a DeMorgan class with a negation operator. We will denote

$$\bar{c}(x,y) = f_c^{-1}\left(\frac{1}{2}(f_c(x) + f_c(y))\right) \qquad \bar{d}(x,y) = f_d^{-1}\left(\frac{1}{2}(f_d(x) + f_d(y))\right)$$

the corresponding mean operators.

Table 1. Truth table for classical implication

p	q	p → q
0	0	1
0	1	1
1	0	0
1	1	1

2   Fuzzy Implication and Its Properties

Here, we will use the definition proposed by Baczyński [2,15]. Then, we will summarise the properties and definitions associated with the implication operator.

Definition 3. A function $i : [0,1]^2 \rightarrow [0,1]$ is called a fuzzy implication if it satisfies, for all $x, x_1, x_2, y, y_1, y_2 \in [0,1]$, the following conditions:

$$\text{if } x_1 \leq x_2, \text{then } i(x_1,y) \geq i(x_2,y), \text{ i.e., } i(\cdot,y) \quad \text{is} \quad \text{decreasing,} \qquad (\text{i1})$$

$$\text{if } y_1 \leq y_2, \text{then } i(x,y_1) \leq i(x,y_2), \text{ i.e., } i(x,\cdot) \quad \text{is} \quad \text{increasing,} \qquad (\text{i2})$$

$$i(0,0) = 1, \qquad (\text{i3})$$

$$i(1,1) = 1, \qquad (\text{i4})$$

$$i(1,0) = 0. \qquad (\text{i5})$$

The set of all fuzzy implications will be denoted by $\mathscr{I}$.

Remark 1. From Definition 3, we may deduce that each fuzzy implication i is constant for $x = 0$ and for $y = 1$, i.e. i satisfies the following properties, called the left and right boundary conditions, respectively:

$$i(0,y) = 1, \quad y \in [0,1], \quad i(x,1) = 1, \quad x \in [0,1].$$

Therefore, i also satisfies the normality condition

$$i(0,1) = 1 \tag{i6}$$

Consequently, every fuzzy implication restricted to the set $[0,1]^2$ coincides with the classical implication, and hence fulfils the binary implication truth table provided in Table 1.

It is also apparent that the fuzzy implication i (a function of two real variables) is continuous.

Additional properties of fuzzy implications were postulated by theorists (see Trillas and Vaverde, Dubois and Prade, Smeth and Magrez, Fodor and Roubens, Gottwald [12,17,18]).

The fuzzy implication i is said to satisfy
the left neutrality property if

$$i(1,y) = y, \quad y \in [0,1]; \tag{1}$$

the exchange principle, if

$$i(x, i(y,z)) = i(y, i(x,z)), \quad x,y,z \in [0,1]; \tag{2}$$

the identity principle, if

$$i(x,x) = 1, \quad x \in [0,1]; \tag{3}$$

the ordering property, if

$$i(x,y) = 1 \text{ if and only if } x \le y, \quad x,y \in [0,1]. \tag{4}$$

One of the most importatnt tautologies in classical two-valued logic is the law of contraposition.

Let $i \in \mathscr{I}$ and η be a fuzzy negation. i is said to satisfy
The law of contraposition with respect to $\eta(x)$, if

$$i(x,y) = i(\eta(y), \eta(x)), \quad x,y \in [0,1], \tag{5}$$

The law of left contraposition with respect to η, if

$$i(\eta(x), y) = i(\eta(y), x), \quad x,y \in [0,1], \tag{6}$$

The law of right contraposition with respect to η, if

$$i(x, \eta(y)) = i(y, \eta(x)), \quad x,y \in [0,1], \tag{7}$$

The general form of the law of importation is given by

$$i(c(x,y), z) = i(x, i(y,z)), \quad x,y,z \in [0,1], \tag{8}$$

where i is a fuzzy implication and c is a conjunctive operator.

3   Distributivity of Fuzzy Implication

In classical logic, the distributivity of a binary operator over another determines the underlying structure of the algebra imposed by these operators. In
fuzzy logic too, we find that many papers discuss the distributivity of t-norms
over t-conorms (see [3–5,13]), uninorms over t-operators [14,16] and the like.
The following are the four basic distributive equations involving an implication
operator:

$$(x \lor y) \to z \equiv (x \to z) \land (y \to z), \tag{d1}$$

$$(x \land y) \to z \equiv (x \to z) \lor (y \to z), \tag{d2}$$

$$x \to (y \land z) \equiv (x \to y) \land (x \to z), \tag{d3}$$

$$x \to (y \lor z) \equiv (x \to y) \lor (x \to z), \tag{d4}$$

Each of the above equivalences is a tautology in classical logic. Here, we will
concentrate on $(d2)$ and $(d4)$ to find the necessary and sufficient conditions. In
this equation the conjunctive operator is denoted by $\hat{c}(x, y)$ and the disjunctive
operator by $\hat{d}(x, y)$. The properties of $\hat{c}(x, y)$ and $\hat{d}(x, y)$ are given in Definition 2.

To solve the distributivity aspect of the fuzzy implication problem, we have
to find the proper disjunctive operator $\hat{d}(x, y)$ when the definition of the implication is

$$x \to y = \bar{x} \lor y = \hat{d}(x, y) \tag{9}$$

This is the so-called S implication. To find the proper solution, let us turn to
the autodistributivity functional equations (see Aczel [1]):

$$F(x, F(y, z)) = F(F(x, y), F(x, z)) \tag{10}$$
$$F(F(x, y), z) = F(F(x, z), F(y, z))$$

We need a definition of reducibility:

Definition 4. $F(x, y)$ is reducible if

$$F(t, x) \neq F(t, y) \quad F(x, t) \neq F(y, t) \qquad \text{for } x \neq y$$

Proposition 1 [1]. *The most general, continuous, reducible function* $F(x, y) \in$
(a, b) *and* $(x, y \in (a, b))$ *that satisfies the autodistributivity* Eq. (10) *is*

$$F(x, y) = f^{-1}\left((1 - q)f(x) + qf(y)\right), \qquad (q \in (0, 1)) $$

with an arbitrary, continuous monotonic f *(generator function) and constant* q.
If $F(x, y)$ *is commutative, then* $q = \frac{1}{2}$.

$$F(x, y) = f^{-1}\left(\frac{1}{2}(f(x) + f(x))\right) \tag{11}$$

Now we can give the necessary and sufficient conditions for the distributivity of the fuzzy implication operator. The autodistributivity equation is generally solved on $[a, b]$ interval. In our case $[a, b] = [0, 1]$.

Proposition 2. *The distributivity of fuzzy implications $(d2)(d4)$ holds if and only if*

$$\hat{d}(x, y) = \bar{d}(x, y)$$

$$\bar{c}(x, y) = f^{-1}\left(\frac{1}{2}(f_c(x) + f_c(y))\right)$$

$$\bar{d}(x, y) = f^{-1}\left(\frac{1}{2}(f_d(x) + f_d(y))\right)$$

$$\bar{i}(x, y) = f^{-1}\left(\frac{1}{2}(f_d(\eta(x)) + f_d(y))\right)$$

Proof. The proof is based on the autodistributivity functional equation and its solution.

The distributivity of fuzzy implication $(d2)$ and $(d4)$ can be written using the S implication (9)

$$\overline{x \wedge y} \vee z = (\bar{x} \vee \bar{y}) \vee z = (\bar{x} \vee z) \vee (\bar{y} \vee z)$$
$$\bar{x} \vee (y \vee z) = (\bar{x} \vee y) \vee (\bar{x} \vee z),$$

replacing $\bar{x}, \bar{y}$ and $\bar{z}$ by x, y, z and denoting $x \vee y$ by $\hat{d}(x, y)$. Then we get the autodistributivity Eq. (10). Based on Aczel's theorem, the necessary and sufficient conditions for autodistributivity are

$$\hat{d}(x, y) = \bar{d}(x, y) = f^{-1}\left(\frac{1}{2}(f_d(x) + f_d(y))\right).$$

So the fuzzy implication operator is

$$\bar{i}(x, y) = f^{-1}\left(\frac{1}{2}(f(\eta(x)) + f(y))\right).$$

Remark 2. $\bar{i}(x, y)$ is a new type of S implication based on the mean operator $\bar{d}(x, y)$. This new operator also depends on the negation operator and because it is characterized by its fixed point, the corresponding mean implication operator is also characterized by this fixed point and is denoted by $i_\nu(x, y)$. To see its properties we will use a special class of operator system.

4   Pliant Implication Operator

Several different structures appear in the framework of fuzzy theory. One of them is the Pliant concept. In this system we have infinitely many negation operators, and hence infinitely many fixed points can be defined (see e.g. [6,8,9,11]). All Pliant operators are created by using just one generator function.

Definition 5. We call $c(x, y)$ and $d(x, y)$ Pliant operators when the generator functions, given in Definition 2, satisfy

$$f_c(x)f_d(x) = 1$$

Let us denote $f_c(x)$ by $f(x)$. Then the Pliant operators are

$$c(x, y) = f^{-1}(f(x) + f(y)) \qquad \bar{c}(x, y) = f^{-1}\left(\frac{1}{2}(f(x) + f(y))\right)$$

$$d(x, y) = f^{-1}\left(\frac{1}{\frac{1}{f(x)} + \frac{1}{f(y)}}\right) \qquad \bar{d}(x, y) = f^{-1}\left(\frac{2}{\frac{1}{f(x)} + \frac{1}{f(y)}}\right)$$

$$= f^{-1}\left(\frac{f(x)f(y)}{f(x) + f(y)}\right) \qquad = f^{-1}\left(\frac{2f(x)f(y)}{f(x) + f(y)}\right)$$

Here, not only the associative operator is given, but so are the corresponding bisymmetric mean operators.

The corresponding negation operator is

$$\eta_\nu(x) = f^{-1}\left(\frac{f^2(\nu)}{f(x)}\right), \tag{12}$$

where ν is the fixed point of the negation operator. This value can be interpreted as a threshold (decision) level. The modifier operator in the Pliant system has the following form

$$\tau_{\nu_*}(x) = f^{-1}\left(f(\nu)\frac{f(x)}{f(\nu_*)}\right)$$

And if $\nu < \nu_*$ then it is a strengthened unary operator; and if $\nu > \nu_*$ it is a weakened unary operator (see [7,10]).

Remark 3. If $f(x) = \left(\frac{1-x}{x}\right)^\alpha$ when $\alpha \in \mathbb{R}^+$, then we have the Dombi operator system.

Definition 6. A function $i : [0, 1]^2 \to [0, 1]$ is called a (d, η)-implication if there exists a disjunctive operator and a fuzzy negation η such that

$$i_\nu(x, y) = x \to y = \bar{x} \vee y = d(\eta_\nu(x), y)$$

$$\bar{i}_\nu(x, y) = x \dashrightarrow y = \bar{x} \,\bar{\vee}\, y = \bar{d}(\eta_\nu(x), y).$$

$x, y \in [0, 1]$ if η_ν is a strong fuzzy negation. Here the mean implication operator is based on the mean disjunctive operator. The implication operator has a lower ν index which indicates the fixed point of the negation operator.

Proposition 3. *The general form of the Pliant and mean-Pliant implication operators is:*

$$i_\nu(x, y) = f^{-1}\left(\frac{f^2(\nu)f(y)}{f^2(\nu) + f(x)f(y)}\right), \tag{13}$$

$$\bar{i}_\nu(x, y) = f^{-1}\left(\frac{2f^2(\nu)f(y)}{f^2(\nu) + f(x)f(y)}\right). \tag{14}$$

when the implication operator is defined by (10).

Proof. We get the result by direct calculation using the Pliant operators.

In the Dombi operator case, we get the following expressions:

$$i_\nu^{(\alpha)}(x,y) = \frac{\left((1-\nu)^{2\alpha}x^\alpha y^\alpha + \nu^{2\alpha}(1-x)^\alpha(1-y)^\alpha\right)^{\frac{1}{\alpha}}}{\left((1-\nu)^{2\alpha}x^\alpha y^\alpha + \nu^{2\alpha}(1-x)^\alpha(1-y)^\alpha\right)^{\frac{1}{\alpha}} + (1-\nu)^2(1-y)x} \quad (15)$$

$$\bar{i}_\nu^{(\alpha)}(x,y) = \frac{\left((1-\nu)^{2\alpha}x^\alpha y^\alpha + \nu^{2\alpha}(1-x)^\alpha(1-y)^\alpha\right)^{\frac{1}{\alpha}}}{\left((1-\nu)^{2\alpha}x^\alpha y^\alpha + \nu^{2\alpha}(1-x)^\alpha(1-y)^\alpha\right)^{\frac{1}{\alpha}} + 2^{\frac{1}{\alpha}}(1-\nu)^2(1-y)x} \quad (16)$$

If $\nu = \frac{1}{2}$, then

$$i_\nu^{(\alpha)}(x,y) = \frac{\left(x^\alpha y^\alpha + (1-x)^\alpha(1-y)^\alpha\right)^{\frac{1}{\alpha}}}{\left(x^\alpha y^\alpha + (1-x)^\alpha(1-y)^\alpha\right)^{\frac{1}{\alpha}} + (1-y)x} \quad (17)$$

See Figs. 1, 2 and 3

$$\bar{i}_\nu^{(\alpha)}(x,y) = \frac{\left(x^\alpha y^\alpha + (1-x)^\alpha(1-y)^\alpha\right)^{\frac{1}{\alpha}}}{\left(x^\alpha y^\alpha + (1-x)^\alpha(1-y)^\alpha\right)^{\frac{1}{\alpha}} + 2^{\frac{1}{\alpha}}(1-y)x} \quad (18)$$

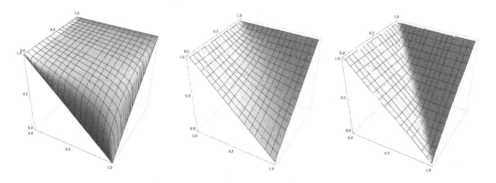

Fig. 1. $\alpha = 0.3$ **Fig. 2.** $\alpha = 1$ **Fig. 3.** $\alpha = 8$

If $\alpha = 1$, then

$$i_\nu(x,y) = i_\nu^{(1)}(x,y) = \frac{(1-\nu)^2 xy + \nu^2(1-x)(1-y)}{(1-\nu)^2 x + \nu^2(1-x)(1-y)}. \quad (19)$$

See Figs. 4, 5 and 6

$$\bar{i}_\nu(x,y) = \bar{i}_\nu^{(1)}(x,y) = \frac{(1-\nu)^2 xy + \nu^2(1-x)(1-y)}{(1-\nu)^2 x(2-y) + \nu^2(1-x)(1-y)}. \quad (20)$$

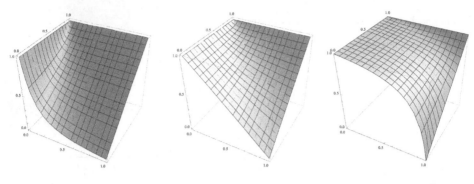

Fig. 4. $\nu = \frac{1}{4}$ **Fig. 5.** $\nu = \frac{1}{2}$ **Fig. 6.** $\nu = \frac{3}{4}$

If $\alpha = 1$ and $\nu = \frac{1}{2}$, then

$$i(x,y) = i_{\frac{1}{2}}(x,y) = \frac{xy + (1-x)(1-y)}{x + (1-x)(1-y)} \tag{21}$$

$$\bar{i}(x,y) = \bar{i}_{\frac{1}{2}}(x,y) = \frac{xy + (1-x)(1-y)}{x(2-y) + (1-x)(1-y)} \tag{22}$$

In Table 2, we summarize the properties of $i_{\nu}(x,y)$ and $\bar{i}_{\nu}(x,y)$.

It is obvious that in the S implication case, the identity principle is not valid. The fixed point of the negation $[0,1]$ interval is divided into two intervals $[0,\nu), [\nu,1]$. Here, this fixed point can be viewed as a threshold. We can weaken the indentity principle so that $\bar{i}_{\nu}(x,x) \geq \nu$.

Proposition 4. *The Pliant implication function* $\bar{i}_{\nu} : [0,1]^2 \to [0,1]$ *satisfies the weak identity principle. That is*

$$\bar{i}_{\nu}(x,x) \geq \nu. \tag{23}$$

Proof. Since

$$\bar{i}(x,y) = f^{-1}\left(\frac{2f^2(\nu)f(x)}{f^2(\nu) + f^2(x)} \right) \geq \nu,$$

it follows that

$$\frac{2f^2(\nu)f(x)}{f^2(\nu) + f^2(x)} \leq 1$$

One of the most important features of the implication operator is the ordering principle. The S implication doesn't have this property. In the next theorem, we will show that a weakened form of it does.

Proposition 5. *For the mean implication operator, the following is valid:*

$$if \quad x \leq y \quad then \quad \bar{i}_{\nu}(x,y) \geq \nu$$

Table 2. Properties of the implication and mean-implication operators

	Property		$i_\nu(x,y)$	$\bar{i}_\nu(x,y)$
1	Continuity	$i(\lim\limits_{n\to\infty} x_n, \lim\limits_{n\to\infty} y_n) = \lim\limits_{n\to\infty} i(x_n, y_n)$	✓	✓
2	Lipschitz	$|F(x_1, y_1) - F(x_2, y_2)| \le c \cdot (|x_1 - x_2| + |y_1 - y_2|)$	✓	✓
3	First place monotonicity	if $x_1 \le x_2$, then $i(x_1, y) \ge i(x_2, y)$	✓	✓
4	Second place isotonicity	if $y_1 \le y_2$, then $i(x, y_1) \le i(x, y_2)$	✓	✓
5.	Compatibility	$i(0,0) = 1$	✓	✓
		$i(1,1) = 1$	✓	✓
		$i(1,0) = 0$	✓	✓
		$i(0,1) = 1$	✓	✓
6	Dominance of falsity	$i(0, x) = 1$	✓	✓
7	Dominance of truth consequent	$i(x, 1) = 1$	✓	✓
8	Left neutrality	$i(1, x) = x$	✓	–
9	The strong negation principle	$i(x, 0) = \eta_\nu(x)$	✓	–
10	Law of contraposition	$i(x, y) = i(\eta_\nu(y), \eta_\nu(x))$	✓	✓
		$i(\eta_\nu(x), y) = i(\eta_\nu(y), x)$	✓	✓
		$i(x, \eta_\nu(y)) = i(y, \eta_\nu(x))$	✓	✓
11	Exchange principle	$i_\nu(x, i_\nu(y, z)) = i_\nu(y, i_\nu(x, z))$	✓	✓
12	Law of importation	$i(c(x, y), z) = i(x, i(y, z))$	✓	✓
13	Identity principle	$i(x, x) = 1$	–	–
14	Ordering property	$i(x, y) = 1$ if and only if $x \le y$	–	–
15	Distributivity property 1	$\bar{i}(c(x, y), z) = \bar{d}(\bar{i}(x, z), \bar{i}(y, z))$	–	✓
16	Distributivity property 2	$\bar{i}(x, \bar{d}(y, z)) = \bar{d}(\bar{i}(x, y), \bar{i}(x, z))$	–	✓

Proof. We use the fact that $x \le y$ so $f(x) \ge f(y)$ and $\frac{f(\nu)}{f(x)} \le \frac{f(\nu)}{f(y)}$ and it is known that

$$2 \le \frac{f(x)}{f(\nu)} + \frac{f(\nu)}{f(x)} \le \frac{f(x)}{f(\nu)} + \frac{f(\nu)}{f(y)}.$$

Hence

$$\frac{2f(\nu)f(y)}{f(x)f(y) + f^2(\nu)} \le 1.$$

Next, multiplying both sides by $f(\nu)$ and applying $f^{-1}(x)$ (a strictly decreasing function), we get the desired result. That is,

$$\frac{2f^2(\nu)f(y)}{f(x)f(y) + f^2(\nu)} \le f(\nu)$$
$$\bar{i}(x, y) \ge \nu.$$

Remark 4. The converse is not true i.e. there exist an x and y such that

$$\bar{i}_\nu(x,y) \geq \nu \text{ and } x \not\leq y \tag{24}$$

5   Implication and Modus Ponens

An implication operator plays an important role in the deductive process of a logic, which is usually realized by some rules of inference. Modus ponens is one such rule of inference, wherein given two classical logic propositions $A \to B$ and from A we infer B. A similar rule of inference in the case of fuzzy propositions is called the generalized modus ponens (GMP), wherein given two fuzzy propositions $A' \to B'$ and from A' we infer B'. A key point of this inference is that even if $A' \neq A$, we will still be able to infer a reasonable conclusion B'.

The classical description of modus ponens is

$$\frac{\begin{aligned} x &= 1 \\ x \to y &= 1 \end{aligned}}{y = 1}$$

Next, we will use the threshold concept and we will show that the following proposition is valid.

Proposition 6. *In the Pliant system, the modus ponens is true using the mean implication. Namely,*

$$\frac{\begin{aligned} x &\geq \nu \\ \bar{i}_\nu(x,y) &\geq \nu \end{aligned}}{y \geq \nu} \tag{25}$$

Proof. So we can write (25) using the fact that $f(x)$ is a strictly decreasing function and using the Pliant operators, we get

$$\frac{\begin{aligned} \frac{f(x)}{f(\nu)} &< 1 \\ \frac{2\frac{f(y)}{f(\nu)}}{1 + \frac{f(x)}{f(\nu)}\frac{f(y)}{f(\nu)}} &< 1 \end{aligned}}{\frac{f(y)}{f(\nu)} < 1} \tag{26}$$

So $1 + \frac{f(x)}{f(\nu)}\frac{f(y)}{f(\nu)} > 2\frac{f(y)}{f(\nu)}$ and we have $1 - \frac{f(y)}{f(\nu)} > \frac{f(y)}{f(\nu)}\left(1 - \frac{f(x)}{f(\nu)}\right)$ because $\frac{f(x)}{f(\nu)} < 1$ so $\frac{f(y)}{f(\nu)}\left(1 - \frac{f(x)}{f(\nu)}\right) > 0$ and $1 - \frac{f(y)}{f(\nu)} > 0$.

The Generalized Modus Ponens (GMP) can be solved by replacing $x \geq \nu$ by $\tau_{\nu_x}(x) \geq \nu$. ν_x is given and we have to find $\tau_{\nu_y}(y)$.

Proposition 7. *In the Pliant system, the solution of the Generalised Modus Ponens is*

$$\frac{\tau_{\nu_x}(x) \geq \nu}{\overline{i}_\nu(x,y) \geq \nu} \quad or \quad \frac{x \geq \nu_x}{\overline{i}_\nu(x,y) \geq \nu}$$
$$\frac{\overline{i}_\nu(x,y) \geq \nu}{\tau_{\nu_y}(y) \geq \nu} \quad \quad \frac{\overline{i}_\nu(x,y) \geq \nu}{y \geq \nu_y}$$

and if $\nu_x \leq f^{-1}(2f(\nu))$, then

$$\nu_y = f^{-1}\left(\frac{f^2(\nu)}{2f(\nu) - f(\nu_x)}\right).$$

Proof. Now we use this proposition and in (26) the inequality relation of the implication operator will have the following form

$$\frac{f(y)}{f(\nu)} \leq \frac{1}{2 - \frac{f(x)}{f(\nu)}}.$$

Here we use the fact that $\nu_x \leq f^{-1}(2f(\nu))$. The larger bound for this inequaility is $x = \nu_x$. So solving for y, we get the desired result $(y = \nu_y)$.

6   Conclusions

Here, a new so-called mean implication operator is introduced. Necessary and sufficient conditions were given for two types of fuzzy distributivity implications. The implication operator has a fixed point as well, which corresponds to the negation operator we need. The fixed point in the negation operator is used as a threshold. Based on this threshold, some weakened properties of the implication operator were proved in the framework of the Pliant concept and modus ponens was also examined. It is an open question whether the general mean implication operator has this property too.

References

1. Aczel, J.: Lectures on Functional Equations and Their Applications, vol. 19 (1966)
2. Baczyński, M., Jayaram, B.: Fuzzy Implications. Springer, Berlin (2008)
3. Bertoluzza, C.: On the distributivity between t-norm and t-conorms. In: Proceedings of the 2nd IEEE International Conference on Fuzzy Systems (FUZZ-IEEE 1993), San Francisco, USA, pp. 140–147, March 1993 (1993)
4. Bertoluzza, C., Doldi, V.: On distributivity between t-norms and t-conors. Fuzzy Sets Syst. **142**, 85–104 (2004)
5. Carbonell, M., Mas, M., Suner, J., Torrens, J.: On distributivity and modularity in De Morgan triplets. Int. J. Uncertain. Fuzziness Knowl.-Based Syst. **4**, 351–368 (1996)
6. Dombi, J.: DeMorgan systems with an infinitely many negations in the strict monotone operator case. Inf. Sci. **181**(8), 1440–1453 (2011)
7. Dombi, J.: Modifiers based on connectives. In: Eleventh International Conference on Fuzzy Set Theory and Applications, p. 1 (2012)

8. Dombi, J., System, P.O., Fodor, J., Ryszard, K., Araujo, C.P.S.: Recent Advances in Intelligent Engineering Systems, pp. 31–58. Springer, Berlin (2012)
9. Dombi, J.: On a certain class of aggregative operators. Inf. Sci. **245**, 313–328 (2013)
10. Dombi, J.: Unary operators. In: Advances in Applied, Pure Mathematics: Proceedings of the International Conference on Pure Matematics, Apllied Mathematics, Computational Methods (PMAMCM), pp. 91–97 (2014)
11. Dombi, J.: On consistent operator systems. In: Baczynski, M., De Baets, B., Mesiar, R. (eds.) Proceedings of the 8th International Summer School on Aggregation Operators (AGOP), Katowice, Poland, pp. 31–32 (2015)
12. Fodor, J., Roubens, M.: Fuzzy Preference Modelling and Multicriteria Decision Suppert. Kluwer Academic Publishers, Dordrecht (1994)
13. Klement, E.P., Mesiar, R., Pap, E.: Triangular Norms. Kluwer, Dordecht (2000)
14. Mas, M., Mayor, G., Torrens, J.: The distributivity condition for uninorms and t-operators. Fuzzy Sets Syst. **128**, 209–225 (2002)
15. Reiser, R.H.S., Bedregal, B., Baczyński, M.: Aggregating fuzzy implications. Inf. Sci. **253**, 126–146 (2013)
16. Ruiz, D., Torrens, J.: Distributive idempotent uninorms. Int. J. Uncertain. Fuzziness Knowl.-Based Syst. **11**, 413–428 (2003)
17. Trillas, E., Alsina, C.: On the law $[p \wedge q \to r] = [(p \to r) \vee (q \to r)]$ in fuzzy logic. IEEE Trans. Fuzzy Syst. **10**(1), 84–88 (2002)
18. Trillas, E., Alsina, C.: Short note: on a conditional-conjunctive law with fuzzy sets. Int. J. Uncertain. Fuzziness Knowl.-Based Syst. **21**(1), 1–8 (2013)

Some Results About Fuzzy Consequence Operators and Fuzzy Preorders Using Conjunctors

Carlos Bejines[1], María Jesús Chasco[1], Jorge Elorza[1], and Susana Montes[2(✉)]

[1] Departamento de Física y Matemática Aplicada, Facultad de Ciencias,
Universidad de Navarra, Pamplona, Spain
{mjchasco,jelorza}@unav.es, cbejines@alumni.unav.es
[2] Departamento de Estadística e Investigación Operativa,
Facultad de Ciencias, Universidad de Oviedo, Oviedo, Spain
montes@uniovi.es

Abstract. The purpose of this paper is to study fuzzy operators induced by fuzzy relations and fuzzy relations induced by fuzzy operators. Many results are obtained about the relationship between $*$-preorders and fuzzy consequences operators for a fixed t-norm $*$. We analyse these properties by considering a semi-copula (generalization of t-norm concept) instead of a t-norm. Moreover, we show that the conditions imposed cannot be relaxed. We have been able to prove some important results about the relationships between fuzzy relations and fuzzy operators in this more general context.

1 Introduction

Fuzzy relations and fuzzy operators are essential tools in fuzzy logic. The fuzzy operator induced by a fuzzy relation through Zadeh's compositional rule and the fuzzy relation induced by a fuzzy operator are concepts that have been studied extensively (see for instance [3,4,8–10,16]).

Fuzzy consequence operator notion was introduced by Pavelka in 1979. He extends the notion of consequence operator defined by Tarski to fuzzy set theory field (see [15]). From then, fuzzy consequence operators have been largely studied in the approximate reasoning context using different fuzzy logics (see for instance [2–4,9,10,12,13]).

Since the introduction of Fuzzy Set Theory by Zadeh in 1965, the concept of fuzzy $*$-preorder has been essential on fuzzy logic. Moreover, fuzzy consequence operators are closely related to fuzzy preorders through coherent operators which were introduced in the early nineties by Castro and Trillas in [4].

T-norms are a useful tool in order to define the transitivity for fuzzy relations but it is well-known that these operators are too restrictive in some cases (see for instance [5,6]). Associativity, commutativity and the boundary conditions can be weakened.

© Springer International Publishing AG 2018
V. Torra et al. (eds.), *Aggregation Functions in Theory and in Practice*,
Advances in Intelligent Systems and Computing 581, DOI 10.1007/978-3-319-59306-7_26

This paper is organized as follows: in Sect. 2, we show some basic notions which are required in the sequel. In Sect. 3, we study and characterize the relationship between preorders and fuzzy consequence operators at the general context of conjunctors. Section 4 is devoted to the study of a specific families of fuzzy consequence operators. In Sect. 5, we characterize when an operator generated by a fuzzy relation is a FCO for a finite universe. In fact, we prove that this operator is a FCO if a only if the fuzzy relation is a •-preorder. Finally, we address some conclusions.

2   Preliminaries

In this section we give a brief account of some notions and results which are necessary in the following sections. A binary operation $* : [0,1] \times [0,1] \longrightarrow [0,1]$ is called a t-norm (see [14]) if it verifies the following properties:

1. $(a * b) * c = a * (b * c)$ for all $a, b, c \in [0,1]$.
2. $a * b = b * a$ for all $a, b \in [0,1]$.
3. $1 * a = a$ for all $a \in [0,1]$.
4. If $a_1 \leq a_2$, then $a_1 * b \leq a_2 * b$ for all $a_1, a_2, b \in [0,1]$.

Conjunctor (see [5,6]) is a more general concept which allows us to work with fuzzy relations and operators:

Definition 1. A binary operation $\bullet : [0,1] \times [0,1] \longrightarrow [0,1]$ is called a *conjunctor* if:

1. If $a_1 \leq a_2$ and $b_1 \leq b_2$, then $a_1 \bullet b_1 \leq a_2 \bullet b_2$ for all $a_1, a_2, b_1, b_2 \in [0,1]$.
2. $0 \bullet 1 = 1 \bullet 0 = 0$ y $1 \bullet 1 = 1$.

The conjunctor $\bullet$ is said to be associative if $(a \bullet b) \bullet c = a \bullet (b \bullet c)$ for all a, b, $c \in [0,1]$. Furthermore, if the conjunctor $\bullet$ verifies: $1 \bullet a = a \bullet 1 = a$ for all $a \in [0,1]$, it is called a *semi-copula* (see for instance [1,7]).

Note that the conjunction for classical relations satisfies the two conditions in the previous conditions and that any t-norm is a particular case of conjunctor.

Let X be a non-empty universal set and let R be a fuzzy relation on X:

1. If $R(x,x) = 1$ for all $x \in X$, R is named reflexive relation.
2. If $R(x,y) = R(y,x)$ for all $x, y \in X$, R is named symmetric relation.
3. If $R(x,y) \bullet R(y,z) \leq R(x,z)$ for all $x, y, z \in X$, R is named •-transitive relation.

If R is a fuzzy relation which verifies (1) and (3), R is called •-preorder. If R is a fuzzy relation which verifies (1), (2) and (3), R is called •-indistinguishability (or •-equivalence).

Definition 2. A function $C : [0,1]^X \longrightarrow [0,1]^X$ is called Fuzzy Consequence Operator (FCO, for short) if it fulfils the following axioms:

(C1) $\mu \subset C(\mu)$ for all fuzzy set in $[0,1]^X$ (Inclusion).
(C2) If $\mu_1 \subset \mu_2$, then $C(\mu_1) \subset C(\mu_2)$ for all $\mu_1, \mu_2 \in I^X$ (Monotony).
(C3) $C(C(\mu)) \subset C(\mu)$ for all $\mu \in I^X$ (Idempotence).

The inclusion of fuzzy subsets is given by the pointwise order, that is, $\mu_1 \subset \mu_2$ if $\mu_1(x) \le \mu_2(x)$ for all $x \in X$.

We denote by $\underline{k}$ the constant fuzzy set defined by $\underline{k}(x) = k$ for all $x \in X$.

Remark 1. Note that under the axioms $(C1)$ and $(C2)$, the idempotence axiom $(C3)$ may be written equivalently as $C(C(\mu)) = C(\mu) \forall \mu \in I^X$.

Fuzzy consequence operators have been extensively studied (see for instance [11,17]) when we deal with t-norms. They are also called closure operators, both in the algebraic and in the topological context. The following notion is a generalization for the case of conjunctors. It will be a very important property when we study the relationship between fuzzy consequence operators and fuzzy relations.

Definition 3. Given a conjunctor $\bullet$, an operator C is called $\bullet$-coherent if it verifies for all $x, a \in X$ and $\mu \in [0,1]^X$:

$$\mu(a) \bullet C(\varphi_a)(x) \le C(\mu)(x)$$

where φ_a denotes the singleton a, that is, $\varphi_a(a) = 1$ and $\varphi_a(x) = 0$ for all $x \neq a$.

3   On the Relationship Between $\bullet$-preorders and Fuzzy Consequence Operators

In [9], Elorza and Burillo achieved some results about the relationship of fuzzy preorders and fuzzy consequence operators. Now, we will extend these results to the case we are working with conjunctors instead of just t-norms.

Let us introduce the notion of fuzzy operator induced by a fuzzy relation and viceversa, based on the studies developed in [4].

Definition 4. Let $\bullet$ be a conjunctor and let R be a fuzzy relation, the fuzzy operator induced by R through sup-$\bullet$ Zadeh's compositional rule is given by

$$C_R^\bullet(\mu)(x) = \sup_{w \in X} \{\mu(w) \bullet R(w, x)\}$$

Definition 5. Let $C : [0,1]^X \to [0,1]^X$ be a fuzzy operator, the fuzzy relation induced by C is defined by

$$R_C(x, y) = C(\varphi_x)(y)$$

These two concepts are very related, as we can see at the following result.

Theorem 1. *Let* • *be a semi-copula and let* R *be a fuzzy relation on the universal set* X, *then*

$$R = R_{C_R^\bullet}$$

Proof. Let x, y be two elements in X, we have that

$$R_{C_R^\bullet}(x,y) = C_R^\bullet(\varphi_x)(y) = \sup_{\omega \in X} \{\varphi_x(\omega) \bullet R(\omega, y)\} = 1 \bullet R(x,y) = R(x,y)$$

Thus, R and $R_{C_R^\bullet}$ are the same relation. □

In the particular case that R has some specific properties, the fuzzy operator induced by it also fulfils some extra properties. Indeed,

Theorem 2. *Let* X *be a finite universal set, let* • *be an associative semi-copula, and let* R *be a fuzzy relation on* X. *If* R *is* •*-preorder, then the operator* $C_R^\bullet$ *is a FCO.*

Proof

(C1) $\mu(x) = \mu(x) \bullet R(x,x) \leq \sup_{\omega \in X} \{\mu(\omega) \bullet R(\omega, x)\} = C_R^\bullet(\mu)(x)$ for all $x \in X$.

(C2) If $\mu_1 \subset \mu_2$, then we have:

$$C_R^\bullet(\mu_1)(x) = \sup_{\omega \in X} \{\mu_1(\omega) \bullet R(\omega, x)\} \leq \sup_{\omega \in X} \{\mu_2(\omega) \bullet R(\omega, x)\} = C_R^\bullet(\mu_2)(x)$$

for any $x \in X$.

(C3) Let x be any element in X, then

$$
\begin{aligned}
C_R^\bullet(C_R^\bullet(\mu))(x) &= \sup_{\omega \in X} \{\sup_{\tau \in X} \{\mu(\tau) \bullet R(\tau, \omega)\} \bullet R(\omega, x)\} \\
&= \sup_{\omega \in X} \{\sup_{\tau \in X} \{(\mu(\tau) \bullet R(\tau, \omega)) \bullet R(\omega, x)\}\} \\
&= \sup_{\omega \in X} \{\sup_{\tau \in X} \{\mu(\tau) \bullet (R(\tau, \omega) \bullet R(\omega, x))\}\} \\
&\leq \sup_{\omega \in X} \{\sup_{\tau \in X} \{\mu(\tau) \bullet R(\tau, x)\}\} = \sup_{\tau \in X} \{\mu(\tau) \bullet R(\tau, x)\} = C_R^\bullet(\mu)(x)
\end{aligned}
$$

Therefore, $C_R^\bullet$ fulfils all the requirement in Definition 2, that is, it is a fuzzy consequence operator. □

Note that the existence of the unity for the conjunctor cannot be removed in Theorems 1 and 2 as it is shown in the following example.

Example 1. Let X be an universal set and let • be the conjunctor defined by: $1 \bullet 1 = 1$ and $x \bullet y = 0$ otherwise.

- Let R be a •-fuzzy preorder such that there exist $x, y \in X$ with $0 < R(x,y) < 1$. Then,

$$R_{C_R^\bullet}(x,y) = C_R^\bullet(\varphi_x)(y) = \sup_{\omega \in X} \{\varphi_x(\omega) \bullet R(\omega, y)\} = 1 \bullet R(x,y) = 0 \neq R(x,y)$$

  that is, the equality for Theorem 1 is not fulfilled.

- Let μ be the fuzzy set such that $\mu(x) = \frac{1}{2}$ for all $x \in X$. Then,

$$\mu(x) = \frac{1}{2} > 0 = \sup_{w \in X} \left\{ \frac{1}{2} \bullet R(w,x) \right\} = \sup_{w \in X} \{ \mu(w) \bullet R(w,x) \} = C_R^\bullet(\mu)(x)$$

Since $C_R^\bullet$ does not satisfy the axiom $(C1)$ in Definition 2, it is not fuzzy consequence operator for any $\bullet$-preorder R. Thus, we can see the existence of the unity for the conjunctor is a necessary condition in Theorem 2.

We can see from Theorem 2 as the commutative property is not necessary for proving that preorder induce fuzzy consequence operators. Thus, we do not need to work with t-norms and it is enough to deal with associative semi-copula. Indeed, we are going to see that $C_R^\bullet$ is even $\bullet$-coherent, even in the case the conjunctor is not associative.

Theorem 3. *Let $\bullet$ be a semi-copula and let R be a fuzzy relation on an universal set X, then $C_R^\bullet$ is $\bullet$-coherent.*

Proof. Let consider any $x, a \in X$ and any $\mu \in [0,1]^X$, then

$$\mu(a) \bullet C_R^\bullet(\varphi_a)(x) = \mu(a) \bullet \sup_{w \in X} \{ \varphi_a(w) \bullet R(w,x) \} = \mu(a) \bullet (1 \bullet R(a,x))$$

$$= \mu(a) \bullet R(a,x) \leq \sup_{w \in X} \{ \mu(w) \bullet R(w,x) \} = C_R^\bullet(\mu)(x)$$

that is, $C_R^\bullet$ is $\bullet$-coherent. $\qquad\qquad \square$

Conversely, we are going to see the properties inherit by the fuzzy relation induced by a fuzzy consequence operator.

Theorem 4. *Let $\bullet$ be a conjunctor and let $C : [0,1]^X \to [0,1]^X$ be a $\bullet$-coherent FCO, then the fuzzy relation R_C induced by C is a $\bullet$-preorder.*

Proof. $(R1)$ $R_C(x,x) = C(\varphi_x)(x) \geq \varphi_x(x) = 1 \ \forall x \in X$ by the axiom $(C1)$ in Definition 2.
$(R2)$ Since C is $\bullet$-coherent and a fuzzy consequence operator,

$$R_C(x,y) \bullet R_C(y,z) = C(\varphi_x)(y) \bullet C(\varphi_y)(z) \leq C(C(\varphi_x))(z) = C(\varphi_x)(z) = R_C(x,z)$$

for any $x, y, z \in X$. $\qquad\qquad \square$

4 Families of Fuzzy Consequence Operators Using Conjunctors

Let S be a non-empty family of fuzzy subsets of X, an operator C_S was defined in [9] as

$$C_S : [0,1]^X \longrightarrow [0,1]^X$$
$$\mu \longmapsto C_S(\mu)$$

where

$$C_S(\mu)(x) = \inf_{\nu \in S, \mu \subset \nu} \{ \nu(x) \}$$

It was also proven that C_S is a fuzzy consequence operator which satisfies:

1. If $\mu \in S$, then $C_S(\mu) = \mu$.
2. If $S = \{\mu\}$, then

$$C_S(\rho) = \begin{cases} \mu & if \rho \subset \mu \\ X & otherwise \end{cases}$$

The use of $*$-coherence property in fuzzy consequence operators setting is a key concept, so it is interesting to ask when C_S will be $\bullet$-coherent for a conjunctor $\bullet$. The following theorem shows a sufficient condition in order to get it.

Let Φ_1 be the set of fuzzy singletons, that is, $\Phi_1 = \{\varphi_x \mid x \in X\}$.

Theorem 5. *If $\Phi_1 \subset S$, then C_S is $\bullet$-coherent for every semi-copula $\bullet$.*

Proof. Since $a \bullet 1 = a$, we have $a \bullet b \leq a \bullet 1 = a$. In the same way, $1 \bullet b = b$ and $a \bullet b \leq b$.

Thus,

$$\mu(a) \bullet C_S(\varphi_a)(x) = \mu(a) \bullet \varphi_a(x) \leq \min\{\mu(a), \varphi_a(x)\} \leq \mu(x) \leq C_S(\mu)(x)$$

for any $x, a \in X$. $\qquad\square$

The chain $\Omega_p^* \subset \tilde{\Omega}^* \subset \Omega_{rt}^* \subset \Omega \subset \Omega'$ was proven in the case of t-norms (see [9]), where $\Omega' = \{C \mid C : I^X \longrightarrow I^X\}$, $\Omega = \{C \mid C \text{ is } FCO\}$, $\Omega_{rt}^* = \{C \in \Omega \mid R_C \text{ is } * \text{-preorder}\}$, $\tilde{\Omega}^* = \{C \in \Omega \mid C \text{ is } * \text{-coherent}\}$ and $\Omega_p^* = \{C \in \Omega \mid \exists R \; * \text{-preorder with } C = C_R^*\}$. There is not problem to extend this chain to semi-copulas in a natural way:

$$\Omega_p^{\bullet} \underset{(1)}{\subset} \tilde{\Omega}^{\bullet} \underset{(2)}{\subset} \Omega_{rt}^{\bullet} \underset{(3)}{\subset} \Omega \underset{(4)}{\subset} \Omega'$$

In fact, (1) is due to Theorem 3, (2) is due to Theorem 4 and (3) and (4) are obvious.

Moreover, the inclusions in this chain are strict, like the following examples show.

Example 2 ($\Omega_p^{\bullet} \neq \tilde{\Omega}^{\bullet}$). Let $\bullet$ a semi-copula and let S be a family of fuzzy sets of X such that $\Phi_1 \subset S$, if C_S is not the identity operator, then C_S is $\bullet$-coherent but C_S is not in $\Omega_p^{\bullet}$.

In fact, by Theorem 5, $Cs \in \tilde{\Omega}^{\bullet}$. If C_S is induced by a $\bullet$-preorder R, then by Theorem 1, we have that C_S is equal to $C_{R_{C_S}^{\bullet}}$ for the $\bullet$-preorder R. Then, we have: for all $x \in X$

$$C_{R_{C_R}^{\bullet}}(\mu)(x) = \sup_{w \in X}\{\mu(w) \bullet C_S(\varphi_w)(x)\} = \sup_{w \in X}\{\mu(w) \bullet \varphi_w(x)\} = \mu(x) \bullet 1 = \mu(x)$$

Hence, $\mu(x) = C_S(\mu)(x)$ for all $x \in X$, but there exists some $x \in X$, $\mu(x) \neq C_S(\mu)(x)$ because C_S is not the identity operator.

Example 3 ($\tilde{\Omega}^{\bullet} \neq \Omega_{rt}^{\bullet}$). Let $\bullet$ a semi-copula and let X be a set with two or more elements. If $a \neq b$ in X, μ a fuzzy set such that $\mu(x) < 1$ for all $x \in X$ and $\mu(a) > \mu(b)$, $S = \{\mu\}$. Then $C_S \in \Omega_{rt}^{\bullet}$, but C_S is not $\bullet$-coherent.

In fact, since C_S fulfils $(C1)$, R_{C_S} is a reflexive relation. For each φ_x, we ha⋯
that μ does not contain φ_x, then for all $x, y, z \in X$:

$$R_{C_S}(x, z) = C_S(\varphi_x)(z) = 1 \geq R_{C_S}(x, y) \bullet R_{C_S}(y, z)$$

Therefore, R_{C_S} is a $\bullet$-preorder. Now, since $\mu(a) > \mu(b)$, we have:

$$\mu(a) \bullet C_S(\varphi_a)(b) = \mu(a) \bullet 1 = \mu(a) > \mu(b) = C_S(\mu)(b)$$

Therefore, C_S is not $\bullet$-coherent.

Example 4 ($\Omega_{rt}^\bullet \neq \Omega$). Let $\bullet$ a semi-copula and let X be a set with three or more elements, let $x, y, z \in X$ different elements and let μ be a fuzzy set such that $\mu(x) = 1 \neq \mu(y) > \mu(z)$. If $S = \{\mu\}$, then $C_S \in \Omega$ but $C_S \notin \Omega_{rt}^\bullet$.
In fact, we know that C_S is fuzzy consequence operator. Now, since $\varphi_x \subset \mu$ and $\varphi_y \not\subset \mu$, we have:

$$C_S(\varphi_x)(z) = \mu(z) < \mu(y) = \mu(y) \bullet 1 = C_S(\varphi_x)(y) \bullet C_S(\varphi_y)(z)$$

Hence, $R_{C_S}(x, z) < R_{C_S}(x, y) \bullet R_{C_S}(y, z)$. We conclude R_{C_S} is not $\bullet$-preorder.

5 Operators Induced by Relations Using Conjunctors

The purpose of this section is to show when the $\bullet$-preorder obtained in Theorem 4 from the fuzzy consequence operator obtained from a $\bullet$-preorder remains equal to the initial $\bullet$-preorder.

In order to do that, two new axioms for fuzzy consequence operators have to be introduced:

$(C4)$ $C(\mu \vee \nu) = C(\mu) \vee C(\nu)$ for all $\mu, \nu \in [0, 1]^X$
$(C5)$ $C(\underline{k} \wedge \varphi_x) = \underline{k} \bullet C(\varphi_x)$ for all $\varphi_x \in \Phi_1$ and all constant fuzzy set $\underline{k}$.

For any finite universe X, Esteva et al. (see [12]) proved the two following results:

- Let X be a finite universal set, let $*$ be a t-norm and let $C : [0, 1]^X \longrightarrow [0, 1]^X$ be a fuzzy operator, then C satisfies $(C1), (C4)$ and $(C5)$ if and only if there exists a reflexive relation R such that $C = C_R^*$.
- Let X be a finite universal set, let $*$ be a t-norm and let $C : [0, 1]^X \longrightarrow [0, 1]^X$ be a fuzzy operator, then C satisfies $(C1), (C3), (C4)$ and $(C5)$ if and only if there exists a $*$-preorder R such that $C = C_R^*$.

Both results are extended the natural way to semi-copulas:

Theorem 6. *Let X be a finite universal set, let $\bullet$ be a semi-copula and let $C : [0, 1]^X \longrightarrow [0, 1]^X$ be an operator, then C satisfies $(C1), (C4)$ and $(C5)$ if and only if there exists a reflexive relation R such that $C = C_R^\bullet$.*

Proof. Suppose R is a reflexive relation such that $C = C_R^\bullet$. We have

(C1)
$$C_R^\bullet(\mu)(x) = \sup_{\omega \in X} \{\mu(\omega) \bullet R(\omega, x)\} \geq \mu(x) \bullet R(x, x) = \mu(x)$$

(C4)
$$\begin{aligned}
C_R^\bullet(\mu \vee \nu)(x) &= \sup_{\omega \in X} \{\mu \vee \nu(\omega) \bullet R(\omega, x)\} \\
&= \sup_{\omega \in X} \{(\mu(\omega) \bullet R(\omega, x)) \vee (\nu(\omega) \bullet R(\omega, x))\} \\
&= \sup_{\omega \in X} \{(\mu(\omega) \bullet R(\omega, x))\} \vee \sup_{\omega \in X} \{(\nu(\omega) \bullet R(\omega, x))\} \\
&= C_R^\bullet(\mu)(x) \vee C_R^\bullet(\nu)(x)
\end{aligned}$$

(C5)
$$\begin{aligned}
C_R^\bullet(\underline{k} \wedge \varphi_x)(y) &= \sup_{\omega \in X} \{(\underline{k} \wedge \varphi_x)(\omega) \bullet R(\omega, y)\} \underline{k}(x) \bullet R(x, y) \\
&= \underline{k}(x) \bullet C_R^\bullet(\varphi_x)(y) = k \bullet C_R^\bullet(\varphi_x)(y)
\end{aligned}$$

Conversely, suppose the fuzzy operator C satisfies $(C1), (C4)$ and $(C5)$. We define $R(x, y) := C(\varphi_x)(y)$. It is clear that R is reflexive.

Let us see that $C = C_R^\bullet$:

Any fuzzy set h can be written as $h = \bigvee_{x \in X} \varphi_x \wedge \underline{h(x)}$. Therefore, by $(C4)$ and $(C5)$,

$$C(h) = C(\bigvee_{x \in X} (\varphi_x \wedge \underline{h(x)})) = \bigvee_{x \in X} C(\varphi_x \wedge \underline{h(x)}) = \bigvee_{x \in X} \underline{h(x)} \bullet C(\varphi_x)$$

and thus,

$$C(h)(y) = \sup_{x \in X} \{h(x) \bullet C(\varphi_x)(y)\} = \sup_{x \in X} \{h(x) \bullet R(x, y)\} = C_R^\bullet(h)(y)$$

where we have taken into account that X is finite. $\qquad\qquad\square$

Theorem 7. *Let X be a finite universal set, let $\bullet$ be an associative semi-copula and let $C : [0, 1]^X \longrightarrow [0, 1]^X$ be a fuzzy operator, then C satisfies $(C1), (C3), (C4)$ and $(C5)$ if and only if there exists a $\bullet$-preorder R such that $C = C_R^\bullet$.*

Proof. If C satisfies $(C1), (C3), (C4), (C5)$, from the previous theorem, there exists a reflexive fuzzy relation R such that $C = C_R^\bullet$.

Prove that R is $\bullet$-transitive. For $x, y, z \in X$, we have

$$R(x, y) \bullet R(y, z) \leq \sup_{\omega \in X} \{R(x, \omega) \bullet R(\omega, z)\} = \sup_{\omega \in X} \{C_R^\bullet(\varphi_x)(\omega) \bullet R(\omega, z)\}$$

$$= C_R^\bullet(C_R^\bullet(\varphi_x))(z) \overset{(C3)}{\leq} C_R^\bullet(\varphi_x)(z) = R(x, z)$$

Conversely, from the previous theorem, we obtain $(C1), (C4)$ and $(C5)$ and from Theorem 2, we obtain $(C3)$. $\qquad\qquad\square$

Some of the following results hold for any universal set X (not necessa finite).

Theorem 8. *Let R be a fuzzy relation on an universal set X and let $\bullet$ be semi-copula. Then:*

(1) $C_R^\bullet$ satisfies (C1) if and only if R is reflexive.
(2) If $C_R^\bullet$ satisfies (C3), then R is $\bullet$-transitive. Moreover if $\bullet$ is associative and X is finite, then the converse is true.

Proof. (1) We suppose that $C_R^\bullet$ is (C1), then

$$R(x,x) = R_{C_R^\bullet}(x,x) = C_R^\bullet(\varphi_x)(x) \geq \varphi_x(x) = 1$$

Conversely, $\mu(x) = \mu(x) \bullet R(x,x) \leq \sup_{w \in X}\{\mu(w) \bullet R(w,x)\} = C_R^\bullet(\mu)(x)$ for all $x \in X$.

(2) We suppose that $C_R^\bullet$ is (C3), since $R = R_{C_R^\bullet}$, we have

$$R(x,y) \bullet R(y,z) \leq \sup_{w \in X}\{R(x,w) \bullet R(w,z)\} = \sup_{w \in X}\{C_R^\bullet(\varphi_x)(w) \bullet R(w,z)\}$$
$$= C_R^\bullet(C_R^\bullet(\varphi_x))(z) \leq C_R^\bullet(\varphi_x)(z) = R(x,z)$$

Conversely, if $\bullet$ is associative and X is finite, we have:

$$
\begin{aligned}
C_R^\bullet(C_R^\bullet(\mu))(x) &= \sup_{w \in X}\{\sup_{\tau \in X}\{\mu(\tau) \bullet R(\tau,w)\} \bullet R(w,x)\}\\
&= \sup_{w \in X}\{\sup_{\tau \in X}\{(\mu(\tau) \bullet R(\tau,w)) \bullet R(w,x)\}\}\\
&= \sup_{w \in X}\{\sup_{\tau \in X}\{\mu(\tau) \bullet (R(\tau,w) \bullet R(w,x))\}\}\\
&\leq \sup_{w \in X}\{\sup_{\tau \in X}\{\mu(\tau) \bullet R(\tau,x)\}\} = \sup_{\tau \in X}\{\mu(\tau) \bullet R(\tau,x)\} = C_R^\bullet(\mu)(x)
\end{aligned}
$$

$\square$

Moreover, it is clear that the operator $C_R^\bullet$ is always (C2) for every relation R. Then:

Corollary 1. *Let R be a fuzzy relation on a finite universal set X and let $\bullet$ be an associative semi-copula. Then, $C_R^\bullet$ is a FCO if and only if R is a $\bullet$-preorder.*

Conclusion

We have studied in this work several relationships between fuzzy relations and fuzzy consequence operators. We have shown that, under the conditions of semi-copulas instead of t-norms, it is still possible to obtain some interesting results.

Acknowledgements. The authors acknowledge the financial support of the Spanish Ministerio de Economía y Competitividad (Grant TIN2014-59543-P and Grant MTM 2016-79422-P) and Carlos Bejines also thanks the support of the Asociación de Amigos of the University of Navarra.

C. Bejines et al.

?ferences

1. Bassan, B., Spizzichino, F.: Relations among univariate aging, bivariate aging and dependence for exchangeable lifetimes. J. Multivar. Anal. **93**(2), 313–339 (2005)
2. Carmona, N., Elorza, J., Recasens, J., Bragard, J.: Permutable fuzzy consequence and interior operators and their connection with fuzzy relations. Inf. Sci. **310**, 36–51 (2015)
3. Castro, J.L., Delgado, M., Trillas, E.: Inducing implication relations. Int. J. Approx. Reason. **10**(3), 235–250 (1994)
4. Castro, J.L., Trillas, E.: Tarski's fuzzy consequences. In: Proceedings of the International Fuzzy Engineering Symposium 1991, vol. 1, pp. 70–81 (1991)
5. Díaz, S., Montes, S., De Baets, B.: On the compositional characterization of complete fuzzy pre-orders. Fuzzy Sets Syst. **159**, 2221–2239 (2008)
6. Díaz, S., Montes, S., De Baets, B.: General results on the decomposition of transitive fuzzy relations. Fuzzy Optim. Decis. Making **9**, 129 (2010)
7. Durante, F., Guiselli Ricci, R.: Supermigrative semi-copulas and triangular norms. Inf. Sci. **179**, 2689–2694 (2009)
8. Elorza, J., Fuentes-González, R., Bragard, J., Burillo, P.: On the relation between fuzzy closing morphological operators, fuzzy consequence operators induced by fuzzy preorders and fuzzy closure and co-closure systems. Fuzzy Sets Syst. **218**, 73–89 (2013)
9. Elorza, J., Burillo, P.: On the relation of fuzzy preorders and fuzzy consequence operators. Int. J. Uncertain. Fuzziness Knowl.-Based Syst. **7**(3), 219–234 (1999)
10. Elorza, J., Burillo, P.: Connecting fuzzy preorders, fuzzy consequence operators and fuzzy closure and co-closure systems. Fuzzy Sets Syst. **139**(3), 601–613 (2003)
11. Esteva, F.: On the form of negations in posets. In: Proceedings of The Eleventh International Symposium on Multiple-Valued Logic (1981)
12. Esteva, F. García, P., Godo, L., Rodríguez, R.O.: Fuzzy approximation relations, modal structures and possibilistic logic. Mathware Soft Comput. V (23) pp. 151–166 (1998)
13. Esteva, F., García, P., Godo, L., Rodríguez, R.O.: On implicative closure operators in approximate reasoning. Int. J. Approx. Reason. **33**, 159184 (2003)
14. Klement, E.P., Mesiar, R., Pap, E.: Triangular Norms. Kluwer, Dordrecht (2000)
15. Pavelka, J.: On fuzzy logic i, Zeitschr. f. Math. Logik und Grundlagen d. Math. Bd. 25, 4552 (1979)
16. Recasens, J.: Indistinguishability Operators. STUDFUZZ, vol. 260. Springer, Heidelberg (2010)
17. Ward, M.: The closure operators of a lattice. Ann. Math. **43**(2), 191–196 (1940)

Author Index

© Springer International Publishing AG 2018

263

V. Torra et al. (eds.), *Aggregation Functions in Theory and in Practice*,
Advances in Intelligent Systems and Computing 581, DOI 10.1007/978-3-319-59306-7

Printed in the United States
By Bookmasters